Towards 4D Bioprinting

Towards 4D Bioprinting

Adrian Neagu

Department of Biophysics, Victor Babes University of Medicine and Pharmacy, Timisoara, Romania;
Department of Physics, University of Missouri, Columbia, MO, United States

Academic Press is an imprint of Elsevier
125 London Wall, London EC2Y 5AS, United Kingdom
525 B Street, Suite 1650, San Diego, CA 92101, United States
50 Hampshire Street, 5th Floor, Cambridge, MA 02139, United States
The Boulevard, Langford Lane, Kidlington, Oxford OX5 1GB, United Kingdom

ISBN: 978-0-12-818653-4

For information on all Academic Press publications visit our website at https://www.elsevier.com/books-and-journals

Publisher: Mara E. Conner
Acquisitions Editor: Carrie L. Bolger
Editorial Project Manager: Kyle Gravel
Production Project Manager: Fizza Fathima
Cover Designer: Christian J. Bilbow

Typeset by TNQ Technologies

Contents

CHAPTER

Introduction

1

As fundamental sciences, engineering, and medicine join forces to provide personalized healthcare, we have the privilege to witness, and contribute to, the birth and growth of disruptive technologies rooted in three-dimensional (3D) printing. The emergence of affordable 3D printers compatible with a variety of 3D printable materials resulted in the widespread use of this technology, way beyond its original purpose of rapid prototyping.

Leveraging the ability of 3D printing to replicate seamless geometric complexity, anatomical models can be fabricated based on the patient's medical imaging. These can be used to plan the operation and discuss it with the patient. Also, patient-specific surgical guides and instruments can be manufactured on the spot to ensure a fast and effective intervention. Implantable medical devices can also be fabricated by 3D printing. In the case of pediatric patients, 4D-printed implants can change in shape and size to match the host's evolving anatomy. Also, people with disabilities can benefit from assistive devices tailored according to their needs using 3D and 4D printing.

Live tissue constructs can be manufactured from the patient's own cells by 3D bioprinting. They can serve as disease models and are useful for testing the efficacy of drugs. Nonetheless, the original motivation of cell printing research was to build organ replacements in the laboratory. Although two decades of research did not bring us to that point, the biofabrication community is confident that, eventually, 3D-bioprinted implantable organs will alleviate donor organ shortage. Finally, 4D bioprinting will provide dynamical tissue constructs capable of recapitulating the response of certain organs to chemical and electrical cues, such as the vasoconstriction caused by caffeine or the motor activity of the gastrointestinal tract governed by nervous influxes.

The scientific literature dedicated to biomedical applications of 3D/4D printing and bioprinting is rapidly expanding. Besides sheer volume, the research output has also grown in diversity. Therefore, experienced investigators often struggle to keep up with the literature outside their own domain of expertise, and newcomers might be discouraged to enter the field. I wrote this book with the aim of providing a coherent overview of 3D/4D bioprinting terminology and technology. Also, I tried to showcase the rich spectrum of applications by selecting representative examples from the primary literature. I thank the authors for their great work and the publishers for the permission to reprint illustrations from the original papers. Since I prefer traveling to cooking, this book is very

Towards 4D Bioprinting. https://doi.org/10.1016/B978-0-12-818653-4.00005-X

much like a tourist guide, as opposed to a collection of recipes. Actually, thinking of the discoveries responsible for the progress of bioprinting over the past two - decades, this book might be considered a treasure map. It provides background knowledge needed for understanding the original papers that shaped the current research landscape. Although the field will likely change between the time of writing and the date of publication, the foundations remain the same, and important milestones may inspire future discoveries.

Treating a vast field within a reasonable size limit required a careful choice of topics and depth of discussion. I apologize to those colleagues whose works are not mentioned in this book, as well as to those specialists who might find their specific topic of interest discussed a bit superficially. For deeper insight, the reader is invited to study the cited literature. With a bibliography of over 500 titles, the text might also serve as a reference guide for novice readers.

This book presents an interdisciplinary topic targeting a diverse audience. Therefore, I did my best trying to stick with straightforward language and avoid excessive jargon. I can only hope this did not result in scientific haiku (although I enjoy haiku, scientific or not).

To understand the text, the reader should only have basic knowledge of biology, chemistry, physics, and mathematics. Chapters are somewhat independent, but connections between them are often mentioned in the text. This book is intended for final-year undergraduate students, graduate students, and researchers willing to expand their horizons.

1. Early milestones of 3D printing

The central idea of 3D printing, that of building a bulky object one layer at a time, has been lingering around for about a century before Charles W. Hull filed his patent application for stereolithography on August 8, 1984 (Hull, 1986).

Actually, it did not just linger. The layer-by-layer fabrication of solid models was used extensively by morphologists since 1876, when the anatomist and embryologist Gustav Born, the father of Nobel laureate physicist Max Born (Born, 2002; Born et al., 1950), developed his wax plate method of embryo modeling (Born, 1876). He sectioned the embryos, using a microtome invented a few years earlier by Wilhelm His, and traced the enlarged histological sections onto wax plates. To preserve the aspect ratio of the anatomical structure, he magnified the cross-sectional image by the ratio of wax plate thickness to histological section thickness. Once the sections were drawn, he cut away the excess wax, preserving temporary support bridges to keep disconnected pieces together, and stacked successive layers on top of each other using vertical guides for precise alignment. His method was described in minute detail in a later paper (Born, 1883) and used since then (Gaunt & Gaunt, 1978; Hopwood, 1999).

Hull's patent was granted on March 11, 1986. It proposed to build a 3D object layer-by-layer by focusing an ultraviolet (UV) light beam onto the surface of a liquid photopolymer solution that turns solid when exposed to UV light. In his own words, stereolithography "is a method and apparatus for making solid objects by successively 'printing' thin layers of a curable material, e.g., a UV curable material, one on top of the other" (Hull, 1986).

By 1983, when this idea came to his mind, Charles (Chuck) Hull was working at a company that used UV light to coat furniture with tough plastic veneers (Hickey, 2014; Ponsford & Glass, 2014). Thus, he was familiar with UV curable materials, such as the acrylic-based Potting Compound 363 produced by Locktite Corporation. He was also aware of ongoing research on building 3D objects by selectively curing a fluid medium at prescribed points of intersection of multiple light beams (or one beam targeting sequentially the same point from different directions). In his patent application, Hull pointed out the drawbacks of those approaches—poor resolution and exposure control due to the interactions of electromagnetic waves with the media they cross on the way toward the target point. By contrast, the stereolithography apparatus (SLA) cured the surface layer of the fluid in millimeter-sized focal spots, which swiped the fluid surface under computer control. Once the first layer was cured, the print bed was lowered into the fluid-filled vat and fresh fluid covered the hardened layer. Selective light exposure brought about the hardening of the new layer, which adhered to the previous one (Hull, 1986).

In his initial setup, Chuck Hull employed a mercury short arc lamp whose output was focused onto a UV transmitting fiber optic bundle of 1 mm in diameter. The bundle was fitted into the housing of a quartz lens that focused the UV beam onto the liquid surface, creating a light intensity of about 1 W/cm^2 at the focal spot. To move the spot under computer control, the lens housing was attached to the pen carriage of a plotter. The patent application also mentions a UV laser as a potential light source and points out its major advantages: higher intensity and the feasibility of optical scanning. Remarkably, alternative setups of the SLA are also described (Hull, 1986), which are incorporated in modern-day digital light processing printers, as well as in the recently developed Fluid-supported Liquid Interface Polymerization (FLIP) 3D printing technique (Beh et al., 2021).

In the summer of 1984, 3 weeks before Hull filed his patent application for stereolithography, three French scientists, Alain Le Méhauté, Olivier de Witte, and Jean Claude André filed theirs for a similar technique, in which laser light was delivered, via an optical fiber, into a photocurable liquid resin. The free end of the optical fiber was supposed to move, under computer control, within the bulk of the resin and cure nearby portions (Moussion, 2014). Their patent application, however, was abandoned by the Compagnie Générale d'Électricité (Le Méhauté's employer, which became Alcatel-Alsthom in 1991) and the Laser Consortium (CILAS) because their decision-makers did not see the commercial potential of the invention. "I'm not bitter. I am proud of the innovative work we undertook and our efforts to promote technological innovation through the impetus of business and economic growth," declared Le Méhauté looking back at those days and the subsequent development of 3D printing. He also added that "I have great respect for Hull who had the courage to initiate the creation of 3D Systems"—a lesson of fair play and positive thinking (Mendoza, 2015).

The first 3D printing company in the world, 3D Systems, was cofounded by Chuck Hull in 1986 and remained at the forefront of 3D printing industry ever since. One year later, 3D Systems put the first 3D printer, SLA-1, on the market. With Chuck Hull as a Chief Technology Officer, 3D Systems continued to innovate and contributed substantially to the advancement of additive manufacturing (Our Story, n.d.).

Medical applications of 3D printing emerged soon thereafter (Mankovich et al., 1990). In their groundbreaking paper, Nicholas J. Mankovich, Andrew M. Cheeseman, and Noel G. Stoker from the University of California, Los Angeles, used an SLA printer to produce a physical model of a human skull starting from a computed tomography scan. They illustrate the power of the method, discuss the encountered difficulties (e.g., imaging artifacts), and propose a vast agenda for future research. They conclude that "the usefulness of such models extends to surgical planning, radiation therapy, patient education, and physician education" (see Chapter 3 to assess the accuracy of their insight).

In an interview given in 2014 to CNN, being asked what was most surprising to him in the evolution of 3D printing, Chuck Hull said "To me, some of the medical applications. I didn't anticipate that, and as soon as I started working with some of the medical imaging people, it became pretty clear that this was going to work. But, you know, they told me, I didn't tell them" (Ponsford & Glass, 2014).

2. Bioprinting—a form of biofabrication

2.1 The beginnings of bioprinting

The idea of using printers to build 3D tissue constructs emerged at the same time but independently of the SLA's commercialization. It sprang from repurposing inkjet printers and graphics plotters to create patterns of biomolecules on 2D substrates (Klebe, 1988).

Robert J. Klebe from the University of Texas Health Science Center, San Antonio, TX, printed fibronectin patterns on polystyrene films. To prevent cell attachment to fibronectin-free regions, the imprinted film was fixed on the bottom of a Petri dish with paraffin wax and treated for 10 min with a 1% solution of thermally denatured bovine serum albumin dissolved in phosphate-buffered saline (PBS). Then, the film was washed twice with PBS, and cells were seeded on the film. This process has led to the formation of a 2D pattern of live cells anchored to the fibronectin layer. To build 3D tissue-like structures, Klebe prepared collagen gel sheets in molds made of a perforated sheet of graph paper sandwiched between two siliconized microscope slides. The collagen was seeded with cells and multiple layers could be glued together with additional collagen. Cell-type-specific extracellular matrix proteins and monoclonal antibodies were also proposed for positioning cells within individual layers, and stacking them was suggested as a method for building a 3D tissue construct. The take-home message of the paper was that further development of the proposed technology "should aid in the production of artificial tissues which resemble natural tissues and organs" (Klebe, 1988).

Cell sheet engineering became practical 2 years later by coating cell culture dishes with a poly(N-isopropylacrylamide) (PNIPAAm) (Yamada et al., 1990). PNIPAAm is a thermoresponsive polymer, which is fully hydrated at room temperature, but at 32°C it becomes hydrophobic and, thereby, it enables cell attachment. (Indeed, cells cannot attach to highly hydrophilic surfaces because that would require displacing tightly bound water molecules.) Yamada and collaborators hypothesized that cells cultured in PNIPAAm-coated polystyrene dishes form confluent layers that can be detached by simply cooling the system below the polymer's transition temperature. This hypothesis proved to be correct and, despite the absence of micropatterning, cell sheet engineering produced clinically relevant applications by the turn of the century (Yang et al., 2007).

Biological cell printing was first accomplished by David J. Odde from the University of Minnesota, Minneapolis, MN, and Michael J. Renn from the Michigan Technological University, Houghton, MI (Odde & Renn, 1999). They used laser-induced optical forces to gently guide and deposit cells on a solid surface. They named their approach laser-guided direct writing (LGDW). The abilities of LGDW were illustrated by depositing chicken embryonic spinal cord cells onto a glass plate in a predefined arrangement. Cells were not harmed by the near-infrared laser beam that propelled them toward the substrate—deposited cells developed neurites, which indicates that they remained viable and functional. The range of LGDW could be extended to several centimeters by guiding the cells within hollow optical fibers. Fiber guiding was preferred over free guiding because (i) the fiber's lumen provided an unperturbed environment for cell movement and (ii) it enabled accurate positioning of cells by pointing the fiber's tip toward selected target points. Moreover, using multiple fibers would allow for printing a variety of cell types. Commenting on the potential applications of LGDW, Odde and Renn wrote that "the ability to organize cells spatially into well-defined 3D arrays that closely mimic the native tissue architecture can potentially help in the fabrication of engineered tissue," and, based on their proof-of-concept experiments, they concluded that LGDW "potentially allows the 3D patterning of cells using multiple cell types with cell placement at arbitrarily selected positions" (Odde & Renn, 1999). The impatient reader might wish to take a look at Chapter 5 for further details on LGDW.

Along a different line of thinking, an important development originated from the work of Rüdiger Landers and Rolf Mülhaupt of the Albert-Ludwigs-Universität Freiburg, Germany (Landers & Mülhaupt, 2000). These authors constructed a 3D plotter capable of computer-controlled pneumatic extrusion of prepolymer solutions and pastes in liquid media to produce solid objects of complex shapes and intricate microarchitecture. They demonstrated the 3D plotting of silicone microdots and strands to build tubes and porous constructs akin to tissue engineering scaffolds.

In their pioneering work, they noticed that 3D plotting does not involve harsh physicochemical factors and, therefore, is suitable to handle biomaterials loaded with live cells (Landers & Mülhaupt, 2000). The team also created agar hydrogel scaffolds by 3D plotting and made them appropriate for cell attachment via surface treatment—soaking in concentrated $CaCl_2$ solution, rinsing with distilled water, and immersion in a diluted solution of hyaluronic acid and alginic acid (Landers et al., 2002). In a later work, dedicated to the fabrication of biodegradable polyurethane scaffolds, Rolf Mülhaupt's team proposed a new name for their technique, 3D bioplotting, even though their plotting material did not incorporate live cells, yet (Pfister et al., 2004). Rightfully so, since they pleaded for including cells and their works inspired the development of today's pneumatic extrusion-based bioprinters.

W. Cris Wilson, Jr., and Thomas Boland revolutionized the additive manufacturing of tissue constructs by adapting commercial inkjet printers to deliver cell suspensions (Wilson and Boland, 2003). This was the first time to use a jet-based instrument for fully automated, unattended printing of live cells. Wilson and Boland designed print heads comprising nine independently operated piezoelectric pumps. Each pump was fed with a cell suspension of its own via a flexible tubing and expelled droplets through a sterile needle. The pumps were controlled by a microchip programmable interface controller via an original software, whereas the print head movement remained under the control of the printer driver (rewritten to handle inks of different viscosities and electrical charges). Hewlett–Packard offered a generous gift to the tissue engineering community by providing the source code of the HP550C printer driver to Wilson and Boland. The modified printer was able to dispense 15 nL droplets of bovine aortal endothelial cell suspension onto a Matrigel substrate, each droplet containing one or two cells. Postprinting cell viability was about 75%, remarkably high for a technology in its incipient phases. The authors suggested that cell damage might have been caused by dehydration since the minuscule droplets evaporated quickly. Hence the hydrogel substrate had a double role in protecting the dispensed cells: it cushioned their landing and kept them hydrated. In striking contrast with the bold title of the paper (the first to mention "organ printing"), its take-home message is careful and (with the hindsight of two decades of progress) realistic: "systematic three-dimensional cellular assemblies may become possible with the use of the ink-jet approach" and "these devices may have many potential applications, ranging from drug screening to tissue engineering" (Wilson and Boland, 2003). The companion paper by Boland and coworkers, which appeared in the same issue of *The Anatomical Record*, provides further evidence that cell printing onto superimposed layers of thermosensitive gels has the potential to create 3D tissue constructs and explores avenues toward printing 3D organs. In particular, it suggests dispensing cell aggregates in successive layers contiguously to allow them to fuse on their own—a scenario that is quite common in developmental biology (Boland et al., 2003). For a deeper insight into using cell aggregates as building blocks of 3D-printed tissues, the team contacted Gabor Forgacs from the University of Missouri–Columbia, MO, who had spent more than a decade investigating the biomechanical properties of cell aggregates and also had a solid experience in developmental biology (Forgacs & Newman, 2005). Their joint paper became the most cited work on organ printing (Mironov et al., 2003).

Today, a search for the term "organ printing" returns over 140,000 hits on Google. Back then, the terminology was so new that even the researchers felt the urge to explore the thin borderline between science and fiction (Jakab et al., 2004). The idea of printing an implantable organ was in utter contrast with the possibilities of the hardware and materials within reach. Indeed, organ printing is still a long-term goal (Ng et al., 2019), but it seems more tangible today, as demonstrated by many examples presented in the forthcoming chapters. As the field matured, its terminology became more nuanced: "3D bioprinting" became the popular term for the additive manufacturing of live structures, whereas the cell-containing materials dispensed by 3D bioprinters are called "bioinks." Bioprinting got embedded into the broader field of biofabrication, which encompasses a vast set of bottom-up technologies of tissue engineering and regenerative medicine. Novel 3D bioprinting techniques mushroomed during the past decade. Chapter 5 presents the most common ones in detail, Chapter 11 discusses recent advances, and new ones will likely emerge by the time this book reaches its audience.

The next section is dedicated to the definitions of the terms "bioprinting" and "biofabrication"—discussed in 8 million and 11 million Google documents, respectively.

2.2 The terminology of biofabrication

The term "bioprinting" was coined at The First International Workshop on Bioprinting and Biopatterning, held in September 2004, in Manchester, UK, organized by Brian Derby from the University of Manchester; Douglas B. Chrisey from the U.S. Naval Research Laboratory, Washington, DC, USA; Richard K. Everett from ONR Global, London, UK; and Nuno Reis from Universidade da Beira Interior, Covilhã, Portugal. At this seminal workshop, bioprinting was defined as "the use of material transfer processes for patterning and assembling biologically relevant materials – molecules, cells, tissues, and biodegradable biomaterials – with a prescribed organization to accomplish one or more biological functions" (Mironov et al., 2006). Technical challenges faced by the new technology were recognized, such as the need for a fluid vehicle "that shortly after printing requires consolidation and should consequently behave as a visco-elastic solid," and "this phase change must occur without damage to the biochemicals, cells, or more complex units within the fluid," establishing important guidance for material development. Despite many uncertainties about the future of the new technology, the participants decided to organize periodic meetings and pondered the opportunity of initiating a professional society and, perhaps, a journal (Mironov et al., 2006).

A dedicated journal, entitled *Biofabrication*, was introduced at the fourth meeting, organized in July 2009, in Bordeaux, France. Published by IOP Science,

Biofabrication grew rapidly in content and prestige. In the meantime, also other high-quality journals emerged, including *Bioprinting* (by Elsevier) and *International Journal of Bioprinting* (by WHIOCE Publishing). The Bordeaux meeting has also provided an updated definition of "bioprinting": "the use of computer-aided transfer processes for patterning and assembling living and non-living materials with a prescribed 2D or 3D organization in order to produce bio-engineered structures serving in regenerative medicine, pharmacokinetic and basic cell biology studies" (Guillemot et al., 2010).

In the first issue of the new journal, the term "biofabrication" was defined as "the production of complex living and nonliving biological products from raw materials such as living cells, molecules, extracellular matrices, and biomaterials" (Mironov et al., 2009). This definition pertains to the field of tissue engineering and regenerative medicine because, since 1994, the term biofabrication has been used in several disciplines, with a variety of meanings (Groll et al., 2016).

The fifth meeting, in October 2010, Philadelphia, USA, marked an important milestone: the International Society for Biofabrication (ISBF) was founded (International Society for Biofabrication (ISBF), 2010). Nowadays, the annual meetings organized by the ISBF, the so-called International Conferences on Biofabrication, attract hundreds of participants from all over the world.

The growing popularity and research production, however, often lead to inconsistent terminology. To fight this tendency, prominent members of the ISBF periodically publish perspective papers to clarify the conceptual framework of this vivid research field (Groll et al., 2019, 2016; Moroni et al., 2018). In particular, the definition of the field itself has been revisited to accommodate novel approaches. Currently, biofabrication is achieved via two main strategies: "bioprinting" and "bioassembly." The latter is defined as "the fabrication of hierarchical constructs with a prescribed 2D or 3D organization through automated assembly of pre-formed cell-containing fabrication units generated via cell-driven self-organization or through preparation of hybrid cell-material building blocks, typically by applying enabling technologies, including microfabricated molds or microfluidics." For bioassembly, one starts from preformed multicellular units such as organoids, cell sheets, or cell fibers. Composed of cells and their extracellular matrix, such units can be obtained through multicellular self-organization in specific environments created by microfluidics and micromolding techniques. A variety of automated methods can be applied to achieve bioassembly depending on the size and geometry of the building blocks (e.g., in the case of cell fibers, one can use methods of textile industry, such as weaving, winding, and knitting). Thus, in the context of tissue engineering and regenerative medicine, according to the currently accepted working definition, biofabrication consists in "the automated generation of biologically functional products with structural organization from living cells, bioactive molecules, biomaterials, cell aggregates such as micro-tissues, or hybrid cell-material constructs, through bioprinting or bioassembly and subsequent tissue maturation processes" (Groll et al., 2016). To assist newcomers and consolidate the professional jargon of active investigators, a vast review article has been published by biofabrication experts, which includes a glossary of terms and a description of the major technologies used in biofabrication (Moroni et al., 2018).

Together with its many (but carefully selected) references, this book is meant to help readers eager to explore the rapidly changing world of bioprinting.

References

Beh, C. W., Yew, D. S., Chai, R. J., Chin, S. Y., Seow, Y., & Hoon, S. S. (2021). A fluid-supported 3D hydrogel bioprinting method. *Biomaterials, 276*, 121034. https://doi.org/10.1016/j.biomaterials.2021.121034

Boland, T., Mironov, V., Gutowska, A., Roth, E. A., & Markwald, R. R. (2003). Cell and organ printing 2: Fusion of cell aggregates in three-dimensional gels. *Anatomical Record Part A Discoveries in Molecular Cellular and Evolutionary Biology, 272*, 497–502. https://doi.org/10.1002/ar.a.10059

Born, G. (1876). Ueber die Nasenhöhlen und den Thränennasengang der Amphibien. *Morphologisches Jahrbuch, 2*, 577–646.

Born, G. (1883). Die plattenmodellirmethode. *Archiv für Mikroskopische Anatomie, 22*, 584–599.

Born, G. V. R. (2002). The wide-ranging family history of Max Born. *Notes and Records of the Royal Society of London, 56*, 219–262. http://www.jstor.org/stable/3557669.

Born, M., Brandt, W., & Born, G. (1950). In memoriam Gustav Born, experimental embryologist. *Acta Anatomica, 10*, 466–475. https://doi.org/10.1159/000140488

Forgacs, G., & Newman, S. A. (2005). *Biological physics of the developing embryo*. Cambridge University Press. https://books.google.ro/books?id=rUyVWQhk7CkC.

Gaunt, W. A., & Gaunt, P. N. (1978). *Three dimensional reconstruction in biology*. University Park Press.

Groll, J., Boland, T., Blunk, T., Burdick, J. A., Cho, D.-W., Dalton, P. D., Derby, B., Forgacs, G., Li, Q., Mironov, V. A., Moroni, L., Nakamura, M., Shu, W., Takeuchi, S., Vozzi, G., Woodfield, T. B. F., Xu, T., Yoo, J. J., & Malda, J. (2016). Biofabrication: Reappraising the definition of an evolving field. *Biofabrication, 8*, 013001. https://doi.org/10.1088/1758-5090/8/1/013001

Groll, J., Burdick, J. A., Cho, D. W., Derby, B., Gelinsky, M., Heilshorn, S. C., Jüngst, T., Malda, J., Mironov, V. A., Nakayama, K., Ovsianikov, A., Sun, W., Takeuchi, S., Yoo, J. J., & Woodfield, T. B. F. (2019). A definition of bioinks and their distinction from biomaterial inks. *Biofabrication, 11*, 013001. https://doi.org/10.1088/1758-5090/aaec52

Guillemot, F., Mironov, V., & Nakamura, M. (2010). Bioprinting is coming of age: Report from the international conference on bioprinting and biofabrication in Bordeaux (3B'09). *Biofabrication, 2*, 010201. https://doi.org/10.1088/1758-5082/2/1/010201

Hickey, S. (June 22, 2014). *Chuck Hull: The father of 3D printing who shaped technology*. The Guardian. www.theguardian.com/business/2014/jun/22/chuck-hull-father-3d-printing-shaped-technology.

Hopwood, N. (1999). "Giving body" to embryos. Modeling, mechanism, and the microtome in late nineteenth-century anatomy. *Isis, 90*, 462–496. https://doi.org/10.1086/384412

Hull, C. W. (March 11, 1986). Apparatus for production of three-dimensional objects by stereolithography. *United States Patent 4,575,330*. https://patentimages.storage.googleapis.com/5c/a0/27/e49642dab99cf6/US4575330.pdf.

International Society for Biofabrication (ISBF). (2010). https://biofabricationsociety.org/about/.

Jakab, K., Neagu, A., Mironov, V., & Forgacs, G. (2004). Organ printing: Fiction or science. *Biorheology, 41*, 371–375.

Klebe, R. J. (1988). Cytoscribing: A method for micropositioning cells and the construction of two- and three-dimensional synthetic tissues. *Experimental Cell Research, 179*, 362–373. https://doi.org/10.1016/0014-4827(88)90275-3

Landers, R., Hübner, U., Schmelzeisen, R., & Mülhaupt, R. (2002). Rapid prototyping of scaffolds derived from thermoreversible hydrogels and tailored for applications in tissue engineering. *Biomaterials, 23*, 4437–4447. https://doi.org/10.1016/S0142-9612(02)00139-4

Landers, R., & Mülhaupt, R. (2000). Desktop manufacturing of complex objects, prototypes and biomedical scaffolds by means of computer-assisted design combined with computer-guided 3D plotting of polymers and reactive oligomers. *Macromolecular Materials and Engineering, 282*, 17–21. https://doi.org/10.1002/1439-2054(20001001)282:1<17::AID-MAME17>3.0.CO;2-8

Mankovich, N. J., Cheeseman, A. M., & Stoker, N. G. (1990). The display of three-dimensional anatomy with stereolithographic models. *Journal of Digital Imaging, 3*, 200. https://doi.org/10.1007/BF03167610

Mendoza, H. R. (May 15, 2015). *Alain Le Méhauté, the man who submitted patent for SLA 3D printing before Chuck Hull*. https://3dprint.com/65466/reflections-alain-le-mehaute/.

Mironov, V., Boland, T., Trusk, T., Forgacs, G., & Markwald, R. R. (2003). Organ printing: Computer-aided jet-based 3D tissue engineering. *Trends in Biotechnology, 21*, 157–161. http://www.sciencedirect.com/science/article/pii/S0167779903000337.

Mironov, V., Reis, N., & Derby, B. (2006). Review: Bioprinting: A beginning. *Tissue Engineering, 12*, 631–634. https://doi.org/10.1089/ten.2006.12.631

Mironov, V., Trusk, T., Kasyanov, V., Little, S., Swaja, R., & Markwald, R. (2009). Biofabrication: A 21st century manufacturing paradigm. *Biofabrication, 1*, 022001. https://doi.org/10.1088/1758-5082/1/2/022001

Moroni, L., Boland, T., Burdick, J. A., De Maria, C., Derby, B., Forgacs, G., Groll, J., Li, Q., Malda, J., Mironov, V. A., Mota, C., Nakamura, M., Shu, W., Takeuchi, S., Woodfield, T. B. F., Xu, T., Yoo, J. J., & Vozzi, G. (2018). Biofabrication: A guide to technology and terminology. *Trends in Biotechnology, 36*, 384–402. https://doi.org/10.1016/j.tibtech.2017.10.015

Moussion, A. (September 17, 2014). *Rencontre avec Alain Le Méhauté, l'un des pères de l'impression 3D! (Meeting Alain Le Méhauté, one of the fathers of 3D printing!*. www.primante3d.com/inventeur/.

Ng, W. L., Chua, C. K., & Shen, Y.-F. (2019). Print me an organ! Why we are not there yet. *Progress in Polymer Science, 97*, 101145. https://doi.org/10.1016/j.progpolymsci.2019.101145

Odde, D. J., & Renn, M. J. (1999). Laser-guided direct writing for applications in biotechnology. *Trends in Biotechnology, 17*, 385–389. https://doi.org/10.1016/S0167-7799(99)01355-4

Our Story. (n.d.). 3D systems. Retrieved from www.3dsystems.com/our-story.

Pfister, A., Landers, R., Laib, A., Hübner, U., Schmelzeisen, R., & Mülhaupt, R. (2004). Biofunctional rapid prototyping for tissue-engineering applications: 3D bioplotting versus 3D printing. *Journal of Polymer Science Part A: Polymer Chemistry, 42*, 624–638. https://doi.org/10.1002/pola.10807

Ponsford, M., & Glass, N. (February 14, 2014). *The night I invented 3D printing*. CNN. https://edition.cnn.com/2014/02/13/tech/innovation/the-night-i-invented-3d-printing-chuck-hall/index.html.

Wilson, W. C., Jr., & Boland, T. (2003). Cell and organ printing 1: Protein and cell printers. *The Anatomical Record Part A: Discoveries in Molecular, Cellular, and Evolutionary Biology, 272A*, 491–496. https://doi.org/10.1002/ar.a.10057

Yamada, N., Okano, T., Sakai, H., Karikusa, F., Sawasaki, Y., & Sakurai, Y. (1990). Thermo-responsive polymeric surfaces; control of attachment and detachment of cultured cells. *Makromolekulare Chemie, Rapid Communications, 11*, 571–576. https://doi.org/10.1002/marc.1990.030111109

Yang, J., Yamato, M., Shimizu, T., Sekine, H., Ohashi, K., Kanzaki, M., Ohki, T., Nishida, K., & Okano, T. (2007). Reconstruction of functional tissues with cell sheet engineering. *Biomaterials, 28*, 5033–5043. https://doi.org/10.1016/j.biomaterials.2007.07.052

CHAPTER

4D printing: definition, smart materials, and applications

2

The term "4D printing" has been coined by Skylar Tibbits in his TED talk delivered on April 4, 2013. It was formally defined and illustrated in research articles (Raviv et al., 2014; Tibbits, 2014), as well as in a visionary technical report (Campbell et al., 2014, pp. 1−15).

As a part of a vast research field known as "programmable matter" 4D printing is a technology based on multimaterial 3D printing in which the fourth dimension entails the capability of the 3D-printed object to predictably change shape and/or function when exposed to a certain stimulus (e.g., light, heat, pressure, electric current, or change in medium). Certain authors simply state that the fourth dimension is time, hinting toward the ability of the 3D-printed system to evolve. While the simplicity of such a definition is appealing and resembles the elegance of the formulation of Einstein's special theory of relativity in Minkowski's 4D space-time, it misses two essential aspects of 4D printing: (i) the postprinting evolution should occur in an intentional, programmable way and (ii) it is supposed to be triggered by a predetermined stimulus.

This chapter presents current applications of 4D printing and discusses its perspectives. Although 4D printing might well grow into a disruptive technology, able to reshape several industries, with important geopolitical consequences (Campbell et al., 2014, pp. 1−15), our discussion will mainly focus on the implications of 4D printing in biomedical sciences.

1. A technology inspired by life

The central idea of 4D printing is to use a 3D printer to fabricate an object capable to respond in a programmed fashion to certain stimuli. The desirable conformation or functionality of the object emerges autonomously, without external intervention.

Such a scenario is common in living systems, whose active parts are mainly composed of proteins. Proteins are macromolecules with diverse roles in living organisms: they are enzymes (the catalysts of biochemical reactions); they compose the cytoskeleton (which confers structural integrity to the cell); they are molecular motors (assuring intracellular transport and changes in cell shape); they are channels, pumps, or carriers (responsible for the transport of ions and small molecules through cell membranes), and they are involved in recognition and signaling

Towards 4D Bioprinting. https://doi.org/10.1016/B978-0-12-818653-4.00002-4

pathways. These functions, and several others, are assured by a wide variety of proteins, but they are made of merely 20 types of amino acids (plus two, less common ones, present in certain life forms).

For a protein to be synthesized, first a DNA sequence is rewritten in the form of messenger RNA (mRNA), in a process known as transcription, which takes place in the cell nucleus. Then, the mRNA is transferred into the cytoplasm. Here, ribosomes and transfer RNAs (tRNAs) decode the mRNA and facilitate the formation of covalent bonds (peptide bonds) between the amino acids according to the sequence specified in the mRNA (Fig. 2.1) (Fowler et al., 2013).

The result of the translation process shown in Fig. 2.1 is a linear sequence of amino acids, also known as a polypeptide. This product, however, is not yet biologically active (hence, it is not called a protein). As soon as the polypeptide is released in its physiological medium (cytoplasm or cell membrane), it undergoes a spontaneous folding process that gives rise to the native structure of the protein.

Experiments show that the native structure of a protein is the thermodynamically stable one. It depends exclusively on the amino acid sequence and on the solution that baths the protein and not on the folding route. If the protein is placed in a

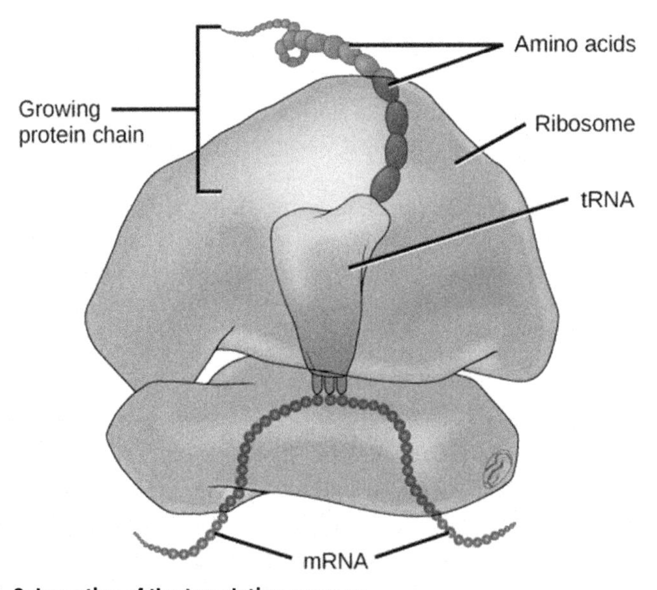

FIGURE 2.1 Schematics of the translation process.

Schematic representation of protein synthesis performed by a ribosome (*blue*) and tRNA molecules (*yellow*) according to the genetic code specified by the mRNA molecule. The tRNA molecules transport amino acids, whereas the ribosome catalyzes peptide bond formation between successive amino acids. For interpretation of the references to color in this figure legend, please refer online version of this title.

Fowler, S., Roush, R., & Wise, J. (2013). Concepts of biology. Houston, Texas: OpenStax. Access for free at https://openstax.org/books/concepts-biology/pages/1-introduction.

test tube and unfolded by raising the temperature, it refolds, assuming the native structure that emerged in the cell, right after synthesis (with rare exceptions of proteins whose biologically active form is kinetically trapped—e.g., insulin) (Dill et al., 2008). Protein structure prediction is a major goal of computational biology; it is increasingly successful, but challenging because of the huge number of conformations a protein must explore to find its native form. Although incomplete, our knowledge of the interactions that drive protein folding enabled the design of new proteins and the prediction of their structure (Dill et al., 2008). A major breakthrough in protein folding research occurred in 2020, when the second version of the AlphaFold artificial intelligence software had won by far the 14-th Critical Assessment of protein Structure Prediction (CASP14) contest (Jumper et al., 2021).

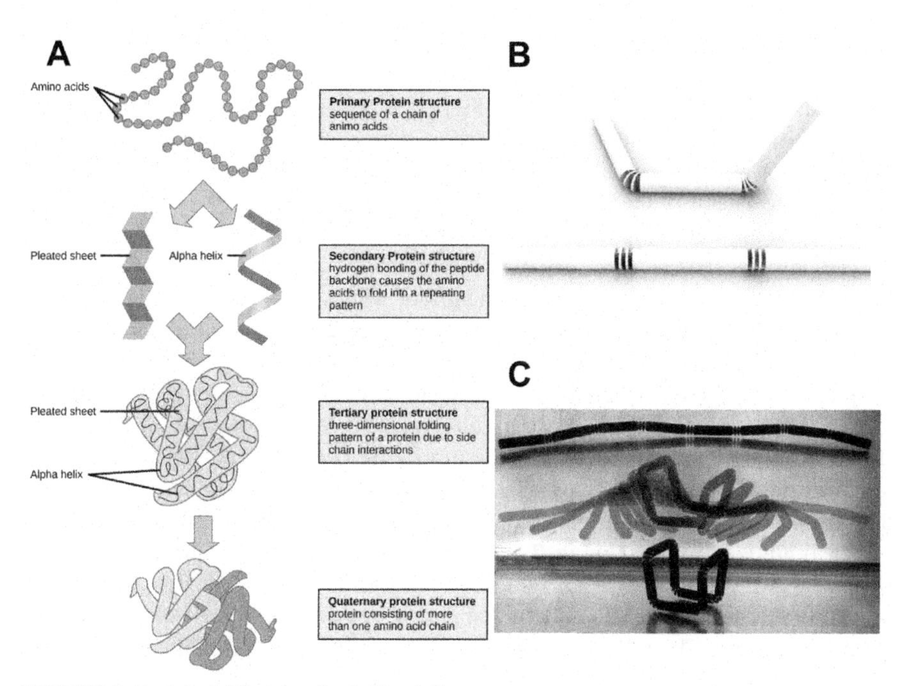

FIGURE 2.2 Protein folding inspired 4D printing.

Stages of protein folding (A). Folding primitives (B) can be fabricated by multimaterial 3D printing. When placed in water, a linear chain of interconnected segments folds into the desired conformation if it includes the right joint designs and materials (C).

With permission from: panel A—Accessible for free at https://openstax.org/books/biology-ap-courses/pages/3-4-proteins; panel B—Raviv, D., Zhao, W., McKnelly, C., Papadopoulou, A., Kadambi, A., Shi, B., ... Tibbits, S. (2014). Active printed materials for complex self-evolving deformations. Scientific Reports, 4, *7422. https://doi.org/10.1038/srep07422; panel C—Tibbits, S. (2014). 4D printing: Multi-material shape change.* Architectural Design, 84, *116–121. https://doi.org/10.1002/ad.1710.*

Fig. 2.2A is a schematic representation of protein folding; it depicts and briefly describes four levels of protein structure (primary, secondary, tertiary, and quaternary).

The fourth dimension of 4D printing resembles protein folding (Fig. 2.2A), whereby a chain of amino acids gradually assumes a secondary structure and a tertiary structure. Certain proteins are composed of multiple amino acid chains; their tertiary structures self-assemble, adopting the biologically functional quaternary structure.

In early examples of 4D printing (Raviv et al., 2014; Tibbits, 2014), self-evolving linear structures were fabricated using a 3D printer. Looking at the analogy with protein synthesis, we notice that, in these examples, the 3D printer played the role of the translation machinery, assembling the linear chain, whereas the postprinting evolution was similar to protein folding. Let us analyze the factors that enable one to predict the folding outcome. To this end, we need to study the structural details of the joints.

The design of the joints depicted in Fig. 2.2B is based on two materials (Raviv et al., 2014): a hydrophilic polymer (red), which expands considerably when immersed in water, and a relatively rigid material (white), which is insensitive to water. A thin strip made of the rigid material should be elastic enough to bend under the action of the adjacent strip of expanding material.

The physical models (Fig. 2.3, *right*), fabricated via inkjet 3D printing of UV-curable polymers, were composed of rigid plastic (*black*) and a hydrophilic polymer with low cross-link density (*light gray*). When exposed to water, the latter turned into a hydrogel, increasing by about 50% in linear size. The elastic modulus of the plastic was 2 GPa, whereas the elastic modulus of the hydrophilic polymer was 40 MPa when dry and 5 MPa when fully hydrated. With a 3D printing resolution of 85 μm in the horizontal plane and 30 μm in the vertical direction, Raviv et al. combined these two materials to print stretching and folding primitives (Fig. 2.3) and objects that responded as planned when immersed in water (Raviv et al., 2014).

The primitives depicted in Fig. 2.3 react predictably to being submerged in water. The linear stretching primitive (Fig. 2.3A) changes its length over time in contact with water because the hydrophilic polymer disks expand from an initial thickness of 0.81−1.25 mm. The relative elongation of the entire construct can be controlled by adjusting the proportion of expanding disks within its structure. If the initial length of the construct is $L_0 = l_0 + l_1$, where l_0 is the length of the portion made of expandable disks and l_1 is the length of the part made of plastic disks, the final length can be expressed as $L = (1 + P_e/100)\, l_0 + l_1$, where P_e is the percent linear expansion of the hydrophilic material when exposed to water ($P_e = 53\%$ in the case of the polymer employed in (Raviv et al., 2014)). Then, the relative elongation of the linear stretching primitive, expressed in percentage, $100\% \times (L - L_0)/L_0$, is given by $P_e/(1 + l_1/l_0)$.

In the ring stretching primitive (Fig. 2.3B), the inner half-rings are made of expanding material, whereas the outer ones are made of plastic. When soaked in water, the inner arcs expand, causing an increase in the radius of curvature of both half-

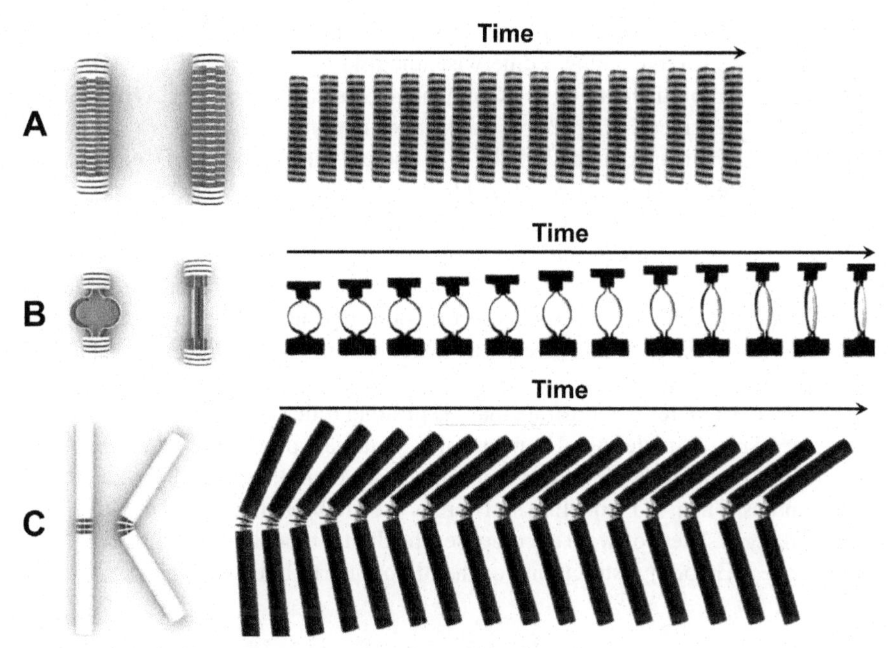

FIGURE 2.3 4D-printed stretching and folding primitives.

Stretching primitives (A and B) and folding primitives (C), depicted as models in the initial state (*left*), models in the final state (*middle*), and as 4D-printed physical objects (*right*), shown at various stages of spontaneous evolution triggered by immersion in water.
Modified from Raviv, D., Zhao, W., McKnelly, C., Papadopoulou, A., Kadambi, A., Shi, B., Hirsch, S., Dikovsky, D., Zyracki, M., Olguin, C., Raskar, R., & Tibbits, S. (2014). Active printed materials for complex self-evolving deformations. Scientific Reports, 4, *7422. https://doi.org/10.1038/srep07422.*

rings (i.e., they gradually approach the fully extended conformation of two parallel bars). The initial length of this construct is $L_0 = 2R + l_1$ where R is the radius of curvature of the dry half-rings, whereas l_1 is the length of the rest of the construct, which does not change as a result of hydration. If the expandable material would have $P_e = (\pi/2 - 1) \times 100\%$, i.e., $P_e \approx 57\%$, the final length of the construct would be $L = \pi R + l_1$, and the relative elongation of the ring stretching primitive would be $100\% \times (\pi/2 - 1)/[1 + l_1/(2R)]$. The slightly smaller value of P_e in the case of the material used by Raviv et al. explains why the fully hydrated half rings did not become precisely straight (Fig. 2.3B, *right*).

The folding primitive includes a strip of expanding material printed over a strip of plastic (Raviv et al., 2014). In the example of Fig. 2.3C, these strips are straight (when dry) and parallel to the vertical plane that is perpendicular to the figure. The expandable strip is on the left, causing the primitive to fold toward the right. Printing the strips the other way around would reverse the folding direction. Making the strips parallel to the plane of the figure would cause the joint to fold toward or away from the reader, depending on the material order. The folding angle is limited by the physical contact between the rigid disks printed perpendicularly to the strips (provided that P_e is large enough; otherwise, the folding angle is limited by the capacity of the hydrophilic material to expand in the presence of water). A typical joint geometry is represented in Fig. 2.4 in the state of maximal folding.

Consider a joint that contains N disks of diameter D and thickness d (see Fig. 2.4; $N = 2$), connected by strips of equal length L_1 (Raviv et al., 2014). The maximum folding angle of such a joint is given by $\alpha = (N + 1)\alpha_1$, where $\alpha_1 = L_1/(0.5\, D)$ is the angle between two disks that touch each other in one point. Hence, the largest folding angle can be controlled precisely by adjusting the ratio L_1/D and the number of disks per joint. The internal radius of the joint depends on the thickness of the disks, $r = 0.5d/\tan(\alpha_1/2)$, whereas the external one is well approximated by $R \cong r + D$.

The above examples demonstrate that 4D printing relies on at least one stimulus-responsive (smart) material. Knowing the material properties, one employs mathematical modeling to design an object that evolves as expected under the action of the stimulus. Then, 3D modeling software is used to build the digital model of the object. Finally, a slicer software decomposes the digital model into a set of 2D slices and generates (G-code) instructions for the 3D printer to build the object.

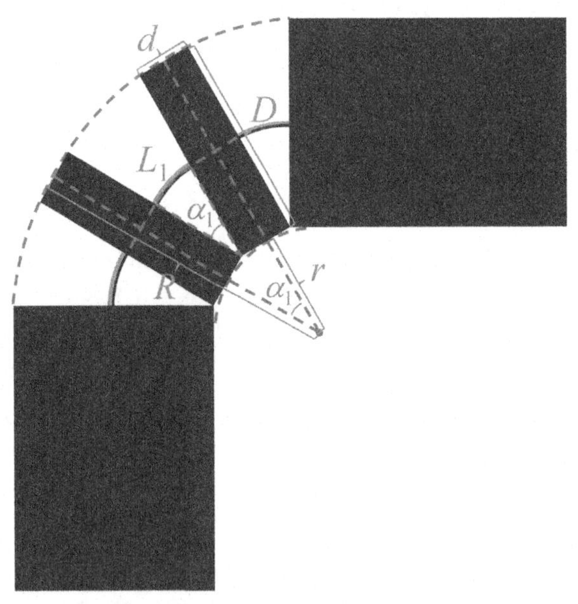

FIGURE 2.4 The geometry of a folding primitive.

The geometric quantities that control the maximum folding angle of a joint. The identical strips that join the bars and disks comprise two layers: one that expands when placed in water (*light gray*) and another one made of plastic (*dark gray*); bars and disks are also made of plastic.

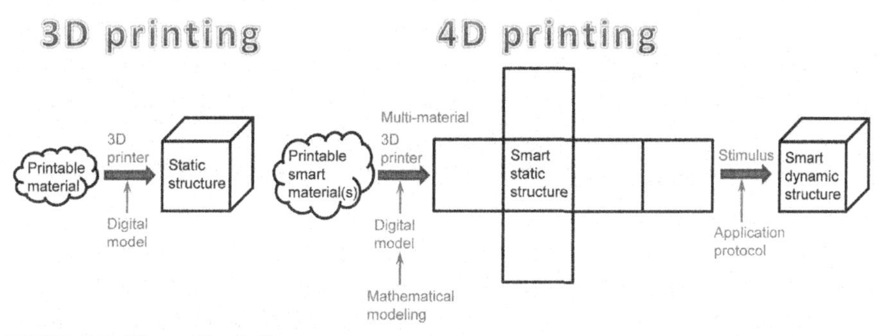

FIGURE 2.5 3D vs. 4D printing.

Diagram of the essential elements of 3D and 4D printing.

The diagrams from Fig. 2.5 put the main features of 3D printing and 4D printing in contrast. In 3D printing, a computer-controlled device (3D printer) manipulates a material of suitable physicochemical properties to turn a digital model into a physical structure. Although there are commercially available and relatively affordable, 3D printers that are able to deliver several materials in a given printing process (multimaterial 3D printers), most applications of 3D printing rely on single-material printers. A 3D-printed object might suffer slight changes in shape and size while the delivered material hardens. These changes, however, are unintentional; they are taken into account during the design phase to assure that the final product has the desired static geometry.

In 4D printing, usually, a multimaterial 3D printer is used to build an object that incorporates at least one smart material (Momeni et al., 2017). The result is a static structure capable of responding to an environmental change (i.e., it is smart). To make the response predictable, the results of mathematical and/or computational modeling are taken into account while building the digital model of the object. The latter specifies the architecture of the object and the material distribution within its bulk, whereas the preliminary step of theoretical modeling establishes the connection between the desired final shape, material properties, and stimulus properties. Once a stimulus is applied, the structure becomes dynamic, gradually adopting the expected shape, property, or functionality. For certain materials, however, the desired evolution of the printed structure is not achieved by simply exposing it to the stimulus. Instead, the stimulus needs to be applied according to a precise protocol (e.g., heat followed by light), or the printed object needs to be conditioned by applying mechanical load and changing the temperature, known as thermomechanical programming. Examples of this kind will be discussed in the next section, dedicated to smart materials.

In conclusion, a 4D-printed object responds to a predetermined stimulus in a programmed fashion, dictated by the geometry of the initial (3D printed) conformation of the object and by the material properties of the voxels that compose it. Hence, a

4D-printed object is physically programmed to adopt a certain form and/or function as a result of a change in the environment.

2. Stimulus-responsive materials developed for 4D printing

Although the examples analyzed in the previous section were chosen to clarify the definition of 4D printing, they also illustrate the need for 3D printable materials that are capable to respond to diverse stimuli. To date, most 4D printing applications took advantage of two classes of smart materials: shape memory polymers and hydrogels.

2.1 Shape memory polymers

Shape memory is a material property. An object made of a shape memory material displays the shape memory effect: it recovers its original size and shape when heated above a certain phase transformation temperature, known as switching temperature, T_{SW}—a glass transition temperature or melting temperature.

To be more specific, let us look into how has the shape been altered. First, a mechanical load was applied at a temperature higher than T_{SW}. Then, the temperature was lowered below T_{SW} while keeping the mechanical load on. Finally, as the load was removed, the object preserved its altered shape. This three-step procedure (load-cool-unload), called thermomechanical programming, renders the object stimulus-responsive (Qi et al., 2008). The object preserves its temporary shape until it is stimulated by raising the temperature above T_{SW} when it recovers its permanent shape (Fig. 2.6). Shape memory materials allow for repeated programming-recovery cycles.

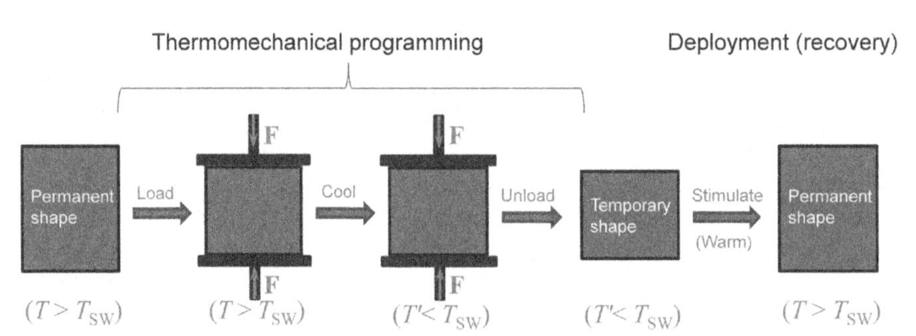

FIGURE 2.6 Thermomechanical programming and deployment of objects made of shape memory materials.

An object made of a shape memory material can be programmed by applying a mechanical load (exerting a force, F), while its temperature is above the switching temperature (T_{sw}—a material constant), lowering its temperature below T_{sw}, and unloading. The structure preserves its deformed (temporary) shape until its temperature remains below T_{sw}. As soon as the temperature becomes larger than T_{sw}, the structure regains its permanent shape.

It might seem strange that inanimate matter has memory, but it is not as mystical as it sounds. Imagine yourself on a cold winter afternoon, building a stack out of muddy hay. The permanent shape of such a construct will be the familiar dome-shape of a haystack. In the evening, you place a heavy crate on top of it. The temperature drops below freezing during the night. When you remove the crate in the morning, the top of the stack will remain flat—this is the temporary shape of the stack. During the day, as the temperature climbs above freezing, the mud in the stack will gradually melt, and the stack regains its permanent shape. This is a rather crude example—while the ability of a frozen muddy haystack to maintain its temporary shape is quite good (i.e., it has a 100% shape fixity), its shape recovery upon unfreezing will be incomplete. Nevertheless, this example illustrates a general feature of shape memory materials: they comprise two components. One of them (the set of entangled strands of grass) is responsible for remembering the permanent shape; the other (the mud) fends for retaining the temporary shape.

Common shape memory materials are alloys, ceramics, and polymers. Shape memory polymers (SMPs) elicited the most interest in the field of 4D printing presumably because they are affordable, easy to program, and easy to deploy; moreover, they have low density and high strain recovery (Akbari et al., 2019). Although the first SMPs were discovered by the mid-twentieth century, the SMP research field became truly vivid during the last two decades, leading to a wealth of applications and elucidating the design principles of SMPs.

Generally, an SMP comprises two different phases (Hager et al., 2015): a stable network that enables the material to retain its original shape, and another phase that fixes the temporary shape via crystallization, glass transition, switching between two liquid crystalline phases, or formation of reversible bonds of chemical or physical nature (Fig. 2.7).

The deformation of the hard phase (permanent network) is the driving force of shape recovery, whereas the reversible bonds that form in the switch phase at low temperatures, $T < T_{SW}$, assure the fixity of the temporary shape.

The stiffness of the hard phase is described by the so-called *rubbery modulus* of the SMP, defined as the Young modulus, E_r, of the SMP at $T > T_{SW}$. Another important property of an SMP is its failure strain, ε_f, measured by uniaxial stretching of a stripe made of the given material; it is defined as the ratio of the deformed sample length at failure and the undeformed sample length (multiplied by 100% if one wishes to express the strain as a percentage).

Shape memory behavior can be characterized by quantitative parameters. To define them, consider again the three steps of thermomechanical programming depicted schematically in Fig. 2.6. The first step (load) consists in a deformation of the hot sample, imposing a preliminary strain or prestrain, ϵ_m = *(intial sample height - deformed sample height) / (intial sample height)*. Here, the index *m* reminds us that this strain is going to be maintained during the next step. The second step (cool) is a strain storage procedure in which the temperature of the sample is lowered below T_{SW} while keeping ε_m constant. Due to thermal contraction, the stress needed

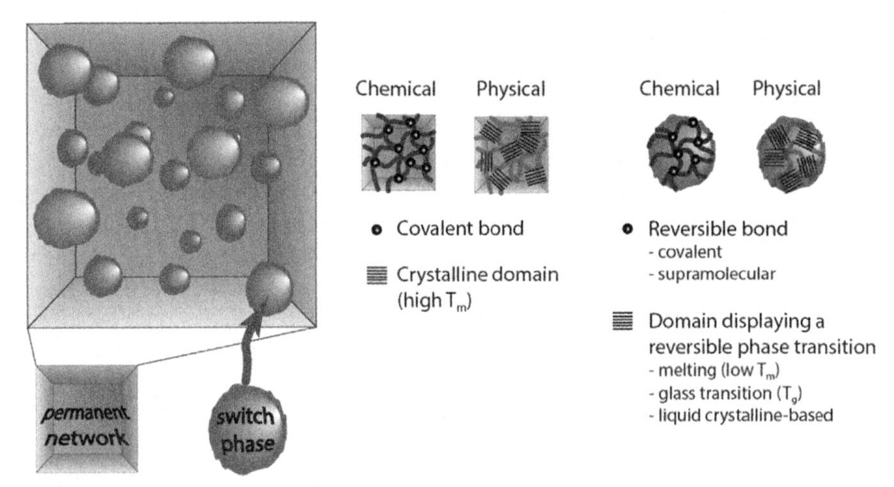

FIGURE 2.7 Generic structure of a shape memory polymer.

A shape memory polymer is composed of a permanent network stabilized by irreversible bonds (as long as the material's temperature is below the melting temperature of the permanent network) and a switch phase that fixes the temporary shape by reversible bonds (as long as the temperature remains below the phase transition temperature of the switch phase).

Reprinted from Hager, M. D., Bode, S., Weber, C., & Schubert, U. S. (2015). Shape memory polymers: Past, present and future developments. Progress in Polymer Science, 49–50, *3–33. https://doi.org/10.1016/j.progpolymsci.2015.04.002, with permission from Elsevier.*

to maintain the desired prestrain decreases, while the sample is cooled down. The third step is a low-temperature unloading process, often accompanied by an elastic rebound ("springback") of the sample (G. Li & Wang, 2016). Because of the springback, the strain of the unloaded sample, ε_u, is slightly smaller than the prestrain, ε_m; their ratio,

$$R_f = \frac{\varepsilon_u}{\varepsilon_m} \tag{2.1}$$

The so-called *shape fixity ratio* characterizes the ability of the switch phase to maintain the deformation imposed during the first step of programming. If the elastic rebound is negligible, $R_f = 1$ or 100%.

Shape recovery is achieved by reheating the sample to a temperature above T_{SW} in the absence of any constraint. This process, also known as free strain recovery, ideally brings the sample back to its initial shape. For many materials, a residual, permanent strain is observed upon reheating: $\epsilon_p =$ *(reheated sample height - intial sample height) / (intial sample height)*. The capability of the material to regain its original shape is characterized by the shape recovery ratio,

$$R_r = \frac{\varepsilon_m - \varepsilon_p}{\varepsilon_m} \tag{2.2}$$

When a polymer block undergoes several thermomechanical programming-reheating cycles, the shape fixity ratio is given by $R_f = \varepsilon_u(N)/\varepsilon_m$, where $\varepsilon_u(N)$ is the strain observed after the unloading step of the N-th cycle.

The shape recovery ratio in the N-th cycle is defined as follows (Li & Wang, 2016):

$$R_r(N) = \frac{\varepsilon_m - \varepsilon_p(N)}{\varepsilon_m - \varepsilon_p(N-1)} \tag{2.3}$$

Phase transitions undergone by an SMP are represented schematically in Fig. 2.8.

SMP fabrication involves an initial processing stage (e.g., extrusion), which takes place at high temperatures—above the melting temperature of the hard phase (high T_m). As the temperature drops below high T_m, the structure assumes its permanent shape Fig. 2.8.

SMPs are classified according to the nature of the phase transition undergone by the soft phase at the switching temperature. The melting transition occurs in semi-crystalline polymers, in physically crosslinked polymers and in chemically cross-linked rubbers. The glass transition is observed in physically crosslinked thermoplastics and in chemically crosslinked thermosets (Hager et al., 2015). Let us take a closer look at a few examples.

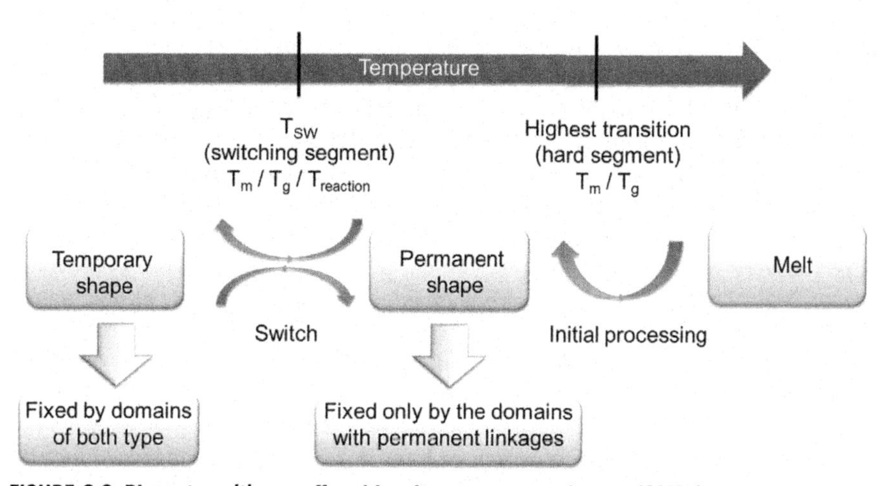

FIGURE 2.8 Phase transitions suffered by shape memory polymers (SMPs).

As the temperature drops below the melting temperature (T_m) or glass transition temperature (T_g) of the hard phase, the permanent shape of the SMP structure emerges. Thermomechanical programming establishes the temporary shape; the latter switches back to the permanent shape as soon as the temperature rises above T_{sw}.

Reprinted from Hager, M. D., Bode, S., Weber, C., & Schubert, U. S. (2015). Shape memory polymers: Past, present and future developments. Progress in Polymer Science, 49–50, *3–33. https://doi.org/10.1016/j.progpolymsci.2015.04.002, with permission from Elsevier.*

In **melting-transition-based SMPs**, the fixity of the temporary shape is assured by the crystallization of the soft phase. Such materials are relatively stiff, and their shape recovery is fast. Examples include polyolefins, polyethers, and polyesters.

Polyolefins, such as low-density polyethylene or an ethylene-1-octene copolymer, have melting temperatures between 60 and 100°C, tunable by adjusting the degree of branching and the crosslinking density. Ultra-high–molecular-weight polyurethane featured a melting temperature of about 150 °C (Hager et al., 2015).

Polyethers have elicited much interest because of their relatively low switching temperatures. For example, ethylene oxide—ethylene terephthalate segmented copolymers have shape recovery temperatures ranging from 44° C to 55 °C. These values were found to depend on the hard phase (poly(ethylene terephthalate)) content (16%–32%) as well as on the molecular weight, M_w, of the soft phase (poly(ethylene oxide)) (4000–10,000 Da). For a given M_w, the smaller the hard phase content the higher the T_m; for a given hard phase content, the larger M_w the higher T_m (Luo et al., 1997). Indeed, long soft segments occupying a high fraction of the copolymer favor the crystallization of the soft phase, thereby rising the shape recovery temperature. Poly(ethylene oxide) served as soft phase in several types of SMPs with hard segments made of polyurethanes, polymethacrylate, or poly(p-dioxanone)-poly(ethylene glycol) (Hager et al., 2015).

Polyesters are widely used materials for the soft phase of SMPs. Among them, a popular choice is poly(ε-caprolactone) or polycaprolactone (PCL), which enabled the fabrication of SMPs whose switching temperatures lie in the range of physiological temperatures. Moreover, PCL is biocompatible, and often used as scaffold material in tissue engineering. It is also biodegradable, making SMPs with PCL content particularly suitable for biological applications. PCL is the soft block in a variety of SMPs, combined with hard phase materials such as polyurethane, polyamides, and polyaramides (Hager et al., 2015). For example, affordable and versatile SMPs resulted from blending PCL with an elastomer—the triblock copolymer poly(styrene-butadiene-styrene), SBS. This material displayed excellent shape fixity and shape recovery when the PCL content (weight percent concentration) ranged from 30% to 70% (Zhang et al., 2009). AFM imaging demonstrated that, in this range of concentrations, both the hard and the soft phases were continuous. SBS and PCL are immiscible. At low concentrations, PCL is present in the form of droplets embedded in the SBS matrix (Fig. 2.9A). Therefore, it is unable to fix the temporary shape.

Starting from a concentration of about 30%, PCL forms a continuous phase. It crystallizes as the slab is cooled down, opposing the relaxation of SBS; thereby, it maintains the temporary shape—i.e., assures shape fixity (Fig. 2.9B). As the PCL concentration exceeds 50%, it becomes the matrix, but the SBS phase remains continuous up to about 70% PCL, ensuring the recovery of the permanent shape as soon as the temperature exceeds the melting temperature of PCL (Fig. 2.9C). At even higher PCL concentrations, at about 90% PCL, the SBS phase becomes disconnected, and shape recovery is impaired (Fig. 2.9D).

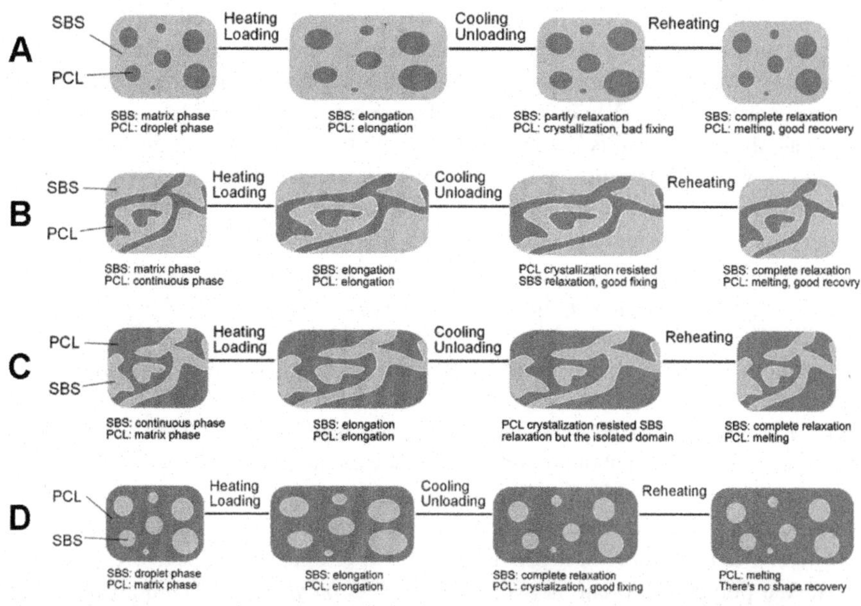

FIGURE 2.9 Shape recovery and shape fixity vs. shape memory polymer (SMP) composition.

The microstructure, thermomechanical programming, and shape recovery of a slab of SBS/polycaprolactone (PCL) blend for different weight percent concentrations of PCL: 20% (A), 30% (B), 70% (C), and 90% (D).

Reprinted from Zhang, H., Wang, H., Zhong, W., & Du, Q. (2009). A novel type of shape memory polymer blend and the shape memory mechanism. Polymer, 50*(6), 1596–1601. https://doi.org/10.1016/j.polymer.2009.01.011, with permission from Elsevier.*

The switching temperature of the SBS/PCL blend ranges between 56 and 58 °C—roughly equal to the melting temperature of PCL. It does not change much with PCL content, presumably due to the immiscibility of SBS and PCL (Zhang et al., 2009).

Glass-transition-based SMPs contain amorphous polymeric materials as soft units. The chains of such polymers are mobile at high temperatures and, therefore, do not resist deformation. As an amorphous polymer is cooled down, in a certain range of temperatures, it gradually adopts a glassy state characterized by limited chain mobility. The glass transition temperature, T_g, is the center of the relatively broad interval of temperatures (of the order of 10 °C) along, which the glass transition occurs. Since glass transition is a second-order phase transition, T_g-based SMPs display a slower shape recovery than T_m-based SMPs. This feature, however, does not hamper biomedical applications: slow shape recovery is desirable, for instance, in orthodontics, and in clinical applications based on self-deploying medical devices that assume their functional shape upon being inserted into the patient's organism (Hu et al., 2012).

Owing to the viscoplasticity of amorphous polymers, in T_g-based SMPs, the temporary shape can also be established by deforming the material within the glass transition zone (warm programming) or even below it (cold programming). While conventional thermomechanical (hot) programming enables one to easily set the temporary shape and results in excellent shape fixity, it takes more time and energy than cold or warm programming. Experiments indicate that free shape recovery does not depend much on the programming temperature, but stress recovery does. Stress recovery is observed when the sample is warmed up above the glass transition zone, while shape recovery is not allowed. Differences between stress recovery patterns originate from different driving forces and differences in the extent of material damage inflicted during programming (Li & Wang, 2016).

Biocompatible SMPs have been synthesized from polyurethane outfitted with amorphous copolyester soft segments made of oligo(*rac*-lactide)-co-glycolide. With a T_g ranging from 50 to 65 °C, the samples made of this material recovered their permanent shape in about 5 min when they were maintained at 70 °C. Composite materials made of polyesters—poly(mannitol sebacate)—and cellulose nanocrystals displayed a shape memory effect with excellent shape fixity and recovery at switching temperatures between 15 and 45 °C (Hager et al., 2015). Covalently crosslinked polymethacrylate networks, as well as poly(glycerol-co-dodecanoate), were also used to synthesize T_g-based biocompatible SMPs (Hu et al., 2012). Photo-curable, methacrylate-based copolymer networks demonstrated tunable thermomechanical properties (rubbery modulus from 1 to 100 MPa, T_g from −50 to 180 °C) depending on the chemical nature and concentrations of their constituents. More importantly, their large failure strain (of up to 300%) enabled ample deformations during programming-recovery cycles. These materials were prepared from benzyl methacrylate (BMA) as a linear chain builder and di(ethylene glycol) dimethacrylate or poly(ethylene glycol) dimethacrylate (PEGDMA) or bisphenol A ethoxylate dimethacrylate as cross-linkers (Ge et al., 2016). One of these materials is especially promising from the point of view of biomedical applications: mixing BMA with PEGDMA in a ratio of 70:30, and adding a photoinitiator, 2,4,6-trimethylbenzoyl diphenylphosphine oxide (TPO) nanoparticles in a concentration of 5% by weight, as well as a photo absorber, Sudan I, in a concentration of 0.05% by weight, resulted in a UV curable solution that could be turned into SMP constructs via digital light processing (DLP) printing. Importantly, the T_g of this methacrylate-based SMP was 30 °C, so the constructs built from this material could be programmed into a compact shape for delivery and recovered their permanent shape once they were warmed up to body temperature (Ge et al., 2021).

In certain T_g-based SMPs, shape recovery can also be triggered by immersion in solvents. For example, in an ether-based polyurethane SMP, the adsorbed water weakens the hydrogen bonding between N—H and CO groups. As a result, the glass transition temperature decreases, triggering shape recovery at room temperature. Solvent-induced shape recovery was observed also in poly(vinyl alcohol)-based materials in the presence of several organic solvents. Nevertheless, the shape recovery of objects made of solvent-sensitive SMPs depends on their size, and the recovery stress is relatively small (Hu et al., 2012).

The SMPs discussed so far have one major drawback: the stimulus elicits the recovery of the permanent shape, and the user needs to conduct a new programming procedure if she/he wishes to get back the temporary shape. A more desirable behavior would be the so-called **two-way shape memory effect**, which amounts to adopting the permanent shape as long as the stimulus is on and rebuilding the temporary shape once the stimulus is turned off. Although still rare, such materials have been synthesized based on liquid crystal elastomers (LCEs). In such polymers, the transition from the anisotropic phase to the isotropic phase usually causes a contraction of the material, but it will expand again when the temperature drops below the transition temperature (Hager et al., 2015). Since LCE-based polymers only display volume changes, bilayer systems have been devised (e.g., a polystyrene layer combined with a nematic LCE layer) to achieve more practical shape changes such as reversible bending. A carefully designed polymer film, comprising actuator and passive domains, displayed complex, reversible shape changes (Agrawal et al., 2014).

Two-way SME has also been achieved in certain polymer systems maintained under constant stress. For example, when constant tensile stress was applied to a sample of cross-linked poly(cyclooctene), it elongated upon cooling and contracted upon heating. Similar behavior was observed also for PCL samples (Hager et al., 2015).

Linking two separate structural units at the molecular level enabled the fabrication of polymers that exhibited two-way SME even in free-standing samples (Behl et al., 2013). In this material, poly(ω-pentadecalactone) (PPD) segments act as the geometry determining (skeleton) domains, whereas PCL segments (75% by weight) play the role of actuator domains. One of the shapes (A) was programmed by imposing a deformation at 100 °C (larger than both $T_{mPPD} = 64$ °C and $T_{mPCL} = 34$ °C), cooling down the sample to 10 °C and subsequently warming it up to 50 °C (situated between the two melting temperatures). On cooling down to 10 °C, the sample shifted to shape B due to the crystallization of the oriented PCL domains. Heating to 50 °C resulted in shape A again, and the shape-shifting cycles could be repeated precisely by the corresponding changes in temperature (Behl et al., 2013).

2.2 Hydrogels

A gel consists of a 3D network of loosely cross-linked polymer filaments embedded in a liquid medium. The liquid prevents the network from collapsing as it keeps the filaments apart, whereas the network keeps the liquid from flowing away. The consistency of a gel depends on its chemical composition and cross-link density; typically, it lies between viscous fluids and quite rigid solids. Therefore, gels can be prepared to match the mechanical properties of most biological tissues. Laser scattering experiments and theoretical modeling have shed light on the interactions that determine a gel's physical properties. These explain why, under certain conditions, a gel can swell or shrink by orders of magnitude as a result of a slight change in temperature or liquid composition (Tanaka, 1981).

When a gel undergoes a phase transition, the process is characterized by the ratio between the final volume and the initial volume, known as the *swelling ratio*.

It is important to keep in mind, however, that a large swelling ratio does not mean that the corresponding process is fast. For instance, phase transitions triggered by changes in the composition of the liquid component of the gel are typically slow because it takes time for the new solvent to penetrate the gel by diffusion. For example, at 22 °C, a centimeter-sized poly(acrylamide) gel soaked in an acetone–water mixture undergoes a spectacular shrinking when the acetone concentration exceeds 42%, but the process takes several days. The time required to reach a new equilibrium volume is proportional to the square of the linear size of the gel. Thin layers or cylinders of gel whose smallest dimension is of the order of 1 μm would swell in milliseconds (Tanaka, 1981).

Hydrogels are gels whose liquid component is an aqueous solution. A vast body of literature demonstrates that biocompatible hydrogels can be synthesized with tunable physical and biochemical properties. They can mimic the natural extracellular matrix of biological cells, enabling fundamental research on cell-substrate interactions as well as a variety of applications in tissue engineering and regenerative medicine. When it comes to their use in 3D/4D printing, hydrogels are selected by also taking into account their processability, mechanical stability, gelation mechanism, and ability to respond to certain stimuli (Li et al., 2020).

Hydrogels can be prepared from natural polymers (biopolymers) or synthetic polymers. While natural polymers are typically bioactive and enable cell adhesion, they are limited in tunability. For example, they can be rendered stiffer by raising the polymer concentration, but this also implies an increase in the concentration of the bioactive molecules that reside on the polymer filaments. Synthetic polymers are more customizable; when properly tailored, they provide a well-controlled, biomimetic environment for live cells. The properties and applications of several classes of hydrogels are summarized in excellent review articles (Li et al., 2020).

Biomedical sciences pushed polymer research forward by their demand for anisotropic hydrogel-based materials of high mechanical strength. Such materials are needed, for example, in the engineering of load-bearing tissue constructs, such as cartilage, tendon, and bone. To meet such demands, composite hydrogels have been developed. One strategy was to prepare blends of two polymers, natural or synthetic. As a result, hydrogels emerged that incorporate interpenetrating polymer networks. Some of them are prepared sequentially (i.e., the second network is polymerized and cross-linked in the presence of a swollen single-network hydrogel), and others are synthesized simultaneously from a mixture of monomers and cross-linking agents of both networks. A remarkable subclass of sequentially prepared interpenetrating polymer network hydrogels comprises the mechanically reinforced double-network hydrogels. Despite their high water content (of about 90% by weight), they are tough and wear-resistant. Their first network, a heavily cross-linked ionic hydrogel, is the minor component that serves as a rigid, brittle skeleton;

it is entangled with the second network, a loosely cross-linked neutral hydrogel—the soft and ductile major component (Li et al., 2020).

Ge and co-workers employed multimaterial 3D printing, via the DLP technique, to fabricate hybrid structures made of a highly stretchable hydrogel covalently bonded with a variety of UV curable polymers, including elastomers, rigid polymers, and methacrylate-based SMPs (Ge et al., 2021). For this purpose, these investigators synthesized an acrylamide-PEGDA (AP) hydrogel of high water content, which could be stretched by an order of magnitude. To prepare the AP hydrogel precursor solution, acrylamide powder was mixed with PEGDA at a ratio of 100:0.625. Then, photoinitiator nanoparticles made of 2,4,6-trimethylbenzoyl diphenylphosphine oxide (TPO) were added in a concentration of 5% by weight. Finally, the entire mixture was dissolved in water until water represented about 80% by weight. The key component of this precursor solution is the water-soluble TPO photoinitiator: it renders the solution UV curable and, thereby, suitable for DLP printing. Since commercially available TPO powders are hydrophobic, Ge et al. devised a procedure to enclose them into surfactant micelles along with a crystallization inhibitor, polyvinylpyrrolidone; these micelles are water-soluble nanoparticles because they expose their hydrophilic head group toward the aqueous medium (Ge et al., 2021).

Nanocomposites made of organic macromolecules (protein filaments) and inorganic nano/microparticles (minerals) are ubiquitous components of hard tissues. Attempting to replicate the properties of certain natural materials, nanocomposite hydrogels have been prepared by incorporating inorganic nanoparticles into hydrogels. As shown in Fig. 2.10, the uniform distribution of nanoparticles within the gel matrix was achieved via five different strategies (Thoniyot et al., 2015).

Nanocomposite hydrogels have been prepared using a wealth of nanoparticles (NPs), including mineral NPs (hydroxyapatite, laponite), metal NPs (Au, Ag), silica NPs (mesoporous silica), carbon NPs (graphene, graphene oxide), magnetic NPs (iron oxide), and polymer NPs (nanocellulose)—see (Li et al., 2020) for a review of their properties and biomedical applications.

In terms of versatility and functionality, nanocomposite hydrogels go far beyond the biological structures that inspired their fabrication. Besides satisfactory mechanical properties, nanocomposite hydrogels developed so far exhibit sensitivity to diverse stimuli (optical, electrical, magnetic).

The next section will present the uses of smart materials in practical applications of 4D printing.

3. Applications of 4D printing

While the literature on 4D printing is rapidly expanding, any selection of research findings is subjective and results in incomplete coverage of the topic. Hence, this section aims at illustrating the potential of 4D printing rather than providing a comprehensive account of the landscape of applicative research achievements. The previous sections of this chapter attempted to provide the background

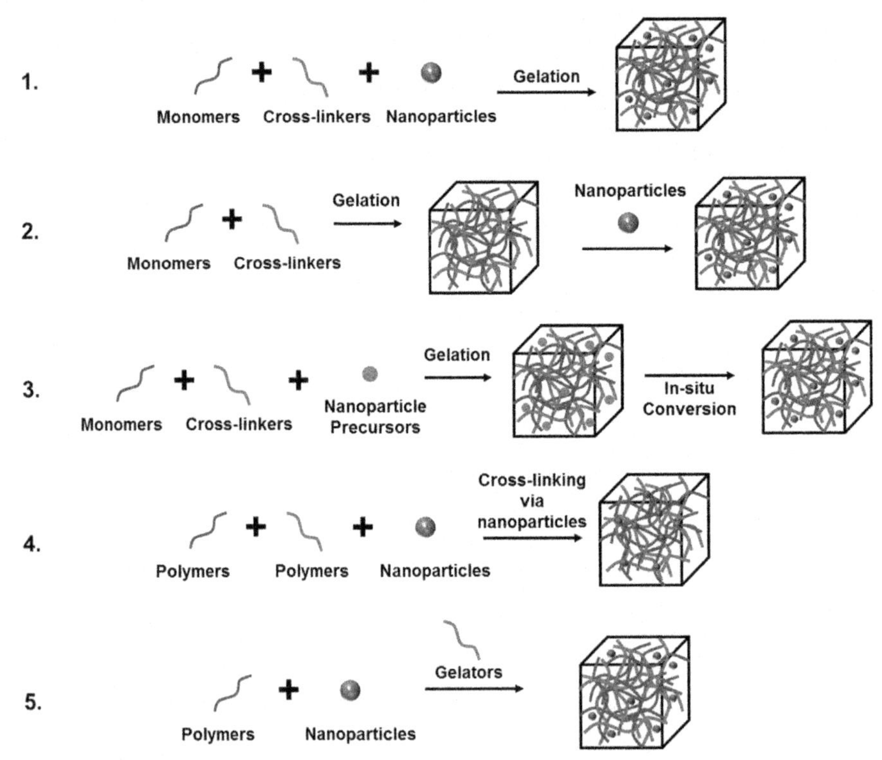

FIGURE 2.10 Fabrication strategies for obtaining composite hydrogels with uniform distribution of nanoparticles.

Hydrogel synthesis within the nanoparticle suspension (1), mechanical incorporation of nanoparticles into the preformed hydrogel (2), nanoparticle synthesis within the hydrogel (3), cross-linking of polymer filaments by nanoparticles (4), and hydrogel formation using polymers, nanoparticles, and gelators (5).

From Thoniyot, P., Tan, M. J., Karim, A. A., Young, D. J., & Loh, X. J. (2015). Nanoparticle–hydrogel composites: Concept, design, and applications of these promising, multi-functional materials. Advanced Science, *2(1–2), 1400010. Reprinted under the terms of the Creative Commons Attribution 4.0 International License (https://creativecommons.org/licenses/by/4.0/legalcode). https://doi.org/10.1002/advs.201400010.*

knowledge needed for understanding the primary literature (original papers); this section seeks to inspire the reader with the success stories of 4D printing research.

3.1 Endoluminal devices

Deployable medical devices are one class of products that might benefit from the autonomous reshaping ability inherent to 4D printing. They can be inserted in body cavities in a compact, temporary shape. This can be done noninvasively or via minimally invasive surgery. Then, in the target area, they adopt their functional shape under the action of stimuli provided by the host organism (such as heat or moisture) or generated externally (e.g., by an electromagnetic field).

Using high-resolution projection microstereolithography (PμSL), Ge et al. printed cardiovascular stent-like structures from methacrylate-based SMPs (Q. Ge et al., 2016). Stent diameters ranged between 2 and 5 mm and their microarchitecture was not limited by geometric complexity. One such structure is depicted in Fig. 2.11.

Finite element modeling of the strain field generated during thermomechanical programming is an essential component of the feasibility analysis because it avoids trial and error experiments. In the case of the stent model shown in Fig. 2.11, the color-coded image of the strain field depicted on the right shows that the strain did not exceed 50%, which is well below the failure strain of the methacrylate-based copolymer material the stent was made of (Ge et al., 2016). Fig. 2.12 shows a more challenging programming procedure: the straightening of a spring.

As revealed by the finite element modeling (Fig. 2.12C), the complete stretching of the spring involves local strains of up to 100% along a helicoidal line that runs

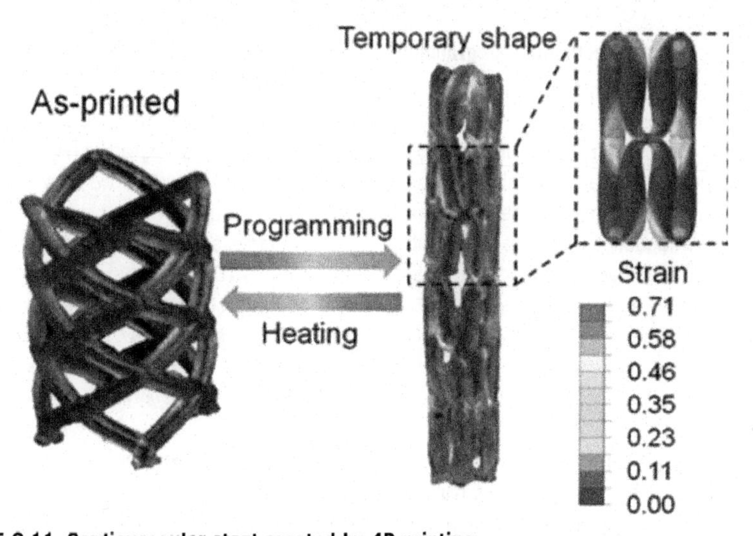

FIGURE 2.11 Cardiovascular stent created by 4D printing.

The permanent shape of the stent is the 3D-printed structure (*left*). Hot programming established the temporary shape that is smaller in diameter (*middle*), creating an inhomogeneous strain distribution within the material, as visualized by finite element modeling (*right*). Modeling was performed using ABAQUS (Simulia, Providence, RI, USA).

From Ge, Q., Sakhaei, A. H., Lee, H., Dunn, C. K., Fang, N. X., & Dunn, M. L. (2016). Multimaterial 4D printing with tailorable shape memory polymers. Scientific Reports, 6, *31110. https://doi.org/10.1038/srep31110.*

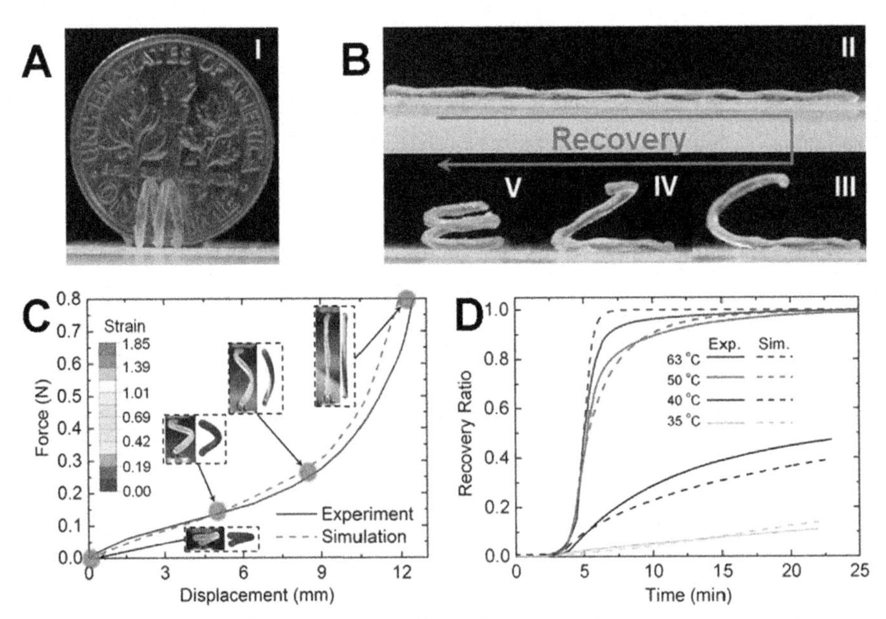

FIGURE 2.12 Thermomechanical programming and recovery of a shape memory spring.

The permanent shape of the spring (A) was obtained via 3D printing. Hot programming gave rise to the temporary shape (B, II), which gradually approached the permanent shape as the temperature exceeded the material's glass transition temperature (B, III–V). The force needed to stretch the spring was measured experimentally (C, *solid curve*) and simulated by finite element modeling (C, *dashed line*); the simulation also provided the strain field that accompanied the deformation (C, *color-coded* snapshots) and the time course of shape recovery observed at different temperatures (D).

From Ge, Q., Sakhaei, A. H., Lee, H., Dunn, C. K., Fang, N. X., & Dunn, M. L. (2016). Multimaterial 4D printing with tailorable shape memory polymers. Scientific Reports, 6, *31110. https://doi.org/10.1038/srep31110.* *Reprinted under the terms of the Creative Commons Attribution 4.0 International License (https://creativecommons.org/licenses/by/4.0/legalcode).*

along the straight strand. Hence, the simulation indicates the smallest acceptable failure strain of the SMP needed for a given application.

Stretching a spring-shaped construct of shape memory poly(vinyl alcohol) (SM-PVA) proved to be essential for intravesical drug delivery. Melocchi et al. developed a proof of concept model of a 4D-printed indwelling device for prolonged release of drugs in the bladder (Melocchi, Uboldi, et al., 2019). The temporary shape (a straight strand) enables noninvasive deployment via a catheter. Once inserted in the bladder, the water-induced decrease in the material's T_g triggers the recovery of the permanent shape of the device (a spring), assuring its retention. The incorporated drugs are released progressively, and the shape memory polymer is eliminated as it is gradually dissolved in the urine.

Besides water-responsiveness, pharmaceutical-grade SM-PVA offers the advantage of printability via fused deposition modeling—a relatively affordable 3D printing technique.

The above strategy proved successful also in the fabrication of gastroretentive drug delivery systems (Melocchi, Uboldi, et al., 2019). This work evaluated prototypes of allopurinol-loaded SM-PVA devices whose permanent shape prevented their evacuation through the pylorus. Thermomechanical programming, accomplished with the help of specific templates, gave rise to compact temporary configurations suitable for oral administration. When placed in 0.1 HCl solution at physiological temperature, each prototype recovered its permanent shape within minutes. Under these conditions, gradual drug release was observed.

Vascular stent-like constructs have been fabricated also by direct ink writing of UV-cross-linkable poly(lactic acid) (c-PLA) loaded with iron oxide nanoparticles—see Fig. 2.13 (Wei et al., 2017). The composite ink was prepared from Fe_3O_4 nanoparticles, benzophenone, PLA, and dichloromethane (the solvent) in a weight ratio of 0.25:0.1:1:3, respectively. When the ink strand was delivered at a velocity of 0.4 mm/s, through a nozzle of 100 μm in inner diameter, by applying a pressure of 1.75 MPa, about half of the solvent evaporated within 5 min. The small diameter of the printed filament facilitated solvent evaporation, ensuring the rigidity of the printed structure. Nevertheless, solvent evaporation had a negligible effect on the size of the printed structure (the strand diameter shrunk by less than 3%).

The original shape (S_O) of the structure (A) was printed according to the digital model (B), resulting in a square grid of superimposed filaments (C) forming a stripe of 2 mm by 30 mm. The deformed shape (S_D) was obtained via thermomechanical programming (D), by manually winding the hot (80 °C) stripe on a wire 1 mm in diameter and subsequent cooling. Shape recovery within a restrictive tube was hypothesized to cause an increase in the spiral structure's diameter (E), and experiments confirmed this hypothesis (F). Hence, the construct works as an intravascular stent (H).

An interesting feature of the composite ink used in the fabrication of the stent construct shown in Fig. 2.13 is its magnetoresponsiveness. The iron oxide nanoparticles suspended in c-PLA enable the magnetic manipulation of any object made of this material. When placed in a nonuniform magnetic field, Fe_3O_4 nanoparticles experience a force that is proportional to the gradient of the magnetic field intensity. They are attracted to the region where the magnetic field is most intense. Hence, stents made of such a material can be guided by externally applied magnetic fields. Moreover, their shape recovery can be triggered externally, by exposing the patient to an alternating electromagnetic field with a frequency of 30 kHz. In such a field, the magnetic moments of the nanoparticles reorient periodically, causing heat dissipation—a phenomenon known as magnetic hyperthermia. Despite the relatively high glass transition temperature of the c-PLA/Fe_3O_4 composite ink ($T_g = 53$ °C), shape recovery might be elicited within the body because the stent's material is warmed up internally, and only a small part of the generated heat is expected to reach the surrounding tissue (Wei et al., 2017).

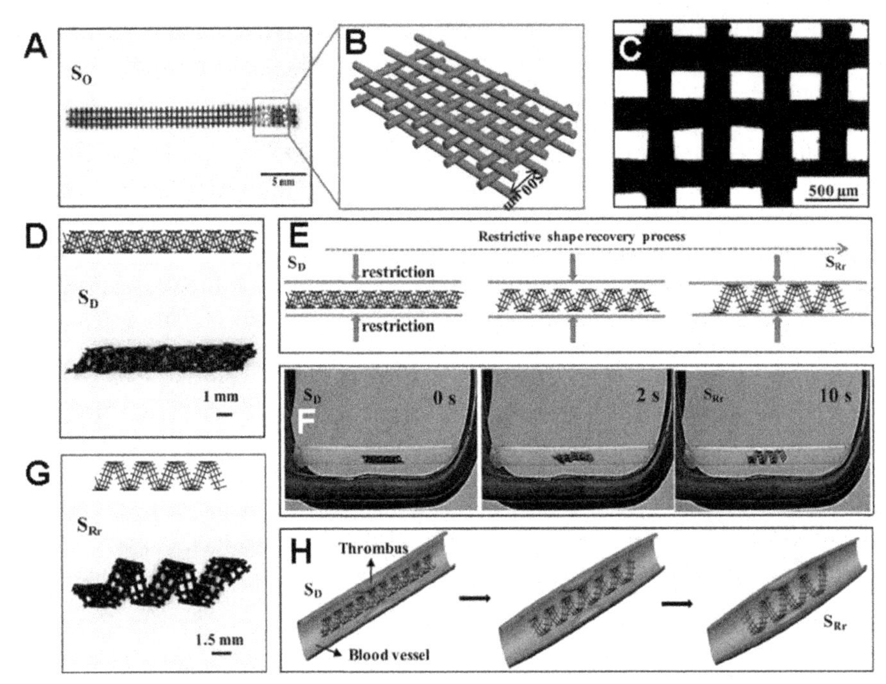

FIGURE 2.13 Direct ink writing of stent-like constructs.

The permanent shape of the construct, visualized by optical microscopy (A); the digital model of the construct (B); the zoomed-in top view of the construct; (C); the temporary shape of the construct (D), shown as a schematic drawing (*top*) and as an optical microscopy image (*bottom*); a schematic cartoon of the restricted shape recovery process (E); experimental snapshots of restricted shape recovery (F); and schematic representation of the potential use of the construct as an intravascular stent (H).

Reprinted with permission from Wei, H., Zhang, Q., Yao, Y., Liu, L., Liu, Y., & Leng, J. (2017). Direct-write fabrication of 4D active shape-changing structures based on a shape memory polymer and its nanocomposite. 9, 876–883. https://doi.org/10.1021/acsami.6b12824. Copyright 2017 American Chemical Society.

Lin et al. used shape memory PLA loaded with Fe_3O_4 nanoparticles (mPLA) to fabricate noninvasively deployable occluders for treating atrial septal defects (Fig. 2.14A). The occluder had an hourglass-shaped frame composed of 3, 4, or 6 arms made of mPLA, illustrated schematically in Fig. 2.14B. The frame was covered by thin occluding membranes made of shape memory PLA via electrospinning. Shown in Fig. 2.14C are photographs of actual 4D-printed occluders.

The occluder was thermomechanically programmed into a cylindrical temporary shape (Fig. 2.14A, right), suitable to be delivered through a catheter of 4.7 mm in diameter. When the frame contained at least 10 %wt magnetic nanoparticles, shape recovery could be induced by an alternating magnetic field of 4 kA/m strength and 27.5 kHz frequency. The 4D-printed occluders had optimal mechanical properties and good histocompatibility. They favored cell adhesion and proliferation, thereby

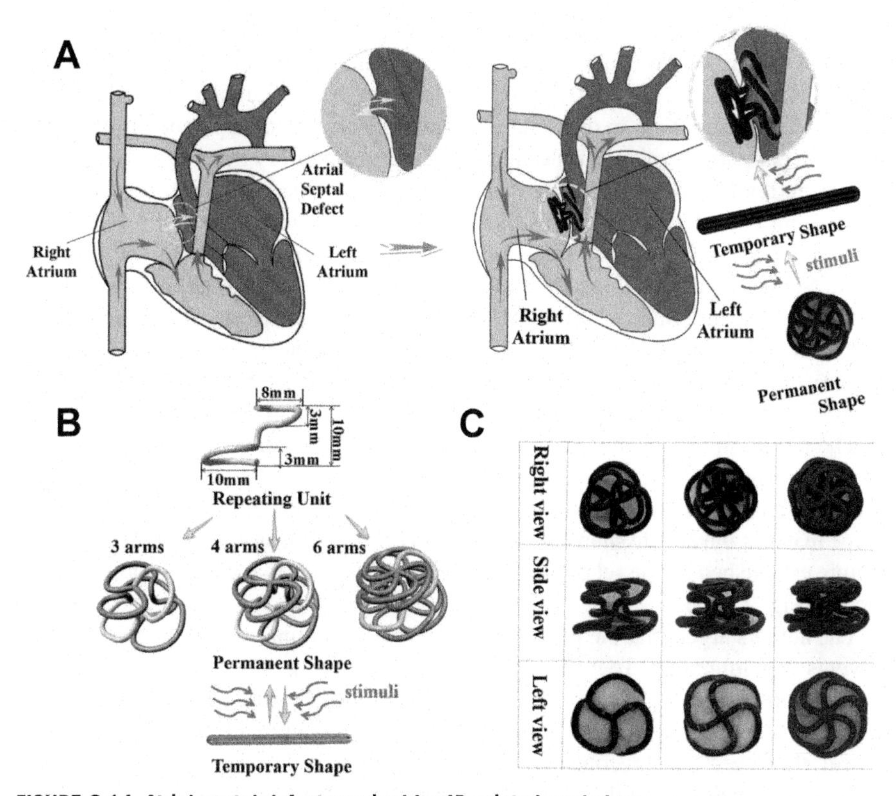

FIGURE 2.14 Atrial septal defect repaired by 4D-printed occluders.

(A) Schematic representation of an atrial sept defect before and after treatment (*left* and *right*, respectively); (B) schematic drawing of the permanent shapes of occluder frames with 3, 4, or 6 arms; the temporary shape consists of closely packed straight filaments (*bottom*); (C) pictures of the permanent shapes of occluders composed of electrospun shape memory polylactic acid (PLA) membranes (*white*) fastened onto 3D-printed mPLA frames (*black*).

Reprinted from Lin, C., Lv, J., Li, Y., Zhang, F., Li, J., Liu, Y., Liu, L. & Leng, J. (2019). 4D-printed biodegradable and remotely controllable shape memory occlusion devices. Advanced Functional Materials, 29, *1906569. https://doi.org/10.1002/adfm.201906569, with permission from John Wiley and Sons.*

promoting fast endothelialization. In vitro tests, performed on excised porcine hearts, demonstrated successful deployment within 16 s and effective occlusion (Lin et al., 2019). A similar strategy proved to be fruitful also for building left atrial appendage occluders (Lin et al., 2021).

Further research needs to address the sole weak point of the 4D-printed occluders developed so far: the relatively high glass transition temperature of the shape memory PLA-based magnetic nanocomposites, ranging from 63 (Lin et al., 2021) to 67 °C (Lin et al., 2019).

Zarek et al. proposed a strategy to fabricate personalized endoluminal devices via melt DLP printing of shape memory polymers and applied it to build a tracheal stent prototype (Zarek et al., 2017). The trachea is a flexible tubular structure about 12 cm in length and 2 cm in diameter (in adult males). When its patency is affected by certain pathologies, a stent is needed to keep the respiratory tract open. Nevertheless, the currently used metallic stents are only recommended as a last resort because they present considerable risks of failure as a result of stent migration or fracture. Besides the loss of function, stent failure is problematic because of the trauma caused by stent removal. Hence, there is a need for anatomically tailored, personalized tracheal stents. Such a prototype was printed from semicrystalline methacrylated PCL. The biodegradability of PCL is not a concern in this application because the lifetime of a PCL construct in the body is of the order of 2 years, whereas a tracheal stent is typically used for a few months (Zarek et al., 2017).

The digital model of the tracheal stent was obtained from MRI imaging. DLP printing, performed at 150 μm slice thickness, took about 6 h to produce a stent of 7 cm in length and wall thickness ranging between 1.5 and 3.5 mm. The stent prototype matches the arcade shape of the trachea (Fig. 2.15A), as well as its microarchitecture established by the set of parallel cartilaginous rings that confer rigidity to the trachea (Fig. 2.15B).

The authors argue that the precise fit to the ridged luminal surface of the trachea prevents stent migration. To further enhance stable fixation, one may scale the digital model to produce a slightly oversized stent (Zarek et al., 2017). These conjectures should be tested in long-term experiments conducted on animal models, and, if successful, they need to be reconfirmed by clinical trials.

An essential feature of the tracheal stent prototype shown in Fig. 2.15 stems from the use of 4D printing. The as-printed, permanent shape of the stent is the functional one (Fig. 2.15B). Deformation into a compact temporary shape was done at a temperature $T > T_m = 47$ °C and the construct was subsequently cooled to physiological temperature (Fig. 2.15C and D, leftmost pictures). The temporary shape enables nontraumatic stent deployment.

For a degree of methacrylation of 88%, the construct displayed a shape fixity ratio of 99%. It recovered its permanent shape within 14 s when the temperature was raised above T_m (Fig. 2.15C and D), matching the target tissue anatomy, with a shape recovery ratio of 98%. For the experimental prototype shown in Fig. 2.15B, shape recovery was achieved in a custom-made thermal chamber. Nevertheless, several options exist for the thermal actuation of endoluminal devices (Zarek et al., 2017), including resistive heating, exposure to infrared radiation generated by diode lasers, or magnetic hyperthermia. The latter, employed in the case of the stent shown in Fig. 2.13, is especially appealing, but DLP, a 3D printing technique based on photocuring, does not work with opaque materials.

Intraluminal stents would benefit from the development of materials with two-way shape memory behavior (see, e.g., (Behl et al., 2013)). Such a material would enable a thermally controlled, nontraumatic removal of the stent.

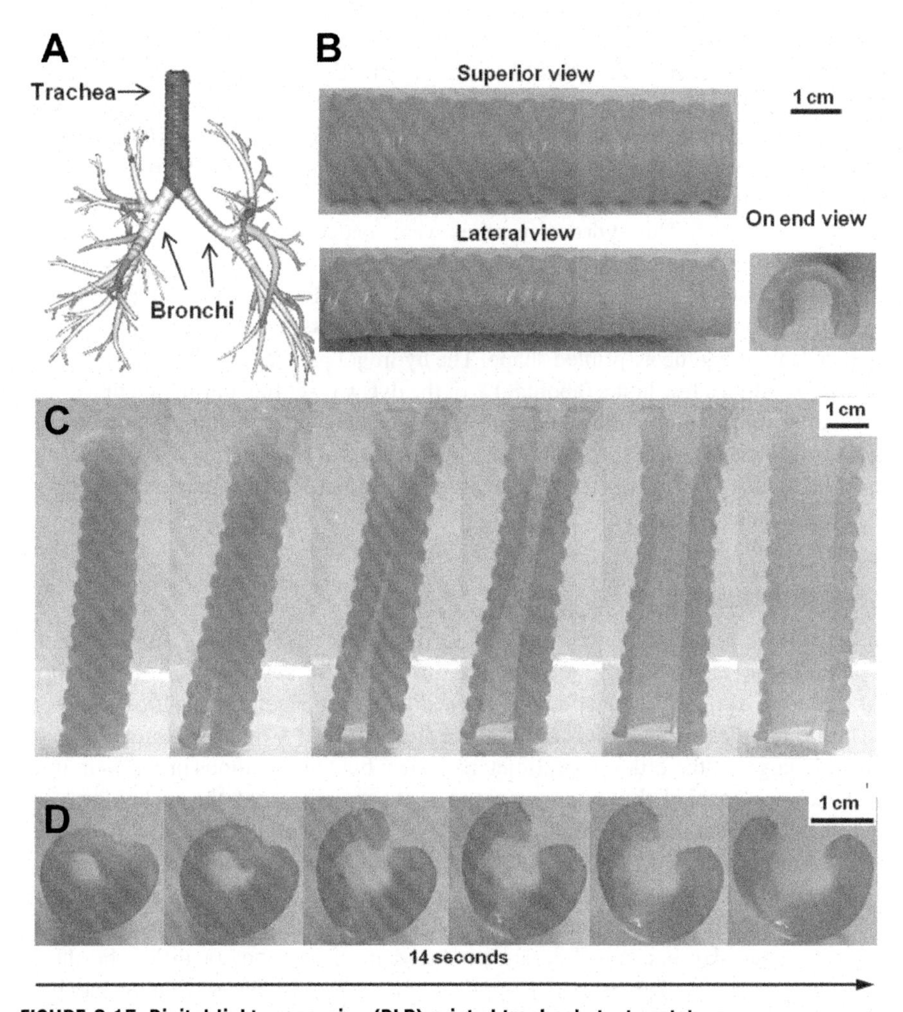

FIGURE 2.15 Digital light processing (DLP)-printed tracheal stent prototype.

The digital model of the tracheobronchial tree of a male subject (A) was derived from a magnetic resonance imaging (MRI) scan. The model was processed using several software packages (Autocad Inventor, Meshmixer, and netFabb) to cut the portion of interest, to convert it into a shell, and remove the dorsal wall. The resulting model was printed (B) using a Freeform Pico 2 DLP printer (Asiga, Australia) outfitted with a custom heating reservoir. Thermomechanical programming established the temporary shape (C and D, *left*) and subsequent heating above the melting temperature of the soft phase triggered shape recovery, shown as subsequent snapshots (C and D, *left* to *right*).

From Zarek, M., Mansour, N., Shapira, S., & Cohn, D. (2017). 4D printing of shape memory-based personalized endoluminal medical devices. Macromolecular Rapid Communications, 38, *1600628. https://doi.org/10.1002/marc.201600628. Reprinted with permission from John Wiley and Sons.*

Remarkable progress was made by using a combination of two DLP printable materials: a methacrylate-based SPM with $T_g = 30$ °C and an acrylamide-PEGDA (AP) hydrogel (Ge et al., 2021). These authors have built a thermoresponsive intravascular stent similar in shape to the one shown in Fig. 2.11 but with hydrogel patches incorporated into the rods that connect adjacent vertices. More precisely, the multimaterial 3D printer included a 200 μm-thick skin of hydrogel in the central portion of each rod. The hydrogel patches were loaded with a red dye, as a drug model. The stent was deformed into a compact, small-diameter cylinder at 37 °C and cooled down to 20 °C. The vitrification of the SMP fixed the compact shape. As soon as the stent was deployed in a model blood vessel maintained at 37 °C, it expanded, attaining the as-printed shape. The hydrogel patches released the dye progressively, within a few hours (about 80% of the dye was set free during the first 2 h). In real drug-release applications, the rate of drug release can be controlled by tuning the hydrogel mesh size, using composite hydrogels that incorporate drug-releasing particles, or using hydrogels that release drugs in response to environmental changes (e.g., a pH change) (Ge et al., 2021).

3.2 Soft actuators

An actuator is a component of a machine responsible for moving something in response to an input. In a robot arm, for instance, the actuators are the motors that make the arm move. Most actuators used today are electromechanical devices that turn an electrical signal into mechanical action. Common actuators are relatively large, of the order of centimeters in size, but the enormous progress in microelectronics enabled the emergence and rapid expansion of the technology of micro-electro-mechanical systems (MEMS). This technology seeks to fabricate miniaturized sensors and actuators and integrate them with printed circuit boards. MEMS devices comprise cantilevers, springs, joints, gears, and rotary motors.

Albeit versatile and functional, MEMS devices also have limitations. They are relatively expensive because they are produced in clean-room facilities. Furthermore, they are rigid, whereas certain applications, such as implantable and wearable devices, require flexibility, and compliance. To address these needs, actuators made of soft and stretchable materials have attracted much attention. They can be actuated through diverse stimuli, such as heat, electric field, magnetic field, or electromagnetic radiation. Such soft actuators are produced by diverse techniques, including 3D printing (El-Atab et al., 2020). Since they are inherently stimulus-responsive, 3D-printed devices that incorporate them count as 4D-printed structures.

Early attempts to build soft actuators by additive manufacturing were made at the advent of 4D printing (Ge et al., 2013). In this work, Dunn et al. created printed active composites (PACs) consisting of an elastomeric matrix reinforced by a set of glassy polymer fibers (Fig. 2.16A).

The shape memory of the polymer fibers brings about complex shape changes as the PAC stripe undergoes thermomechanical programming by being heated to a temperature $T_H = 60$ °C, larger than the glass transition temperature, $T_g \approx 35$ °C, of the

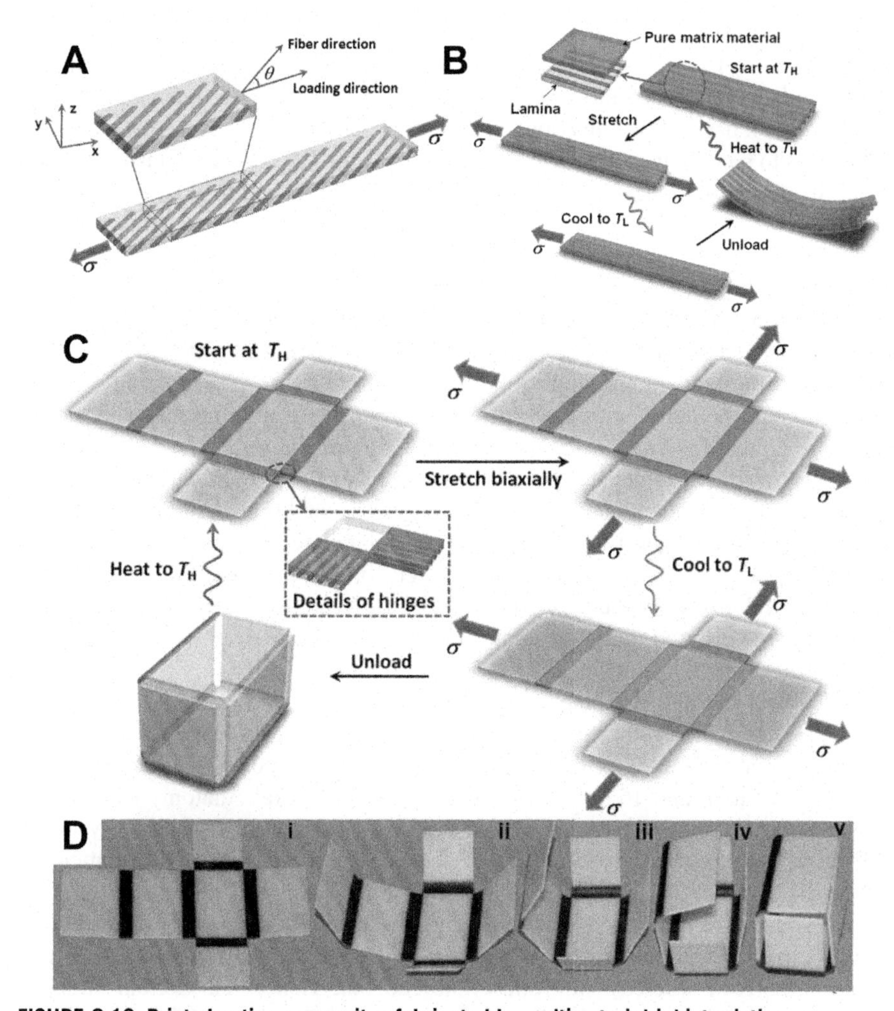

FIGURE 2.16 Printed active composites fabricated by multimaterial inkjet printing.

Schematic representation of a printed active composite (PAC) lamina made of a rubbery elastomeric matrix (*green*, light gray in print) reinforced by parallel SMP strands (*red*, dark gray in print), oriented at an angle from the direction of the uniaxial loading applied during thermomechanical programming (A). A thermoresponsive hinge (soft actuator) is obtained when a pure matrix material layer is printed on top of a PAC lamina. The glass transition temperature of the SMP fibers, T_g. Heating to T_H larger than T_g, stretching, cooling to T_L lower than T_g, and releasing the two-layered laminate results in spontaneous bending as the rubbery matrix tends to recover its original length, while the vitrified SMP fibers remain elongated. The joint straightens as soon as the system is warmed above T_g (B). Schematic of a self-folding box made of stiff plastic plates connected via two-layered PAC hinges. Thermomechanical programming of the box is achieved by biaxial stretching at T_H, cooling to T_L, and unloading (C). Time-lapse photographs show the autonomous folding of the box upon releasing the mechanical loads at T_L (D, *left* to *right*).

Reprinted from Ge, Qi, Qi, H. J., & Dunn, M. L. (2013). Active materials by four-dimension printing. Applied Physics Letters, 103, *131901. https://doi.org/10.1063/1.4819837, with the permission of AIP Publishing.*

SMP the fibers are made of. To program the stripe, one needs to stretch it at T_H, cool it down to $T_L = 15\ ^\circ C$, and release it (see Fig. 2.16B for an example). The elastomeric matrix used for the fabrication of the PACs shown in Fig. 2.16 is in a rubbery state because it has a glass transition temperature of about $-5\ ^\circ C$. Hence, the matrix layer tends to recover its permanent shape, whereas the PAC layer retains the strain imposed during mechanical loading because the fibers are in a glassy state below T_g.

When two-layered hinges are incorporated in structures made of stiff plastic, a smart origami is obtained, which folds autonomously in a preprogrammed fashion (Fig. 2.16C and D).

Self-folding, 4D-printed boxes might be employed for gentle manipulation of tissue-engineered constructs. For this purpose, one would need SMP fibers with a glass transition temperature slightly higher than 37 °C. The programming of the structure could be done, within a sterile hood, using a simple mechanical device to impose a prescribed strain in a thermostated bath maintained at T_H and transferring the system into another sterile bath kept at T_L. By carefully sliding the open box below the tissue construct, and unloading the hinges, conditions will be created for the box to enclose the construct. One can freely manipulate the closed box using mechanical tweezers and release its content at will by briefly warming the hinges above T_g. To this end, one would ideally rely on magnetic hyperthermia or on exposure to visible light or infrared radiation. Materials of this kind are available to date, and many of them are 3D printable with various techniques.

Soft joints capable of remote actuation have been fabricated by embedding strands of digital SMPs of different colors (Veroyellow and Verocyan) in a rubbery transparent matrix (Tango+). To this end, Jeong et al. used a polyjet printer of 50 μm in-plane resolution and 15 μm layer thickness—J750 Digital Anatomy (Stratasys, Edina, MN, USA) (Jeong et al., 2020). The SMP strands had a shape fixity ratio of 99% and a shape recovery ratio of 92%. Such a joint is represented schematically in Fig. 2.17A.

The strands are positioned such that a light beam directed normally to the composite stripe reaches both types. When red light is shined on the joint, the blue strands absorb it, whereas the yellow ones do not. Hence, the bottom strands warm up, and, within half a minute, they exceed the glass transition temperature of Verocyan; as they recover their original length, the joint becomes curved because the top strands remain elongated (Fig. 2.17B, *bottom-right* and *middle*). If the system is subsequently exposed to blue light, preferentially absorbed by the yellow strands, the joint assumes its permanent shape (Fig. 2.17B, *bottom-left*). After thermomechanical programming, this multilayer composite stripe can be reused for another cycle of remote actuation. The programming strain controls the extent of bending.

Fig. 2.18 illustrates a variety of responses observed in composite joints made of a Tango+ layer outfitted with two different sets of SMP fibers—one near the bottom and the other near the top (Jeong et al., 2020). In these applications, three choices of SMPs have been used: Veroyellow ($T_g = 63.2\ ^\circ C$), Verocyan ($Tg = 66.7\ ^\circ C$), and Veroclear ($Tg = 63.5\ ^\circ C$). For the joint shown schematically in Fig. 2.17A, when red light was applied first, the joint assumed an n-shape because the Verocyan fibers

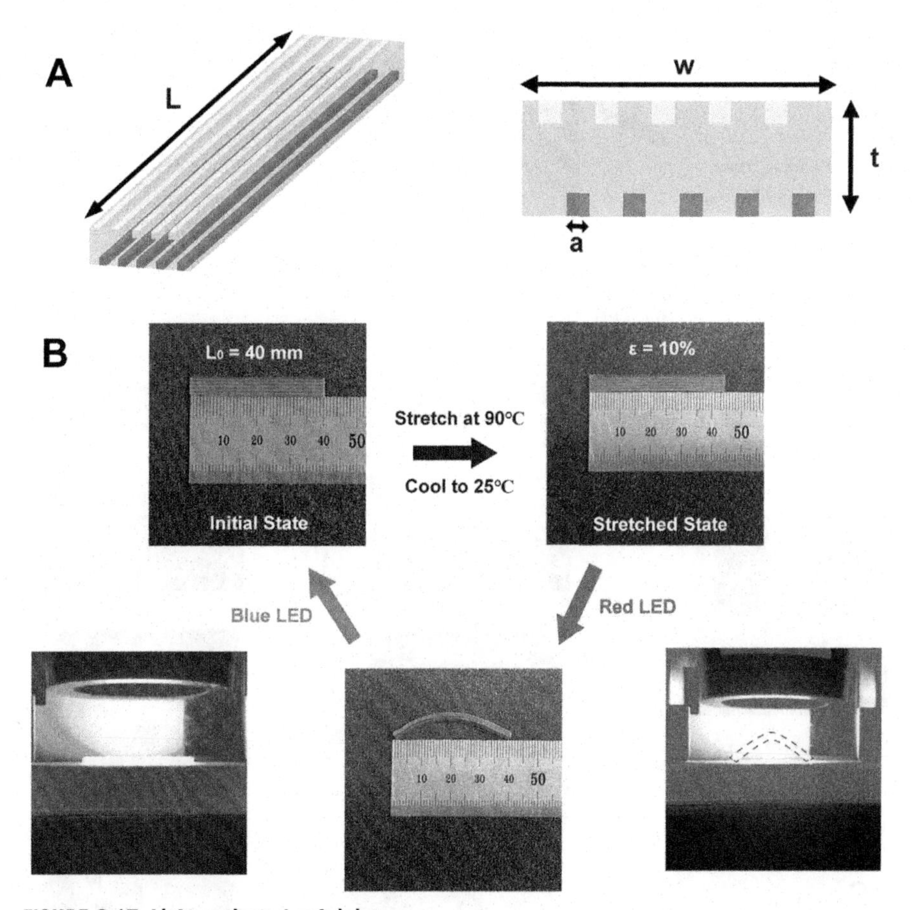

FIGURE 2.17 Light-activated soft joint.

Schematic representation of a soft joint, $L = 4$ cm long, $w = 5.5$ mm wide, and $t = 2$ mm thick, which includes two sets of parallel strands whose cross-section is a square of $a = 0.4$ mm side length (A). Thermomechanical programming of the 3D-printed joint (B, *left*) consists of stretching it by 10% in hot water, cooling it down, and unloading it. The vitrified strands maintain their temporary shape, keeping the structure straight and elongated (B, *right*). Red light is absorbed by the blue strands, which recover their original length, bending the joint as soon as their temperature rises above the glass transition temperature of the material (B, *bottom line, middle*, and *right*). Subsequent exposure to blue light restores the original shape (B, *bottom line, left*).

From Jeong, H. Y., Woo, B. H., Kim, N., & Jun, Y. C. (2020). Multicolor 4D printing of shape-memory polymers for light-induced selective heating and remote actuation. Scientific Reports, 10, *6258. https://doi.org/10.1038/s41598-020-63020-9. Reprinted under the terms of the Creative Commons Attribution 4.0 International License (https://creativecommons.org/licenses/by/4.0/legalcode).*

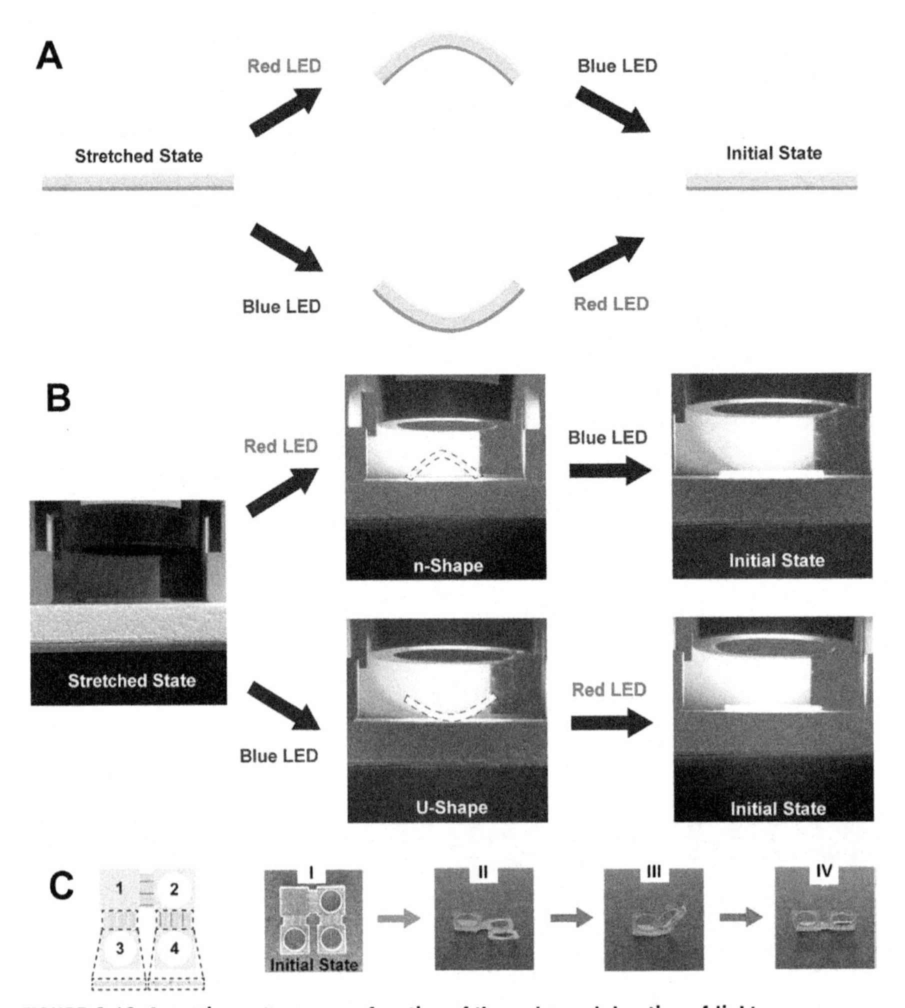

FIGURE 2.18 Actuation patterns as a function of the order and duration of light exposure.

Schematic drawing of light-induced shape changes of the soft joint made of a transparent rubbery matrix incorporating yellow SMP fibers at the *top* and blue ones at the *bottom* (A). Experimental photographs of the joint as a result of sequential exposure to *red* and *blue* light supplied by light-emitting diodes (LEDs) (B). The joint assumes an *n*-shape when *red* light is shined on it first, because it is preferentially absorbed by the *blue* SMP fibers, causing their shape recovery (shortening). A *u*-shape is obtained when *blue* light is applied first because it is absorbed by the yellow SMP fibers. Schematic of a 3D printed construct with two types of hinges, shown in cross-section at the bottom (C). Hinge one to three, which connects the rigid plastic plates 1 and 3, contains Veroyellow fibers at the *top* and Veroclear fibers at the *bottom*; hinges one to two and two to four contain Verocyan fibers at the *top* and Veroclear fibers at the *bottom*. In (C), panels I—IV are experimental snapshots of the initial state of the construct (I) and various states obtained as a result of placed at the bottom softened and recovered their initial length, whereas the Veroyellow fibers at the top remained elongated, in a glassy state. Blue light applied later enabled a complete shape recovery. Reverting the order of illumination caused an opposite bending of the joint, as shown in the schematic representation of Fig. 2.18A, and in the sequence of experimental snapshots of Fig. 2.18B.

A 3D-printed structure outfitted with two different types of joints is shown schematically on the left side of Fig. 2.18C. This multicolor hinged structure comprises square-shaped rigid plastic plates of 5 mm side length (plates 1–4) connected with joints that contain SMP fibers of 2.5 mm in length. The bottom part of the scheme represents the joints in cross-section, revealing that joint one to three is made of yellow SMP fibers at the top and clear SMP fibers at the bottom, whereas joints one to two and two to four contain blue SMP fibers at the top and clear ones at the bottom. The pictures from panel (C) represent the as-printed state of the structure (I), and its sequential shape changes depending on the color of light and the duration of illumination (II–IV). For thermomechanical programming, all the hinges were stretched in hot water and cooled to room temperature. When blue LED light was focused onto the structure, plate 3 rose within 10 s (II). A brief (5 s) exposure to red light made the other two hinges bend, as well (III). The additional action of red light (for 18 s), however, straightened joint one to two because of heat conduction from the blue SMP fibers to the transparent SMP fibers. Despite prolonged red light exposure, joint two to four remained bent because it was tilted with respect to the horizontal, receiving less light from the LED placed above the structure. The experiments depicted in Fig. 2.18 demonstrate that composite soft actuators enable complex shape changes in 4D-printed structures, and these changes can be controlled by the color of light and the duration of illumination.

3.3 Soft micromachines

Microorganisms navigate their aqueous habitat with remarkable agility. To do so, they adapt their shape and propagation strategy to the specifics of their environment. Biomimetic research has led to the fabrication of microswimmers, opening the quest for the development of multifunctional micromachines able to reach remote parts of our body and perform minimally invasive interventions (Jin et al., 2020).

For instance, the *E. coli* bacterial flagellum served as a blueprint for engineering helical magnetic micromachines outfitted with a basket-like cargo holder (Tottori et al., 2012). The inner diameter of the basket was 8.5 μm, whereas the external

light exposure: 10 s in *blue* light (II), followed by 5 s in *red* light (III), and an additional 18 s in *red* light (IV).

From Jeong, H. Y., Woo, B. H., Kim, N., & Jun, Y. C. (2020). Multicolor 4D printing of shape-memory polymers for light-induced selective heating and remote actuation. Scientific Reports, 10, *6258. https://doi.org/10.1038/s41598-020-63020-9.*

one was 10 μm, matching the diameter of the helical body. The micromachine frames were 3D printed from a rigid photoresist by direct laser writing; then, a Ni/Ti bilayer was deposited on them by thin-film evaporation. Under the action of a rotating magnetic field (of about 1.5 mT flux density and 40 Hz frequency) such micromachines displayed a corkscrew-like motion and advanced at speeds of the order of 100 μm/s. Moreover, mouse myoblasts adhered to the Ni/Ti coating and proliferated on it, demonstrating the cytocompatibility of these artificial structures (Tottori et al., 2012). Although such micromachines replicate several traits of mobile unicellular organisms, they lack flexibility and adaptability.

Huang et al. used smart materials to fabricate reconfigurable mobile micromachines composed of a monolayer helical tail and a bilayer tubular head. They took advantage of photolithographic patterning of thermoresponsive hydrogels to build micromachines capable of shape changes due to near-infrared light exposure (Huang et al., 2016). To synthesize a smart hydrogel capable of thermally induced swelling at a critical solution temperature of 42 °C, they mixed N-isopropylacrylamide (NIPAAm) and acrylamide (AAm) monomers with poly(ethylene glycol) diacrylate (PEGDA) crosslinker, in a NIPAAm:AAm:PEDGA molar ratio of 85:15:1, respectively. Ten grams of this mixture was dissolved in 7 g ethyl lactate, and 0.865 g of Fe_3O_4 nanoparticles were added, as well as 0.3 g of 2, 2-dimethoxy-2 phenylacetophenone (DMPA), a photoinitiator needed for ultraviolet light curing (see Chapter 5). Bilayers consisted of a layer of this hydrogel deposited on a PEGDA support layer. During photopolymerization, the nanoparticles were aligned by a magnetic field oriented normally to the hydrogel layer; thus, upon spontaneous bending, the cylindrical head of the micromachine was endowed with radial magnetization. Such a machine rotated along its long axis in the presence of a rotating magnetic field. Also, when heated by near-infrared light, the micromachine switched from a long slender form, akin to a swimming bacterium, to a compact configuration, in which the tail was wrapped around the head. These thermally induced conformational changes were reversible (Huang et al., 2016).

The above-mentioned studies paved the way toward engineering 4D-printed, magnetically driven micromachines capable of responding to multiple stimuli (Lee et al., 2021). In their proof-of-concept study, Lee et al. employed high-resolution 3D printing by two-photon polymerization to fabricate spherical microrollers and double-helical microscrews made of pNIPAAm-AAc, a copolymer of NIPAAm and acrylic acid (AAc), loaded with 4.2 mg/mL iron oxide nanoparticles.

The reversible thermoresponsiveness of the hydrogel prepared by Lee et al. is assured by the pNIPAAm fraction. PNIPAAm features both hydrophilic amide (—CONH—) groups and hydrophobic isopropyl ($-CH(CH_3)_2$) groups. Below pNIPAAm's lower critical solution temperature, $T_c \approx 32$ °C, the polymer filaments are strongly hydrated due to hydrogen bonds formed between the amide groups and water molecules (Lanzalaco & Armelin, 2017). When the temperature is raised above T_c, the hydrophobic interactions between the isopropyl groups overwhelm the hydrogen bonding, water is expelled, and, consequently, the hydrogel shrinks (Fig. 2.19).

The carboxyl (−COOH) groups of the acrylic acid monomers are sensitive to pH and to the presence of divalent cations. In an aqueous solution with pH > pK_a (of about 4.2), the carboxyl groups become deprotonated, leaving behind $-COO^-$ side chains. The negatively charged polymer filaments repel each other and absorb water, boosting the network size (Fig. 2.19B). This effect can be annihilated by adding Ca^{2+} ions to the aqueous phase of the hydrogel because the negative charges of the $-COO^-$ groups are screened by the calcium ions (Lee et al., 2021).

How do 3D-printed pNIPAAm-AAc micromachines respond to all these stimuli? Fig. 2.20A illustrates the shrinking of microscrews observed as the temperature climbed over T_c. Subsequent cooling rendered the polymer filaments hydrophilic again, restoring the initial state. Several heating/cooling cycles were performed with precisely the same response (Fig. 2.20B). The swelling induced by increasing pH is shown in Fig. 2.20C for both microrollers and microscrews. Finally, the extent of the response was maximized by combining all three stimuli: room temperature and alkaline pH acted synergically to bring about swelling, whereas high temperature and electrostatic screening caused shrinking (Fig. 2.20D).

Thanks to the incorporated superparamagnetic iron oxide nanoparticles, the hydrogel micromachines fabricated by Lee et al. could be driven by rotating magnetic fields. Microrollers, for example, acquired a speed of about 2 μm/s under the action of a rotating uniform magnetic field of 10 mT flux density when the rotation frequency was 1.5 Hz. The microscrews attained velocities of about 2.7 μm/s at the optimal actuation frequency of 0.8 Hz.

Magnetically steerable, two-way stimulus-responsive micromachines made of biocompatible materials can be useful as control units in microfluidic chips, such as organ-on-chip devices. Also, they might serve for the chemoembolization of tumors in arterioles. The magnetic steering helps to target tumors precisely, without affecting nearby healthy organs. A cool microroller can be guided along a large vessel and, right before entering a narrower one, it can be heated remotely; it shrinks and eventually blocks the vessel at a distal location, where it narrows down. The shrinking microroller can also release a chemotherapy drug previously loaded into it.

3.4 Artificial tissues

The cooperative behavior of cells in biological tissues is interesting from both scientific and technological points of view. Synthetic systems capable of replicating the controlled communication of live cells could be used to restore the function of damaged tissues (such as the electric conduction in broken nerves). One such system consists of lipid-wrapped aqueous droplets in oil, which adhere to each other forming lipid bilayers akin to cell membranes. Despite structural differences, the rheological properties of such droplet assemblies closely mimic the viscoelastic behavior of biological tissues. Moreover, droplets can be functionalized with membrane proteins to enable electrical communication along desired routes, created by 3D printing.

Right before the birth of the 4D printing concept, Villar et al. printed 3D networks of picoliter aqueous droplets capable of sophisticated collective behavior as

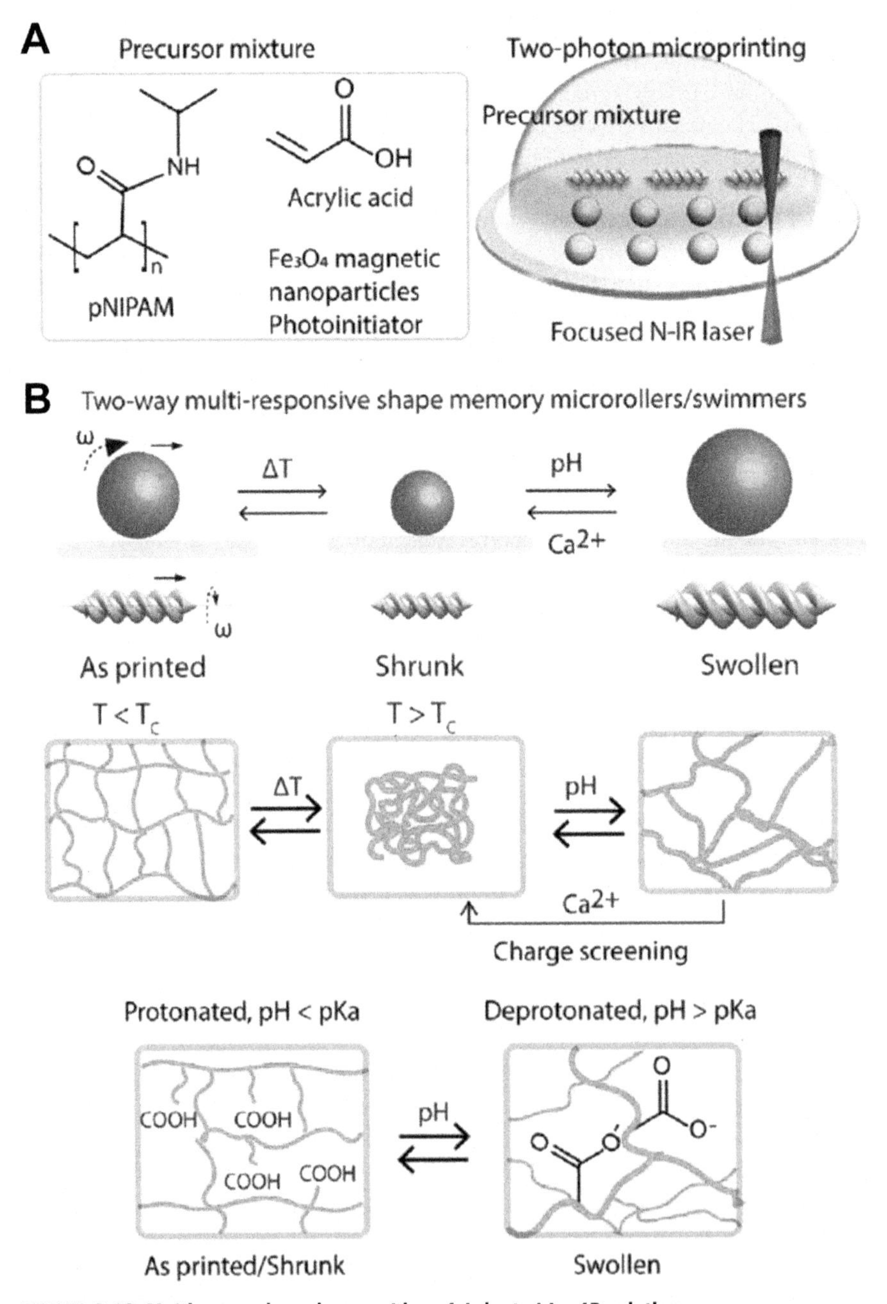

FIGURE 2.19 Multiresponsive micromachines fabricated by 4D printing.

(A) Components of the prepolymer solution (*left*) and 3D printing by direct laser writing via two-photon polymerization (*right*); (B) schematic drawings of reversible modifications in well as of anticipated shape changes driven by differences in osmolarity (Villar et al., 2013). In present-day terminology, they relied on 4D printing to build tissue-like constructs whose postprinting shape morphing resulted in complex geometries.

The 3D printing of lipid-coated aqueous droplets took place in lipid-containing bulk oil or in a drop of such oil hanging in an aqueous solution; in the latter case, the drop was embraced along its equator by a metal ring mounted on a micromanipulator. The movement of the oil drop was synchronized with the ejection, by two nozzles, of small droplets of aqueous solutions, of about 65 pL. Once the printing was completed, the excess oil was removed through one of the nozzles.

To create a functional construct, Villar et al. printed a 3D network with part of the droplets loaded with α-hemolysin (αHL), a protein responsible for the virulence of the bacterium *Staphylococcus aureus*. It is known that 7 αHL monomers form a pre-pore in a lipid bilayer, which subsequently transitions into a transmembrane pore 1.4 nm in diameter. This pore allows the transport of ions and small molecules. Hence, certain droplets were electrically connected, whereas the ones lacking αHL formed an insulating network. A steady resistive current was observed along the printed conductive pathway, whereas the insulating droplets displayed transient capacitive currents.

Combining droplets of high and low osmolarity in a well-defined arrangement, the 3D-printed network suffered an anticipated macroscopic shape change. Water crossed the bilayers, with a permeability coefficient of 27 ± 5 μm/s; therefore, droplets of high osmolarity swelled and the ones of low osmolarity shrank until their osmolarities became equal. As a result, the network suffered predictable, spontaneous deformation as long as the contact between adjacent droplets was maintained. For example, two strips of droplets of different osmolarities on top of each other gradually bent, turning into a closed ring as soon as the ends of the strips met and formed phospholipid bilayers. A lotus flower-like construct made of a layer of droplets of low osmolarity covering a layer of high osmolarity closed spontaneously, forming a hollow sphere, as water permeated from the top layer into the bottom layer (Villar et al., 2013).

Synthetic tissues outfitted with additional membrane proteins are expected to respond to diverse stimuli. Moreover, stimulus-responsive osmolytes will provide further control of postprinting shape morphing. Smart droplet clusters are promising tools for drug delivery. If challenges related to the integration of droplet networks into biological tissues will prove to be surmountable, such 4D-printed synthetic

the size of 3D printed micromachines as a result of temperature, pH, or Ca^{2+} ion concentration changes.

From Lee, Y. W., Ceylan, H. & Yasa, I. C. (2021). 3D-printed multi-stimuli-responsive mobile micromachines. ACS Applied Materials and Interfaces, 13, *12759–12766. https://doi.org/10.1021/acsami.0c18221.*

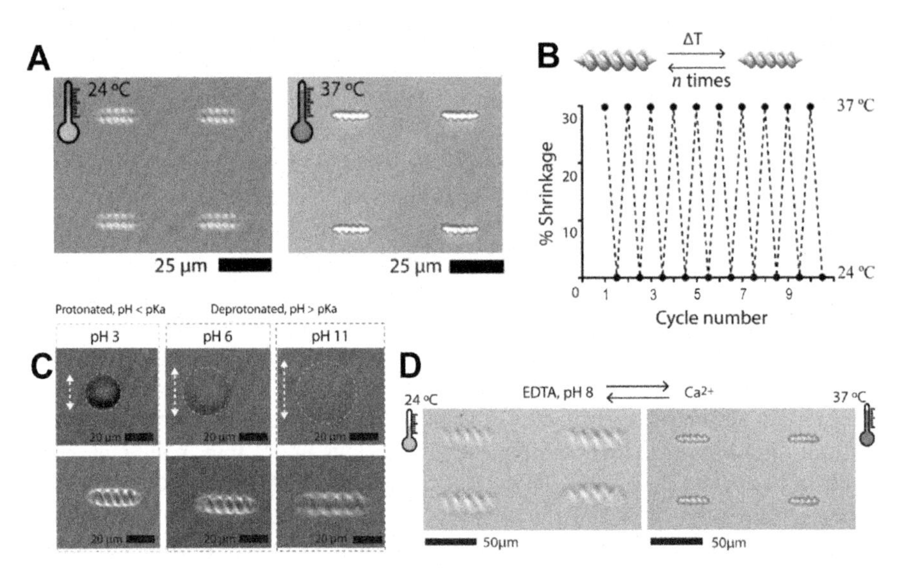

FIGURE 2.20 Optical microscopy images of stimulus-responsive microrollers and microscrews created by 3D printing.

(A) Microscrews below and above the temperature; (B) the percent shrinkage of the length of microscrews during repeated warming/cooling cycles; (C) pH-sensitivity of microrollers (top) and microscrews (bottom); (D) microscrews at room temperature, at basic pH and in the absence of Ca^{2+} ions—ethylenediaminetetraacetic acid (EDTA) is a tetravalent molecule that is capable of forming a stable complex with divalent ions.

From Lee, Y. W., Ceylan, H. & Yasa, I. C. (2021). 3D-printed multi-stimuli-responsive mobile micromachines. ACS Applied Materials and Interfaces, 13, *12759–12766. https://doi.org/10.1021/acsami.0c18221.*

systems will have the potential to restore the function of the host tissue. For example, they could provide a temporary electrical connection in the case of nerve damage.

References

Agrawal, A., Yun, T. H., Pesek, S., Chapman, W., & Verduzco, R. (2014). Shape-responsive liquid crystal elastomer bilayers. *Soft Matter, 10*, 1411–1415. https://doi.org/10.1039/c3sm51654g

Akbari, S., Zhang, Y.-F., Wang, D., & Ge, Q. (2019). 4D printing and its biomedical applications. In *3D and 4D printing in biomedical applications* (pp. 343–372). Wiley—VCH Verlag. https://doi.org/10.1002/9783527813704.ch14

Behl, M., Kratz, K., Zotzmann, J., Nöchel, U., & Lendlein, A. (2013). Reversible bidirectional shape-memory polymers. *Advanced Materials, 25*(32), 4466–4469. https://doi.org/10.1002/adma.201300880

Campbell, T. A., Tibbits, S., & Garrett, B. (2014). *The next wave: 4D printing and programming the material world* (pp. 1–15). https://www.atlanticcouncil.org/in-depth-research-reports/report/the-next-wave-4d-printing-and-programming-the-material-world/

Dill, K. A., Ozkan, S. B., Shell, M. S., & Weikl, T. R. (2008). The protein folding problem. *Annual Review of Biophysics, 37*, 289–316. https://doi.org/10.1146/annurev.biophys.37.092707.153558

El-Atab, N., Mishra, R. B., Al-Modaf, F., Joharji, L., Alsharif, A. A., Alamoudi, H., Diaz, M., Qaiser, N., & Hussain, M. M. (2020). Soft actuators for soft robotic applications: A review. *Advanced Intelligent Systems, 2*(10), 2000128. https://doi.org/10.1002/aisy.202000128

Fowler, S., Roush, R., & Wise, J. (2013). *Concepts of biology.* OpenStax. https://openstax.org/books/concepts-biology/pages/1-introduction.

Ge, Qi, Chen, Z., Cheng, J., Zhang, B., Zhang, Y.-F., Li, H., He, X., Yuan, C., Liu, J., Magdassi, S., & Qu, S. (2021). 3D printing of highly stretchable hydrogel with diverse UV curable polymers. *Science Advances, 7*(2), eaba4261. https://doi.org/10.1126/sciadv.aba4261

Ge, Qi, Qi, H. J., & Dunn, M. L. (2013). Active materials by four-dimension printing. *Applied Physics Letters, 103*, 131901. https://doi.org/10.1063/1.4819837

Ge, Q., Sakhaei, A. H., Lee, H., Dunn, C. K., Fang, N. X., & Dunn, M. L. (2016). Multimaterial 4D printing with tailorable shape memory polymers. *Scientific Reports, 6*, 31110. https://doi.org/10.1038/srep31110

Hager, M. D., Bode, S., Weber, C., & Schubert, U. S. (2015). Shape memory polymers: Past, present and future developments. *Progress in Polymer Science, 49*(50), 3–33. https://doi.org/10.1016/j.progpolymsci.2015.04.002

Huang, H.-W., Sakar, M. S., Petruska, A. J., Pané, S., & Nelson, B. J. (2016). Soft micromachines with programmable motility and morphology. *Nature Communications, 7*, 12263. https://doi.org/10.1038/ncomms12263

Hu, J., Zhu, Y., Huang, H., & Lu, J. (2012). Recent advances in shape–memory polymers: Structure, mechanism, functionality, modeling and applications. *Progress in Polymer Science, 37*(12), 1720–1763. https://doi.org/10.1016/j.progpolymsci.2012.06.001

Jeong, H. Y., Woo, B. H., Kim, N., & Jun, Y. C. (2020). Multicolor 4D printing of shape-memory polymers for light-induced selective heating and remote actuation. *Scientific Reports, 10*, 6258. https://doi.org/10.1038/s41598-020-63020-9

Jin, Q., Yang, Y., Jackson, J. A., Yoon, C., & Gracias, D. H. (2020). Untethered single cell grippers for active biopsy. *Nano Letters, 20*, 5383–5390. https://doi.org/10.1021/acs.nanolett.0c01729

Jumper, J., Evans, R., Pritzel, A., et al. (2021). Highly accurate protein structure prediction with AlphaFold. *Nature, 596*, 583–589. https://doi.org/10.1038/s41586-021-03819-2

Lanzalaco, S., & Armelin, E. (2017). Poly(N-isopropylacrylamide) and copolymers: A review on recent progresses in biomedical applications. *Gels, 3*, 36. https://www.mdpi.com/2310-2861/3/4/36

Lee, Y. W., Ceylan, H., & Yasa, I. C. (2021). 3D-printed multi-stimuli-responsive mobile micromachines. https://doi.org/10.1021/acsami.0c18221

Lin, C., Liu, L., Liu, Y., & Leng, J. (2021). 4D printing of bioinspired absorbable left atrial appendage occluders: A proof-of-concept study. *ACS Applied Materials & Interfaces, 13*, 12668–12678. https://doi.org/10.1021/acsami.0c17192

Lin, C., Lv, J., Li, Y., Zhang, F., Li, J., Liu, Y., Liu, L., & Leng, J. (2019). 4D-printed biodegradable and remotely controllable shape memory occlusion devices. *Advanced Functional Materials, 29*, 1906569. https://doi.org/10.1002/adfm.201906569

Li, G., & Wang, A. (2016). Cold, warm, and hot programming of shape memory polymers. *Journal of Polymer Science Part B: Polymer Physics, 54*(14), 1319–1339. https://doi.org/10.1002/polb.24041

Li, J., Wu, C., Chu, P. K., & Gelinsky, M. (2020). 3D printing of hydrogels: Rational design strategies and emerging biomedical applications. *Materials Science and Engineering: R: Reports, 140*, 100543. https://doi.org/10.1016/j.mser.2020.100543

Luo, X., Zhang, X., Wang, M., Ma, D., Xu, M., & Li, F. (1997). Thermally stimulated shape-memory behavior of ethylene oxide-ethylene terephthalate segmented copolymer. *Journal of Applied Polymer Science, 64*(12), 2433–2440. https://doi.org/10.1002/(SICI)1097-4628(19970620)64:12<2433::AID-APP17>3.0.CO;2-1

Melocchi, A., Inverardi, N., Uboldi, M., Baldi, F., Maroni, A., Pandini, S., Briatico-Vangosa, F., Zema, L., & Gazzaniga, A. (2019). Retentive device for intravesical drug delivery based on water-induced shape memory response of poly(vinyl alcohol): Design concept and 4D printing feasibility. *International Journal of Pharmaceutics, 559*, 299–311. https://doi.org/10.1016/j.ijpharm.2019.01.045

Melocchi, Alice, Uboldi, M., Inverardi, N., Briatico-Vangosa, F., Baldi, F., Pandini, S., Scalet, G., Auricchio, F., Cerea, M., Foppoli, A., Maroni, A., Zema, L., & Gazzaniga, A. (2019). Expandable drug delivery system for gastric retention based on shape memory polymers: Development via 4D printing and extrusion. *International Journal of Pharmaceutics, 571*, 118700. https://doi.org/10.1016/j.ijpharm.2019.118700

Momeni, F., M.Mehdi Hassani, N.,S., Liu, X., & Ni, J. (2017). A review of 4D printing. *Materials & Design, 122*, 42–79. https://doi.org/10.1016/j.matdes.2017.02.068

Qi, H. J., Nguyen, T. D., Castro, F., Yakacki, C. M., & Shandas, R. (2008). Finite deformation thermo-mechanical behavior of thermally induced shape memory polymers. *Journal of the Mechanics and Physics of Solids, 56*(5), 1730–1751. https://doi.org/10.1016/j.jmps.2007.12.002

Raviv, D., Zhao, W., McKnelly, C., Papadopoulou, A., Kadambi, A., Shi, B., Hirsch, S., Dikovsky, D., Zyracki, M., Olguin, C., Raskar, R., & Tibbits, S. (2014). Active printed materials for complex self-evolving deformations. *Scientific Reports, 4*, 7422. https://doi.org/10.1038/srep07422

Tanaka, T. (1981). Gels. *Scientific American, 244*, S124–S127. http://www.jstor.org/stable/24964265.

Thoniyot, P., Tan, M. J., Karim, A. A., Young, D. J., & Loh, X. J. (2015). Nanoparticle–hydrogel composites: Concept, design, and applications of these promising, multi-functional materials. *Advanced Science, 2*(1–2), 1400010. https://doi.org/10.1002/advs.201400010

Tibbits, S. (2014). 4D printing: Multi-material shape change. *Architectural Design, 84*, 116–121. https://doi.org/10.1002/ad.1710

Tottori, S., Zhang, L., Qiu, F., Krawczyk, K. K., Franco-Obregón, A., & Nelson, B. J. (2012). Magnetic helical micromachines: Fabrication, controlled swimming, and cargo transport. *Advanced Materials, 24*, 811–816. https://doi.org/10.1002/adma.201103818

Villar, G., Graham Alexander, D., & Bayley, H. (2013). A tissue-like printed material. *Science, 340*, 48–52. https://doi.org/10.1126/science.1229495

Wei, H., Zhang, Q., Yao, Y., Liu, L., Liu, Y., & Leng, J. (2017). Direct-write fabrication of 4D active shape-changing structures based on a shape memory polymer and its nanocomposite. https://doi.org/10.1021/acsami.6b12824.

Zarek, M., Mansour, N., Shapira, S., & Cohn, D. (2017). 4D printing of shape memory-based personalized endoluminal medical devices. *Macromolecular Rapid Communications, 38*, 1600628. https://doi.org/10.1002/marc.201600628

Zhang, H., Wang, H., Zhong, W., & Du, Q. (2009). A novel type of shape memory polymer blend and the shape memory mechanism. *Polymer, 50*(6), 1596−1601. https://doi.org/10.1016/j.polymer.2009.01.011

CHAPTER

3D and 4D printing of medical devices

3

In the acceptance of the Global Harmonization Task Force, a voluntary group of representatives of regulatory authorities and trade associations from Europe, the United States, Canada, Japan, and Australia, "*medical device means any instrument, apparatus, implement, machine, appliance, implant, reagent for in vitro use, software, material or other similar or related article, intended by the manufacturer to be used, alone or in combination, for human beings, for one or more of the specific medical purpose(s) of: (i) diagnosis, prevention, monitoring, treatment or alleviation of disease; (ii) diagnosis, monitoring, treatment, alleviation of or compensation for an injury; (iii) investigation, replacement, modification, or support of the anatomy or of a physiological process; (iv) supporting or sustaining life; (v) control of conception; (vi) disinfection of medical devices; (vii) providing information by means of in vitro examination of specimens derived from the human body; and does not achieve its primary intended action by pharmacological, immunological or metabolic means, in or on the human body, but which may be assisted in its intended function by such means*" (GHTF/SG1/N071:2012 Definition of the Terms 'Medical Device' and 'In Vitro Diagnostic (IVD) Medical Device,' 2012).

According to the above definition, a material on its own can be considered a medical device provided that it is used according to the manufacturer's specifications for a well-defined intended use. Examples include dental resins. The manufacturer of a resin describes its constituents, mixing ratios and curing time, as well as the types of restorations it is suitable for. Following these instructions leads to expected outcomes since the manufacturer has tested the safety and performance of the resin under the specified circumstances. By contrast, a raw 3D printing material is not a medical device because it is not used directly to treat patients and, despite detailed instructions for using it in a given 3D printer type, there is no guarantee that the final device will meet safety and performance requirements—the additive manufacturing process is too complex to allow for proper specifications (Personalized Medical Devices—Regulatory Pathways, 2020). Hence, regulatory bodies do not deal with raw printing materials for unspecified use, but with specific devices built for well-defined clinical applications (Beitler et al., 2022).

The 3D printing of patient-specific medical devices is by far the most mature of all the medical technologies rooted in additive manufacturing. It did not reach adulthood, though, since it is still growing and its evolution is hard to predict, even from the perspective of active players (Chepelev et al., 2017, 2018). It rather resembles a teenager, ready to explore the unknown and to outrun regulatory efforts (Beitler

Towards 4D Bioprinting. https://doi.org/10.1016/B978-0-12-818653-4.00008-5

et al., 2022). Readers willing to dive into this field have crystal-clear review articles and books at their disposal (Dimitrios Mitsouras et al., 2020; Dimitris Mitsouras et al., 2015; Giannopoulos et al., 2016; Rybicki & Grant, 2017). Therefore, this chapter does not aim at painting a detailed portrait of the field; instead, it is meant to be a teaser and an example for the younger fields of 3D and 4D bioprinting.

1. From medical imaging to patient-matched anatomical models and surgical templates

Volumetric medical imaging, mainly based on computed tomography (CT) or magnetic resonance imaging (MRI), enables 3D visualization of anatomic structures on a computer screen. It facilitates the communication within a medical team as well as between caregivers and the patient. Volumetric images can be acquired also by diagnostic ultrasound or by 3D optical scanning. Ultrasound is affordable and noninvasive but provides lower resolution than CT or MRI and is prone to generate artifacts because of refractive edge shadowing. To mitigate artifacts, 3D acquisitions from various angles are recommended (Tutschek, 2018). Optical scanners can achieve very high resolution, of the order of 10 μm, but they only acquire surface topography (George et al., 2017).

Although a 3D rendering displayed on a computer screen can be informative (for trained people), patient-specific anatomical models provide a tactile, hands-on experience. The creation of such models was the first application of 3D printing in medicine (Mankovich et al., 1990) and remained very popular. The majority of 3D-printed medical devices used nowadays stem from the 3D visualization of anatomical structures. Exceptions include certain drug delivery devices, which may be personalized to some extent, but their shape should not necessarily match the recipient's anatomy.

The 3D printing workflow includes several steps. Some of them, such as image acquisition, are common to most applications, others are specific to device categories. Medical imaging instruments generate their output in Digital Imaging and Communications in Medicine (DICOM) format. Volumetric medical images are visualized in 3D and interpreted by a team of physicians to decide whether 3D-printed models are needed or not. If the answer is positive, the DICOM images are processed to delimit regions of interest; this process, known as segmentation, is usually performed in a semiautomatic manner (using software and expert input) and generates a Standard Tessellation Language (STL) file for each region. While the selected region (e.g., tumor) is a collection of image volume elements (voxels), the STL file is a geometric representation of its frontier as a set of triangles (facets) akin to a jigsaw puzzle (Dimitris Mitsouras et al., 2015). If one imagines the selected region as a potato, the STL file describes its peel divided into hundreds of thousands of facets, in a contiguous arrangement, resembling a mosaic from a building designed by the Austrian artist Friedensreich Hundertwasser. The preparation for

3D printing includes minor refinements of the STL files, such as (i) eliminating holes or surface roughness caused by imaging artifacts, (ii) fragmenting the model for optimal visualization, or (iii) adding support structures to keep delicate anatomical features in place. Also, distinct anatomical structures can be colored differently and magnets can be added to fragmented models to keep parts together.

Common uses of 3D printing in medical device manufacturing have been divided into three groups, which require increasingly stringent quality management to ensure patient safety and treatment efficacy: Group I—anatomical models, Group II—modified anatomical models, and Group III—virtual surgical planning with templates (Christensen & Rybicki, 2017).

These groups are in rough correspondence with the medical device classification scheme developed by the United States Food and Drug Administration (FDA) to determine the type of premarket submission needed for FDA's approval for the device to be marketed (Classify Your Medical Device, 2020). Class I devices pose the smallest risk to the user, and therefore, most of them are exempt from the premarket notification requirement. A manual stethoscope, for instance, is a Class I device. Class II devices are associated with moderate risk, so they mainly require premarket notification. Most medical devices (53%) belong to Class II. Class III devices can expose the patient to a high risk, and therefore, they need to pass the premarket approval process. This class includes about 9% of the medical devices regulated by the FDA. For instance, implantable prosthetics and life support systems belong to Class III.

Although 3D printing is a disruptive technology, prone to generate regulatory challenges, the FDA has cleared tens of 3D-printed medical devices, such as dental crowns and bridges, surgical guides, joint replacements, cranial plates, spinal cages, and bronchial splints (Di Prima et al., 2016). Within the FDA, additively manufactured medical devices are regulated by the Center for Device and Radiological Health, pharmaceutical 3D printing is overseen by the Center for Drug Evaluation and Research (CDER), whereas bioprinting is under the attention of the Center for Biologics Evaluation and Research (CBER). To date, there are no FDA-approved products created by 3D bioprinting, but CBER maintains a close contact with the bioprinting research community and is aware of the potential applications of bioprinting (Di Prima et al., 2016).

1.1 Group I. Anatomical models

This group includes anatomical models created for visualization, surgical planning, and physician–patient discussion for securing informed consent.

For example, kidney cancer patients had the opportunity to learn about their disease using personalized 3D-printed models (Bernhard et al., 2016). Seven patients had a preoperative informed consent discussion about kidney physiology, their condition, tumor characteristics, and the risk of surgical complications. CT scan images were presented and explained during this discussion. Then, their learning outcomes were assessed by questionnaires before and after having seen the 3D-printed models (Fig. 3.1).

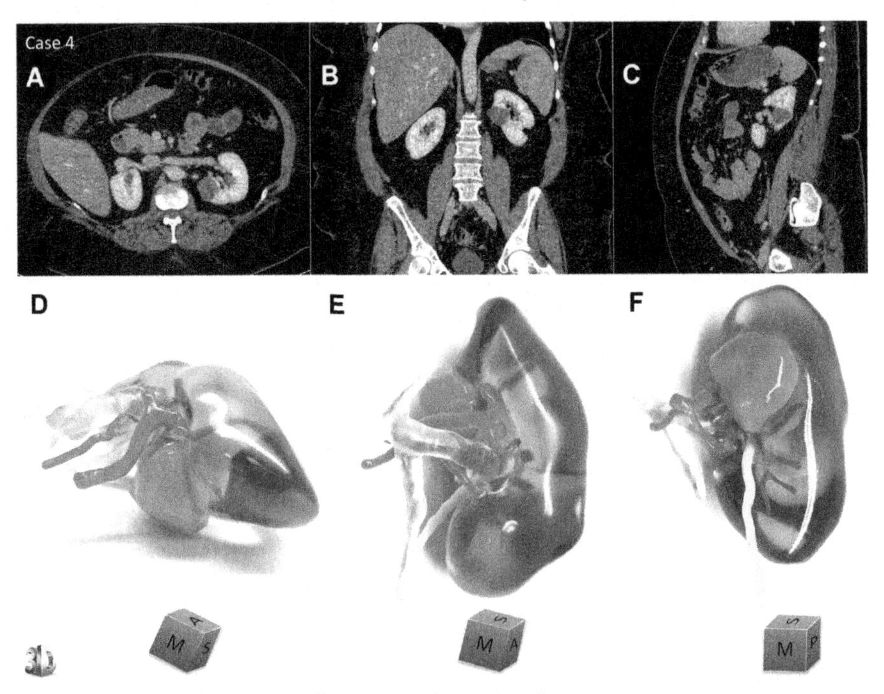

FIGURE 3.1 Anatomical model printed for patient education.

CT scan images of a kidney cancer patient in (A) axial, (B) coronal, and (C) sagittal views; digital photographs of the corresponding 3D-printed kidney model shown in (D) superomedial, (E) anteromedial, and posteromedial (F) views; the cube displayed under each picture shows the physical model's orientation; letters label the cube's faces as follows: inferior (I), superior (S), medial (M), lateral (L), anterior (A), and posterior (P).

Reprinted by permission from Springer Nature from Bernhard, J.-C., Isotani, S., Matsugasumi, T., Duddalwar, V., Hung, A. J., Suer, E., Baco, E., Satkunasivam, R., Djaladat, H., Metcalfe, C., Hu, B., Wong, K., Park, D., Nguyen, M., Hwang, D., Bazargani, S. T., de Castro Abreu, A. L., Aron, M., Ukimura, O. & Gill, I. S. (2016). Personalized 3D printed model of kidney and tumor anatomy: A useful tool for patient education. World Journal of Urology, 34, *337–345. https://doi.org/10.1007/s00345-015-1632-2*

The percentage of correct answers increased from 55% to 93% (mean values of seven respondents) as a result of scrutinizing the physical models. Although test order effects might have contributed to improved results, it is unlikely that they were dominant, since the mean improvements in the percentage of correct answers were topic dependent: 17% for basic kidney physiology, 50% for kidney anatomy, 39% for tumor characteristics, and 45% for the surgical procedure (Bernhard et al., 2016).

Medical students can also benefit from 3D prints of healthy and pathological organs (Garas et al., 2018). Especially in the case of rare malformations, their learning experience is reinforced by being able to "literally grasp the problem" (Tutschek, 2018). Students are also open to this technology, not only as users of 3D-printed learning materials but also as creators. In a study conducted on 429 medical students, 78.6% of them declared to be interested in attending classes or seminars on 3D printing (Wilk et al., 2020).

1.2 Group II. Modified anatomical models

Modified anatomical models are mainly created for enhanced surgical planning. For example, in the case of a cancer patient with a solid tumor, the anatomical model can be altered by removing the tumor to assist reconstructive surgery. Also, in the case of a trauma patient with a unilateral damage, a surgical graft can be designed by mirroring contralateral structures (Fig. 3.2).

The newly designed graft, shown in purple in Fig. 3.2C and D, is meant to replace missing bone. It is exported from the design software as a separate STL file, which can be printed to serve as a guide for autogenous bone harvesting or for manufacturing an alloplastic (synthetic material) graft (Christensen & Rybicki, 2017).

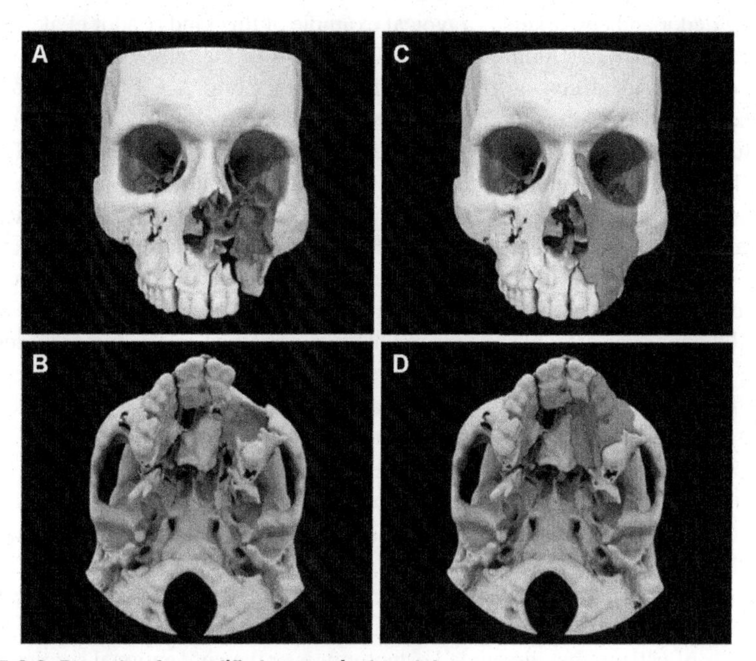

FIGURE 3.2 Example of a modified anatomical model.

3D reconstruction of a trauma patient's maxilla shown in (A) anterior view and (B) inferior view; (C) anterior view of the modified model, which includes a defect-matched graft (purple, dark gray in print) designed by mirroring the bony structures of the intact side; (D) inferior view of the modified model.

From Christensen, A. & Rybicki, F. J. (2017). Maintaining safety and efficacy for 3D printing in medicine. 3D Printing in Medicine, 3, *1. https://doi.org/10.1186/s41205-016-0009-5. Reprinted under the terms of the Creative Commons Attribution 4.0 International License (https://creativecommons.org/licenses/by/4.0/legalcode).*

1.3 Group III. Virtual surgical planning with templates

Group III is composed of tools needed for virtual surgical planning, as well as 3D-printed templates/guides fabricated to aid a surgical intervention. Virtual surgical planning is a digital procedure that complements 3D medical imaging. It is useful, for instance, for simulating the resection of a malignant bone tumor and the subsequent reconstructive surgery (Christensen & Rybicki, 2017). Based on the digital plan, surgical templates are designed using dedicated software and fabricated by 3D printing. When placed over existing anatomical structures, the templates indicate the spots where the surgeon is supposed to make cuts or insert screws/implants to carry out the plan.

Medical devices from this third category were found particularly useful in complex surgical interventions performed in the proximity of essential anatomical structures. Spine surgery is notorious in this respect, and oncological cases are especially challenging because they require accurate tumor resection followed by an elaborate reconstructive operation to restore the mechanical stability of the spine. The case reported by Lador and coworkers is a typical example of this kind (Lador et al., 2020). A 20-year-old man had a minimally invasive surgery for the removal of a suspected osteoid osteoma identified on the fourth lumbar vertebra. A CT scan performed 4 months later revealed the tumor's recurrence, and a biopsy indicated that the tumor was actually an osteoblastoma (benign, but more aggressive than an osteoid osteoma). The medical team opted for a complete tumor resection followed by the cryoablation of the retained part of the diseased vertebra. To aid the entire procedure, they printed the physical model shown in Fig. 3.3 before and after tumor resection (*panels* A and B, respectively). Also, a cryoablation needle insertion guide was designed and fabricated by 3D printing (Fig. 3.3C, *blue*). It is important to wipe out all the abnormal osteoblasts that might have remained in the tumor's vicinity because an osteoblastoma can have severe consequences. It is more fragile than normal bone, so, if it involves the frontal (load-bearing) part of the spine, the affected vertebra can suffer compression fracture.

Fig. 3.3D is a magnetic resonance image of the tumor and its neighborhood (axial view). Imagine being able to manipulate the printouts while planning the details of the surgery as opposed to inspecting cross sections or 3D renderings on a computer screen.

A prospective randomized controlled trial, conducted on 80 patients with end-stage knee osteoarthritis who underwent total knee arthroplasty (TKA), assessed the benefits of patient-specific guided osteotomy over the conventional surgical procedure. More precisely, it tested the hypothesis that 3D-printed guides could assist the intramedullary guide, providing a better femoral rotational alignment—essential for proper knee joint kinematics. On average, the guided surgery took 15 min longer than the conventional one but caused less bleeding and less surgical trauma. The short-term clinical outcomes were satisfactory with both techniques, without statistically significant differences between the two groups (Sun et al., 2020). In another study, TKA was performed on patients with valgus deformity with tibial plateau

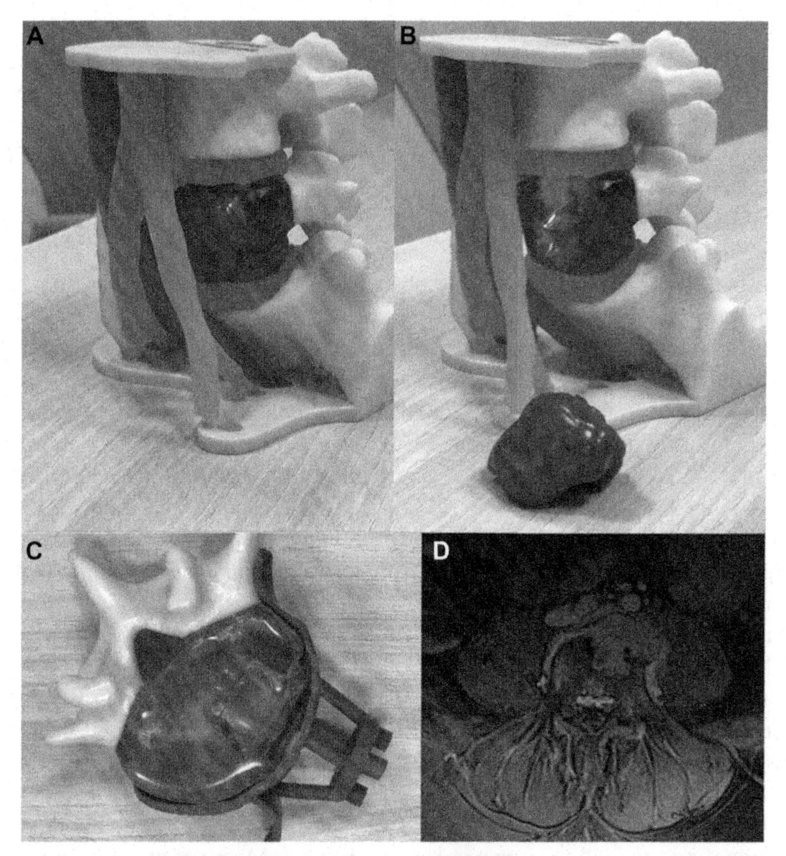

FIGURE 3.3 A representative medical device from Group III.

(A) 3D-printed model representing the anatomical neighborhood of an osteoblastoma (pink) involving the fourth lumbar vertebra; (B) the same model showing the result of tumor resection; (C) a 3D-printed guide for the insertion of cryoablation needles; it is a thermal insulator, so it also protects the adjacent healthy tissue; (D) magnetic resonance image of the vertebra invaded by the tumor. For interpretation of the references to color in this figure legend, please refer online version of this title.

Reprinted from Lador, R., Regev, G., Salame, K., Khashan, M. & Lidar, Z. (2020). Use of 3-dimensional printing technology in complex spine surgeries. World Neurosurgery, 133, *e327–e341. https://doi.org/10.1016/j.wneu.2019.09.002, with permission from Elsevier.*

bone defects and femoral condyle dysplasia (Shen et al., 2019). Ten patients benefitted from personalized osteotomy guide plates, whereas the other 10 patients had conventional TKA. In this case, the guided surgery was completed faster, with smaller intraoperative blood loss, and resulted in significantly higher Knee Society Scores (91.5 vs. 77.3 for the clinical score and 90.3 vs. 77 for the functional score).

Besides more or less modified anatomical models, surgical simulation software, and surgical guides, the spectrum of 3D-printed medical devices also includes patient-specific implants, such as titanium cages meant to replace a broken vertebra. Their endplates can be shaped to match the adjacent vertebrae, and several sizes and lordotic angles can be printed beforehand to address deviations from the preoperative plan (Lador et al., 2020).

Medical devices created by 3D printing have been applied to treat patients in a variety of clinical specialties. Their list is led by orthopedics, but it also includes abdominal surgery, thoracic surgery, plastic surgery, neurosurgery, dental surgery, maxillofacial surgery, gastroenterology, oncology, and ophthalmology. The results, reported in 110 original articles (some of them involving over 400 patients), were generally positive, probably because the applications leveraged the strengths of additive manufacturing and took advantage of diverse printing technologies and materials (Kermavnar et al., 2021; Matsumoto et al., 2015).

When it comes to fabricating personalized implantable devices for pediatric patients, 4D printing plays a fundamental role. The preprogrammed shape morphing can accommodate the patient's growth, as demonstrated already in the early days of 4D printing (Morrison Robert et al., 2015). An audacious and highly rigorous research, conducted by investigators from the University of Michigan, saved the lives of three boys who suffered from tracheobronchomalacia (TBM), a condition that leads to the collapse of the tracheal or bronchial airways during respiration. Their approach was approved by the FDA for emergency use (Di Prima et al., 2016; Morrison Robert et al., 2015).

Infants suffering from severe forms of TBM usually need support during the first 2–3 years of life, until their airway cartilage becomes strong enough to prevent collapse. Therapeutic alternatives available to date are (i) tracheostomy tube placement and mechanical ventilation and (ii) intraluminal airway stent deployment; the first is invasive, whereas the second is risky because stent migration and granulation tissue formation can lead to airway occlusion. Instead, Morrison et al. designed and 3D printed a splint to "embrace" the malacic airway segment and keep it patent by suturing its walls to the splint. The open cylinder design of the splint, resembling the Greek letter Ω in transversal cross section, allowed for facile deployment over the affected airway portion.

Fig. 3.4 illustrates the design stages (*panels* A and B) and a representative product, a poly(ε-caprolactone) (PCL) splint created by selective laser sintering (*panel* C). The entire bioengineering approach was validated in large animal (porcine) model (Hollister et al., 2016).

The 4D-printed pediatric tracheal splints were designed to ensure airway patency by anchoring the airway wall (sutured to the splint) and to become increasingly compliant in the long run (via degradation) to allow for airway growth and mechanically stimulated stiffening without triggering adverse tissue reaction. Besides facilitating tissue growth, a bioresorbable splint material relieves the patient from a second surgical intervention (Hollister et al., 2015; Scott J. Hollister et al., 2016).

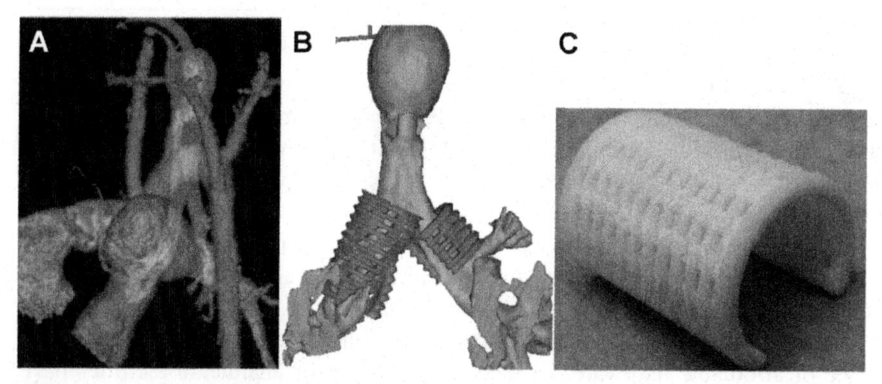

FIGURE 3.4 Tracheal splint created by 4D printing.

(A) 3D rendering of CT scan images indicates the presence of malacic airway portions (asterisks); (B) digital models of splints (red) designed to closely match native geometry in the absence of dynamic airway collapse; (C) poly(ε-caprolactone) (PCL) splint fabricated by 3D printing, with preprogrammed deformability and degradation kinetics.

Reprinted with permission from Hollister, S. J., Flanagan, C. L., Morrison, R. J., Patel, J. J., Wheeler, M. B., Edwards, S. P. & Green, G. E. (2016). Integrating image-based design and 3d biomaterial printing to create patient specific devices within a design control framework for clinical translation. ACS Biomaterials Science & Engineering, 2, *1827–1836. https://doi.org/10.1021/acsbiomaterials.6b00332. Copyright 2016 American Chemical Society.*

The impact of design parameters (splint wall thickness) was investigated by suturing 25-mm-long splints with 14 mm inner diameter and 90 opening over the trachea of 1-month-old pigs and following them for 8 months. Three different splint wall thicknesses (3, 4, and 5 mm) were considered, each being deployed in five animals. The tracheal hydraulic diameter increased by 48% under the thinnest splint, by 37% under the middle one, whereas the 5 mm wall thickness hampered tissue growth, causing stenosis (Hollister et al., 2016).

The long-term shape morphing of the 4D-printed tracheal splint is a result of PCL molecular weight loss with time and the associated loss of mechanical stiffness. A detailed understanding of the kinetics of these processes, as well as of tissue growth, will enable more accurate material programming, and the creation of splints whose shape transition under the stimuli provided by the in vivo environment will precisely match the rate of airway growth.

Finally, please note that, under certain definitions of 4D bioprinting (An et al., 2016), additively manufactured tracheal splints can be considered examples of 4D-bioprinted constructs. One could say they represent the first class of 4D-bioprinted products cleared by the FDA for emergency use. Nonetheless, according to the terminology adopted in this book, they are 4D-printed devices, as called by the authors themselves (Hollister et al., 2016)—see Chapter 9 for a brief account of the debate concerning the definition of 4D bioprinting.

2. The 3D printing of medical devices at the point of care

Historically, medical applications of 3D printing were developed within radiology departments of academic institutions and healthcare facilities (Dimitrios Mitsouras et al., 2020; Dimitris Mitsouras et al., 2015; Matsumoto et al., 2015). Hence, it is unsurprising that the largest interdisciplinary group assembled to deal with standards and quality criteria for 3D printing of medical devices has arisen within the Radiological Society of North America (RSNA); it is called the RSNA Special Interest Group on 3D Printing and plays a pivotal role in devising guidelines for safe and effective use of 3D printing in the clinics (Chepelev et al., 2018).

With the advent of 3D printing as a result of technological progress and the emergence of affordable hardware, an increasing number of healthcare facilities have started to set up their own 3D printing laboratories to build personalized anatomical models and patient-matched surgical tools. Such point-of-care (PoC) 3D printing facilities have proven their positive influence on patient care. Nonetheless, they pose regulatory challenges because they belong to a healthcare provider and play the role of an on-demand medical device manufacturer (Beitler et al., 2022).

From the regulatory perspective, the activity of 3D printing centers at the PoC has been intensely debated for almost a decade by the research community and the FDA (Public Workshop—Additive Manufacturing of Medical Devices: An Interactive Discussion on the Technical Considerations of 3D Printing, 2014), culminating with the recent release of a discussion paper by the FDA (Discussion Paper: 3D Printing Medical Devices at the Point of Care, 2021). The latter is not meant to implement a regulatory policy; it seeks to stimulate discussion for further policy development with the intention of facilitating innovation while assuring a reasonable level of safety and effectiveness of devices 3D printed at the PoC. To this end, the discussion is based on the following principles: (i) the level of FDA oversight needs to correlate with the risks posed both by the device itself and by printing the device at the PoC; (ii) the device needs to meet a set of predetermined specifications regardless of the location of additive manufacturing; (iii) the entities involved in device design and production should understand their responsibilities throughout the device's life cycle; (iv) the healthcare provider should develop capabilities, oversight, training, and experience in 3D printing to mitigate the production risks; and (v) FDA regulations should be implemented via the least burdensome approach, relying, where possible, on existing standards and processes.

As already mentioned, raw printing materials or generic additive manufacturing processes are not considered by regulatory agencies. Instead, the International Medical Device Regulators Forum (IMDRF) proposed to assess the entire *medical device production system (MDPS)*. An MDPS is defined as "*a collection of the raw materials, software and digital files, and main production and post-processing (if applicable) equipment intended to be used by a healthcare provider, or healthcare facility, to produce a specific type of medical device at the point of care, for treating their patients.*" An MDPS also includes the medical device it aims to produce (Personalized Medical Devices—Regulatory Pathways, 2020).

The discussion paper released by the FDA conveys the agency's current view on three scenarios. For each of them, illustrative examples are given and a set of questions are addressed to stakeholders (Discussion Paper: 3D Printing Medical Devices at the Point of Care, 2021).

In the first scenario, a healthcare facility uses an MDPS developed by a traditional manufacturer, who is responsible for complying with all the regulatory requirements. The 3D printing center from the healthcare facility is merely the user of the MDPS—a legally marketed device. For example, a manufacturer could submit a premarket notification (also known as 510(k)) to the FDA for an MDPS that builds patient-matched anatomic models for surgical planning and presurgical training (The 510(k) Program: Evaluating Substantial Equivalence in Premarket Notifications [510(k)], 2014). The MDPS labeling specifies the medical imaging device, design and manufacturing software, design limitations coded into the software, as well as suitable 3D printers and tooling. The 3D printing center uses the cleared MDPS as indicated in its labeling. Nevertheless, concerns remain, especially when the printout needs postprocessing (e.g., machining, mechanical polishing, heat treatment, sterilization). Regulatory implications of postprocessing operations performed at the PoC are still a matter of ongoing debate.

The second scenario considers a traditional manufacturer colocated with the healthcare facility, relieving the latter from all 3D printing activities. The two entities develop a business agreement stipulating that the manufacturer provides the healthcare facility with personalized 3D-printed devices based on the medical imaging data collected by the latter. For example, a traditional manufacturer could secure a 510(k) clearance for an additively manufactured spinal fusion cage. The manufacturer could lease a space on the premises of the healthcare provider and fabricate the cages using its own equipment in accordance with its own quality system. Then, again, the manufacturer would assume the responsibilities for FDA regulatory requirements. Open questions pertain to the quite common patient-specific changes that need to be applied to an already authorized device. While the necessary procedures are in place, further discussion is needed to make sure that manufacturers understand their obligations.

In the third scenario, the healthcare facility assumes all the responsibilities of a traditional manufacturer, including regulatory aspects. Such a scenario is appropriate when all the necessary material and human resources are present at the healthcare facility's 3D printing center, and an internal quality management system is implemented, with well-defined quality standards, and clear procedures for complaint handling and adverse event reporting. For instance, a healthcare provider might want to print patient-specific anatomic models for patient education and surgical planning. It might use image analysis, segmentation software, and a 3D printer to make anatomic models for previous surgeries. Such a retrospective study yields quantitative measures of the accuracy of imaging, printing, and postprocessing (if needed). Then, the PoC 3D printing center might be capable of producing patient-matched anatomic models for treatment planning. The precise regulatory framework for the third scenario is highly nontrivial and requires further discussion.

Until the publication of definitive guidelines by the FDA concerning the activity of PoC 3D printing centers, research institutions and professional organizations share their best practices in the scientific literature. They emphasize the importance of internal quality systems (QS) regulations, personnel training, equipment maintenance procedures, and product fabrication protocols (Beitler et al., 2022). Fig. 3.5 illustrates a protocol for printing a patient-matched anatomical model.

The first two steps from Fig. 3.5 stand for the institutional prerequisites consisting of preparatory activities (e.g., training and material and equipment testing) as well as periodic ones (e.g., hardware calibration and maintenance). It is advisable to document the qualifications and experience of the entire personnel, including radiology technicians, segmentation experts, industrial designers, and engineers. Step 3, carried out by the medical team, involves medical image analysis and weighing the potential benefits of a 3D physical model. Step 4 consists of model segmentation using a locally validated software approved by the FDA for the specific application (e.g., creating anatomical models of maxillofacial bony structures).

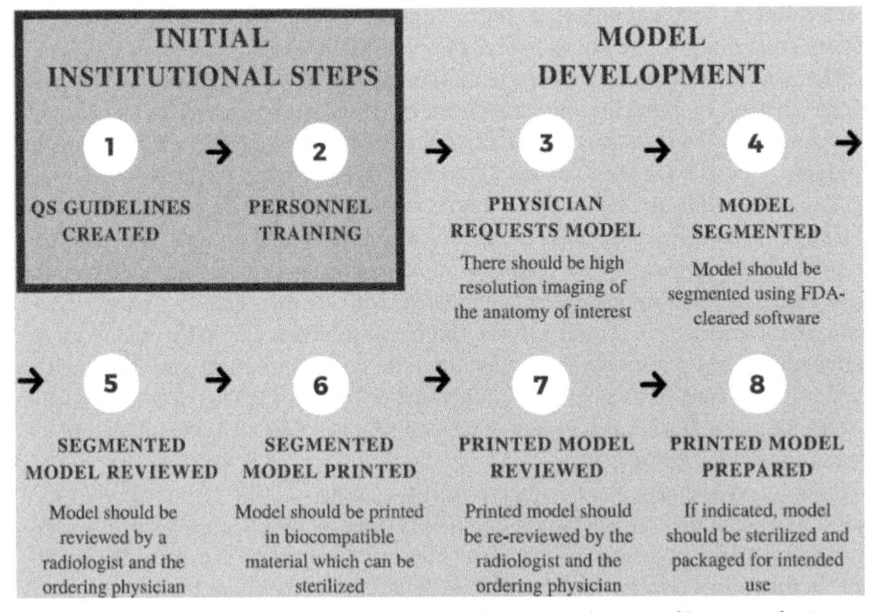

FIGURE 3.5 Additive manufacturing protocol for building a patient-specific anatomical model.

The steps enclosed in the top-left rectangle represent preliminary activities, whereas steps 3-8 describe the workflow for creating 3D-printed, personalized anatomical models.

From Beitler, B. G., Abraham, P. F., Glennon, A. R., Tommasini, S. M., Lattanza, L. L., Morris, J. M. & Wiznia, D. H. (2022). Interpretation of regulatory factors for 3D printing at hospitals and medical centers, or at the point of care. 3D Printing in Medicine, 8, *7. https://doi.org/10.1186/s41205-022-00134-y. Reprinted under the terms of the Creative Commons Attribution 4.0 International License (https://creativecommons.org/licenses/by/4.0/legalcode).*

Once the STL file is refined by eliminating imaging artifacts, the medical team inspects the digital model and discusses material choices, printing parameters (layer height, tool path, etc.), and postprocessing operations (cleaning, sterilization, etc.) with the personnel of the 3D printing center (step 5). Then the model is printed (step 6) and the printout is scrutinized by the medical team (step 7). If deemed appropriate, the printout enters postprocessing, and the resulting final product is packaged and labeled according to the FDA's recommendations (step 8).

Besides the labeling items required for any medical device (such as the product name, serial number, manufacturer's name and address, intended use, etc.) (Labeling: Regulatory Requirements for Medical Devices, 1989), the labeling of patient-specific devices should also include the patient's identifier and the final design version used to print the device. Moreover, special care needs to be taken while establishing the expiration date of a patient-matched device. Instead of a typical shelf-life evaluation, the expiration date should be derived from the date of medical image acquisition and the physician's assessment of the given patient's anatomical stability. It is also recommended to include a precautionary advice in the device labeling that the patient should be examined by the medical team before using the device to make sure that she/he did not suffer clinically relevant anatomical changes since the date of imaging (Indeed, a traumatic event or a rapidly evolving disease might compromise the performance of a device designed a while ago) (Technical Considerations for Additive Manufactured Medical Devices, 2017).

References

An, J., Chua, C. K., & Mironov, V. (2016). *A Perspective on 4D Bioprinting, 2*, 3. https://doi.org/10.18063/ijb.2016.01.003

Beitler, B. G., Abraham, P. F., Glennon, A. R., Tommasini, S. M., Lattanza, L. L., Morris, J. M., & Wiznia, D. H. (2022). Interpretation of regulatory factors for 3D printing at hospitals and medical centers, or at the point of care. *3D Printing in Medicine, 8*, 7. https://doi.org/10.1186/s41205-022-00134-y

Bernhard, J.-C., Isotani, S., Matsugasumi, T., Duddalwar, V., Hung, A. J., Suer, E., Baco, E., Satkunasivam, R., Djaladat, H., Metcalfe, C., Hu, B., Wong, K., Park, D., Nguyen, M., Hwang, D., Bazargani, S. T., de Castro Abreu, A. L., Aron, M., Ukimura, O., & Gill, I. S. (2016). Personalized 3D printed model of kidney and tumor anatomy: A useful tool for patient education. *World Journal of Urology, 34*, 337–345. https://doi.org/10.1007/s00345-015-1632-2

Chepelev, L., Giannopoulos, A., Tang, A., Mitsouras, D., & Rybicki, F. J. (2017). Medical 3D printing: Methods to standardize terminology and report trends. *3D Printing in Medicine, 3*, 4. https://doi.org/10.1186/s41205-017-0012-5

Chepelev, L., Wake, N., Ryan, J., Althobaity, W., Gupta, A., Arribas, E., Santiago, L., Ballard, D. H., Wang, K. C., Weadock, W., Ionita, C. N., Mitsouras, D., Morris, J., Matsumoto, J., Christensen, A., Liacouras, P., Rybicki, F. J., Sheikh, A., & , ... Levitin, A., ... (2018). Radiological Society of North America (RSNA) 3D printing special interest group (SIG): Guidelines for medical 3D printing and appropriateness for clinical scenarios. *3D Printing in Medicine, 4*, 11. https://doi.org/10.1186/s41205-018-0030-y

Christensen, A., & Rybicki, F. J. (2017). Maintaining safety and efficacy for 3D printing in medicine. *3D Printing in Medicine, 3*, 1. https://doi.org/10.1186/s41205-016-0009-5

Classify Your Medical Device. (2020). *United States Food and Drug Administration.*

Di Prima, M., Coburn, J., Hwang, D., Kelly, J., Khairuzzaman, A., & Ricles, L. (2016). Additively manufactured medical products—The FDA perspective. *3D Printing in Medicine, 2*, 1. https://doi.org/10.1186/s41205-016-0005-9

Discussion Paper: 3D Printing Medical Devices at the Point of Care. (December 10, 2021). *United States Food and Drug Administration, Center for Devices and Radiological Health (CDRH).*

Garas, M., Vaccarezza, M., Newland, G., McVay-Doornbusch, K., & Hasani, J. (2018). 3D-Printed specimens as a valuable tool in anatomy education: A pilot study. *Annals of Anatomy—Anatomischer Anzeiger, 219*, 57–64. https://doi.org/10.1016/j.aanat.2018.05.006

George, E., Liacouras, P., Rybicki, F. J., & Mitsouras, D. (2017). Measuring and establishing the accuracy and reproducibility of 3D printed medical models. *RadioGraphics: A Review Publication of the Radiological Society of North America, Inc, 37*, 1424–1450. https://doi.org/10.1148/rg.2017160165

GHTF/SG1/N071:2012 Definition of the Terms 'Medical Device' and 'In Vitro Diagnostic (IVD) Medical Device. (May 16, 2012). *Global Harmonization Task Force (GHTF).*

Giannopoulos, A. A., Mitsouras, D., Yoo, S.-J., Liu, P. P., Chatzizisis, Y. S., & Rybicki, F. J. (2016). Applications of 3D printing in cardiovascular diseases. *Nature Reviews Cardiology, 13*, 701–718. https://doi.org/10.1038/nrcardio.2016.170

Hollister, Scott J., Flanagan, C. L., Morrison, R. J., Patel, J. J., Wheeler, M. B., Edwards, S. P., & Green, G. E. (2016). Integrating image-based design and 3D biomaterial printing to create patient specific devices within a design control framework for clinical translation. *ACS Biomaterials Science & Engineering, 2*, 1827–1836. https://doi.org/10.1021/acsbiomaterials.6b00332

Hollister, S. J., Flanagan, C. L., Zopf, D. A., Morrison, R. J., Nasser, H., Patel, J. J., Ebramzadeh, E., Sangiorgio, S. N., Wheeler, M. B., & Green, G. E. (2015). Design control for clinical translation of 3D printed modular scaffolds. *Annals of Biomedical Engineering, 43*, 774–786. https://doi.org/10.1007/s10439-015-1270-2

Kermavnar, T., Shannon, A., O'Sullivan, K. J., McCarthy, C., Dunne, C. P., & O'Sullivan, L. W. (2021). Three-dimensional printing of medical devices used directly to treat patients: A systematic review. *3D Printing and Additive Manufacturing, 8*, 366–408. https://doi.org/10.1089/3dp.2020.0324

Labeling: Regulatory Requirements for Medical Devices. (August 1989). *United States Food and Drug Administration.*

Lador, R., Regev, G., Salame, K., Khashan, M., & Lidar, Z. (2020). Use of 3-dimensional printing technology in complex spine surgeries. *World Neurosurgery, 133*, e327–e341. https://doi.org/10.1016/j.wneu.2019.09.002

Mankovich, N. J., Cheeseman, A. M., & Stoker, N. G. (1990). The display of three-dimensional anatomy with stereolithographic models. *Journal of Digital Imaging, 3*, 200. https://doi.org/10.1007/BF03167610

Matsumoto, J. S., Morris, J. M., Foley, T. A., Williamson, E. E., Leng, S., McGee, K. P., Kuhlmann, J. L., Nesberg, L. E., & Vrtiska, T. J. (2015). Three-dimensional physical modeling: Applications and experience at mayo clinic. *RadioGraphics, 35*, 1989–2006. https://doi.org/10.1148/rg.2015140260

Mitsouras, Dimitris, Liacouras, P., Imanzadeh, A., Giannopoulos, A. A., Cai, T., Kumamaru, K. K., George, E., Wake, N., Caterson, E. J., Pomahac, B., Ho, V. B., Grant, G. T., & Rybicki, F. J. (2015). Medical 3D printing for the radiologist. *RadioGraphics, 35*, 1965–1988. https://doi.org/10.1148/rg.2015140320

Mitsouras, Dimitrios, Liacouras, P. C., Wake, N., & Rybicki, F. J. (2020). RadioGraphics update: Medical 3D printing for the radiologist. *RadioGraphics, 40*, E21–E23. https://doi.org/10.1148/rg.2020190217

Morrison Robert, J., Hollister Scott, J., Niedner Matthew, F., Mahani Maryam, G., Park Albert, H., Mehta Deepak, K., Ohye Richard, G., & Green Glenn, E. (2015). Mitigation of tracheobronchomalacia with 3D-printed personalized medical devices in pediatric patients. *Science Translational Medicine, 7.* https://doi.org/10.1126/scitranslmed.3010825, 285ra64-285ra64.

Personalized Medical Devices—Regulatory Pathways. (March 18, 2020). *International Medical Device Regulators Forum (IMDRF).*

Public Workshop—Additive Manufacturing of Medical Devices. (October 8, 2014). An interactive discussion on the technical considerations of 3D printing. *United States Food and Drug Administration.* www.fda.gov/MedicalDevices/NewsEvents/WorkshopsConferences/ucm397324.htm.

Rybicki, F. J., & Grant, G. T. (2017). *3D Printing in Medicine: A Practical Guide for Medical Professionals.* Springer. https://doi.org/10.1007/978-3-319-61924-8

Shen, Z., Wang, H., Duan, Y., Wang, J., & Wang, F. (2019). Application of 3D printed osteotomy guide plate-assisted total knee arthroplasty in treatment of valgus knee deformity. *Journal of Orthopaedic Surgery and Research, 14*, 327. https://doi.org/10.1186/s13018-019-1349-9

Sun, M., Zhang, Y., Peng, Y., Fu, D., Fan, H., & He, R. (2020). Accuracy of a novel 3D-printed patient-specific intramedullary guide to control femoral component rotation in total knee arthroplasty. *Orthopaedic Surgery, 12*, 429–441. https://doi.org/10.1111/os.12619

Technical Considerations for Additive Manufactured Medical Devices. (December 5, 2017). *United States Food and Drug Administration.*

The 510(k) Program: Evaluating Substantial Equivalence in Premarket Notifications [510(k)]. (July 28, 2014). *United States Food and Drug Administration.*

Tutschek, B. (2018). 3D prints from ultrasound volumes. *Ultrasound in Obstetrics and Gynecology, 52*, 691–698. https://doi.org/10.1002/uog.20108

Wilk, R., Likus, W., Hudecki, A., Syguła, M., Różycka-Nechoritis, A., & Nechoritis, K. (2020). What would you like to print? Students' opinions on the use of 3D printing technology in medicine. *PLoS One, 15.* https://doi.org/10.1371/journal.pone.0230851. e0230851–e0230851.

CHAPTER

3D and 4D printing of assistive technology

4

We often take pride in our capacity to adapt. Being able to manage unexpected situations through creative thinking and collaboration is one of the strengths of humans. This capacity of ours is truly vital when it comes to helping people in need. Broadly speaking, *assistive technology is "any item, piece of equipment, software program, or product system that is used to increase, maintain, or improve the functional capabilities of persons with disabilities*" (Assistive Technology Industry Association (ATIA), n.d.).

According to the World Intellectual Property Organization (WIPO), about one billion people need assistive technology today, and this number is expected to double in the next 30 years due to improvements in healthcare and the associated increase in average human lifespan (WIPO Technology Trends 2021: Assistive Technology, 2021). Satisfying their needs is not just a humane act; it is required by the Convention on the Rights of Persons with Disabilities, adopted by the United Nations in 2006, signed by 82 states, and entered into force on May 3, 2008. This legally binding document represents the commitment of the signatory states to implement the human rights stipulated in the Convention. Among others, signatory states are obliged "*to undertake or promote research and development of, and to promote the availability and use of new technologies, including information and communications technologies, mobility aids, devices and assistive technologies, suitable for persons with disabilities, giving priority to technologies at an affordable cost*" (Convention on the Rights of Persons with Disabilities and Optional Protocol, 2008).

Based on patenting data, WIPO identified recent trends in assistive technology and pinpointed nine enabling technologies that drive the innovation in this field: (i) the Internet of Things and connectivity, (ii) artificial intelligence, (iii) augmented/virtual reality, (iv) 3D printing, (v) advanced sensors, (vi) autonomous vehicles, (vii) new materials, (viii) advanced robotics, and (ix) brain–computer interface/brain–machine interface. Their synergy has led to the development of "soft" assistive technology for physically disabled people. Here, the adjective "soft" should not be taken literally; it does not refer to objects made of fluffy materials but to comfortable, lightweight, smoothly operating devices (WIPO Technology Trends 2021: Assistive Technology, 2021).

People with physical impairments, whether congenital or acquired, have specific needs, which depend both on their particular anatomy and their degree of disability. Personalization is the key to comfort, function, and appearance. Therefore, it is of no surprise that 3D printing is the rising star of physical medicine and rehabilitation.

Towards 4D Bioprinting. https://doi.org/10.1016/B978-0-12-818653-4.00006-1

Furthermore, 4D printing is being explored as a technique for manufacturing soft robotics (Delda et al., 2021) and shape changing devices (Qamar et al., 2018).

Additive manufacturing has several advantages over conventional manufacturing methods, such as injection molding and machining. First of all, 3D printing enables mass customization—low-cost fabrication of custom products. Second, the product's shape complexity is only limited by the printer's resolution (smallest printable feature size), of 0.1–0.5 mm for most techniques available to date, which is satisfactory for objects comparable in size to human limbs. Third, 3D printing produces little material waste. By comparison, injection molding (the injection of molten plastic into a mold) is cheaper than 3D printing for mass production, but creating individual parts of custom design would require a new mold for each product. The fabrication of molds, however, is costly and time-consuming. Machining (sculpting a part from a block of material using an automated milling machine) enables a certain extent of customization, limited by tool size and path, but, on average, it implies about 10 times higher material costs than 3D printing (Lunsford et al., 2016).

This chapter highlights the progress made so far in harnessing additive manufacturing in the field of assistive technology and introduces the reader to current research aimed at ensuring the efficacy, reliability, and cost-effectiveness of 3D-printed assistive devices. To date, additive manufacturing was found useful in the fabrication of orthoses, prostheses, and a plethora of small devices designed to facilitate the day-to-day activities of people with impairments.

1. Orthoses and prostheses

While an estimated 35–40 million people need prosthetics or orthotics services worldwide, only 5%–15% of the people in need are fortunate enough to have access to them (WHO standards for prosthetics and orthotics, 2017).

According to the International Organization for Standardization (ISO), an *orthosis* is an "*externally applied device used to compensate for impairments of the structure and function of the neuro-muscular and skeletal systems,*" whereas an *external limb prosthesis* is an "*externally applied device used to replace wholly, or in part, an absent or deficient limb segment*" (ISO 8549-1:2020(En) Prosthetics and Orthotics — Vocabulary — Part 1: General Terms for External Limb Prostheses and External Orthoses, 2020).

Orthoses are shell-like structures designed to closely wrap affected body parts to immobilize them, preventing further mechanical damage until they heal. Examples include orthopedic devices meant to replace messy and heavy plaster casts and splints, dynamic splints made to control the spasticity of stroke patients, knee braces, wrist guards, and thoracic braces for scoliosis patients. Most orthoses are Class 1 medical devices according to the classification scheme devised by the United States Food and Drug Administration (FDA) (Classify Your Medical Device, 2020). Further examples are passive dynamic ankle-foot orthoses created to help stroke patients to walk again. Unlike rigid orthoses, these are elastic devices, whose mechanical properties (bending and/or rotational stiffness) ensure the storage and return of mechanical energy during stance in gait, thereby facilitating smooth walking (Faustini et al., 2008). By contrast, active orthoses and exoskeletons are capable of augmenting power at the joints of the wearer by operating in parallel with her/his muscles (Dollar & Herr, 2008). Such motorized orthoses pose higher risks for the patient, and, therefore, they are Class 2 medical devices according to the FDA (Classify Your Medical Device, 2020).

1.1 Static orthoses

When a bone or joint is broken, the injured body part needs to be immobilized until it heals. The lucky readers who are not familiar with the traditional cast and splint application might wish to access the Cleveland Clinic Health Library and enter the search term "casts" (Cleveland Clinic Health Library, n.d.). Although cast immobilization is recommended in many orthopedic conditions, residents receive little training in it because casts are mainly applied by assistants and technicians, whereas residents are preferentially trained in modern surgical techniques. Casting and splinting, however, can lead to serious complications when it is performed by novice practitioners, especially in the case of pediatric patients or patients in coma or under anesthesia (Halanski & Noonan, 2008). Even in the absence of complications, patients might still suffer because conventional casts and splints are heavy and poorly ventilated.

Patient-matched, 3D-printed splints represent an increasingly popular alternative to traditional ones because they are lightweight, well-ventilated, waterproof, and beautiful. To create them, clinicians can take advantage of three cutting-edge technologies: (i) 3D body scanning, (ii) computer-aided design (CAD), and (iii) a variety of 3D printing techniques. The production costs are rapidly decreasing, but the learning curve remains quite steep for the involved clinicians. Therefore, current research efforts focus on streamlining the digital design process (Li & Tanaka, 2018).

To acquire a 3D digital image of the injured limb, one can use a relatively inexpensive 3D optical scanner. Other options are magnetic resonance imaging and computed tomography, as discussed in Chapter 3. Optimal lighting and a smooth scanner path are essential for a quality optical scan. In the example shown in Fig. 4.1A, backlit portions of the patient's wrist were not recorded, resulting in holes in the mesh model generated by the scanner's software Fig. 4.1A, bottom image. Such artifacts are eliminated by interpolation, and the refinement proposed by the visual programming software can be further adjusted by the orthopedist before using the limb's digital model as a template for splint design.

In the course of healing, inflammation is common in fractured limbs, and it is often accompanied by swelling. The 3D-printed splints are made of rigid materials (for effective immobilization), and their components are firmly assembled using rigid caps and screws. Nevertheless, to allow for swelling, the screws can be removed along one of the junctions, and the parts of the splint can be fastened by adjustable Velcro straps.

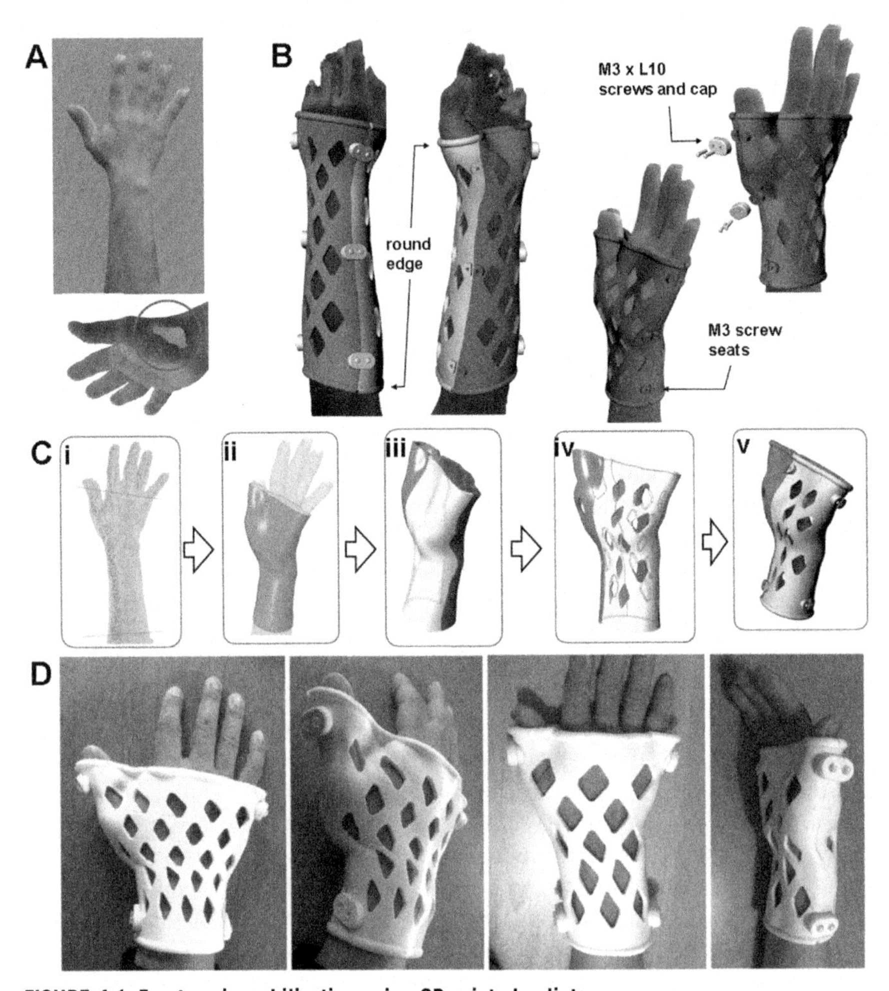

FIGURE 4.1 Fracture immobilization using 3D-printed splints.

(A) The output generated by the 3D optical scanner, showing the patient's forearm (top) and palm (bottom); for part of the wrist, the acquired image is incomplete (encircled); (B) anterior and posterior views of a 3-part forearm splint (left) and a 2-part wrist splint (right); parts are assembled using caps and screws; (C) stages of digital design: (i) load the mesh model and delimit the immobilization area via lines drawn by the clinician; (ii) generate basic covering surface; (iii) divide the model into three independent parts (shown in different colors) and assign wall thickness; (iv) generate diamond lattice structure to reduce weight, enable perspiration, and decorate the product; (v) create rounded edges and add screw seats; (D) pictures of thermoplastic polymer—acrylonitrile butadiene styrene (ABS)—splints fabricated by fused deposition modeling (FDM)—a low-cost 3D printing technology.

From Li, J., & Tanaka, H. (2018). Rapid customization system for 3D-printed splint using programmable modeling technique – a practical approach. 3D Printing in Medicine, 4, *5. https://doi.org/10.1186/s41205-018-0027-6. Reprinted and adapted under the terms of the Creative Commons Attribution 4.0 International License (https://creativecommons.org/licenses/by/4.0/legalcode).*

Optical scanning and splint design by a clinician may take from 20 minutes to 3 hours depending on its complexity and the clinician's CAD proficiency. By contrast, the semi-automatic design workflow proposed by Li and coworkers takes about 3 minutes and does not rely on advanced CAD skills. This spectacular boost in productivity stems from a precompiled customization algorithm, which relieves the clinician from repetitive modeling tasks. It computes the splint wall thickness automatically, generates the lattice pattern, and divides the splint into multiple parts. If several printers are available, each part can be printed on a different printer, which is a great way to save time. Indeed, printing is the major bottleneck of the additive manufacturing approach to fracture immobilization. A typical upper-limb splint can be printed in about 10 hours; printing several parts in parallel can lead to a two to three-fold reduction of the total print duration. By comparison, a traditional cast or splint application takes about 20 minutes. Experts suggest using a provisional immobilization device, while the patient-matched splint is being printed (Li & Tanaka, 2018).

A further drop in the fabrication time of personalized splints has been achieved by (i) eliminating the need for 3D scanning and (ii) 3D printing automatically designed flat splints to be fitted onto the patient's limb via thermoforming (Popescu et al., 2020). Using a caliper, the clinician measures 11 anthropometric parameters (hand width and thickness, wrist width and thickness, arm width and thickness, thumb maximum length and maximum width, palm length, arm length, and thenar distance) and enters them into the web interface of an online application. The clinician selects one of the predefined design variants (hexagonal or diamond lattice, single or double layer) and adjusts the splint thickness. Based on these input parameters, the application updates a predefined CAD model and (within a few seconds) generates the splint's digital model in STL format (Fig. 4.2A) as well as a planar image of the splint in PDF format (Fig. 4.2A). The latter can be printed and tried on to assess the fit before commencing the 3D printing. If the clinician is content with the fit, she/he can download the STL file and feed it to the 3D printer. Since the splint is initially flat (Fig. 4.2A), the printing time is relatively short—0.5–6 hours, depending on the selected design variant and printing parameters (Popescu et al., 2020). Finally, the orthopedist immerses the splint in hot (80°C) water to make it malleable and shapes it to closely wrap the patient's hand. This operation is performed in a warm (40°C) water bath to prevent premature hardening of the thermoplastic material. Once the splint cools and becomes rigid again, it is fastened with Velcro straps (Fig. 4.2A). This approach is feasible due to the thermomechanical properties of poly(lactic acid) (PLA), an affordable (30 USD/kg), and biocompatible material.

For even better comfort, bilayer splints have been fabricated using the online application to generate a mold based on the negative model of the flat splint. The mold (shown in Fig. 4.2B) and the rigid part of the splint (Fig. 4.2B) were clamped together, and a medical-grade, two-component silicone was injected into the mold's channels via small holes left in the splint's walls. After a curing time of about 30 minutes, the mold was removed, leaving behind a rigid PLA structure covered by a soft, skin-friendly silicone padding (Fig. 4.2C, panels iii and iv, purple layer).

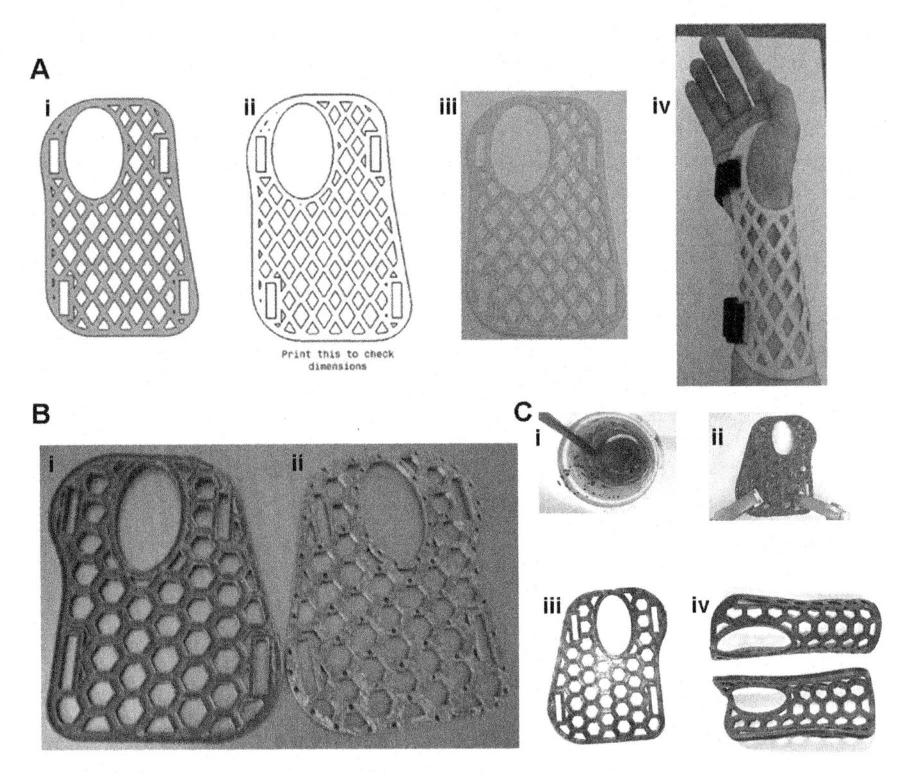

FIGURE 4.2 Fabrication of customized splints by thermoforming 3D-printed flat structures.

(A) Production stages include (i) the automatic generation of the flat 3D splint's digital model, in STL format, (ii) generation of full-scale 2D image of the splint in PDF format, (iii) 3D printing of the poly(lactic acid) (PLA) splint by fused deposition modeling (FDM), and (iv) shaping the splint in warm water to closely match the patient's anatomy; (B) picture of 3D-printed structures needed for bilayer splint fabrication: (i) mold and (ii) rigid layer of the splint, outfitted with holes for injecting silicone into the mold's channels; (C) snapshots of silicone layer preparation: (i) mixing the two components of the silicone, (ii) injecting the silicone into the channels formed underneath the splint while it is fastened over the mold, (iii) removing the mold once the silicone is cured to obtain a flat splint made of a PLA layer (blue) covered by a silicone layer (purple), and (iv) thermoforming the silicone-padded splint. For interpretation of the references to color in this figure legend, please refer online version of this title.

Reprinted with permission from Popescu, D., Zapciu, A., Tarba, C., & Laptoiu, D. (2020). Fast production of customized three-dimensional-printed hand splints. Rapid Prototyping Journal, 26, *134–144. https://doi.org/10.1108/RPJ-01-2019-0009, Emerald Publishing Limited all rights reserved.*

Although the fabrication of a planar thermoformable splint takes at least twice the time needed for traditional splinting, the new technology might gain popularity because the resulting product is customized, comfortable, aesthetic, and affordable. Indeed, the material costs of a 3D-printed bilayer splint are up to 16 USD, about one-third of the price of a conventional splint. Furthermore, with this new technology, the clinician is completely relieved of CAD; she/he just evaluates the patient, uploads anthropometric data, picks one of the available designs, runs the online app, downloads the digital model, sends it to the 3D printer, and deploys the product (Popescu et al., 2020).

Splints have also been used for treating stroke survivors suffering from upper-limb spasticity. Spastic muscles tend to remain contracted for long periods of time. If left untreated, the patient is at risk of developing a clenched fist, with fibrotic remodeling of the flexor muscles of the fingers, which hampers daily activities and proper hygiene. An effective treatment for this condition is to keep the spastic muscles in a stretched position for 6–8 hours per day, for at least 6 months. This can be done by the patient or a physical therapist, but splinting is also an option.

Early studies focused on static splints. They immobilize the hand and wrist to prevent contractures, but they are hard to endure because in moments of intense spasticity the patient experiences pain as the wrist and fingers are pressed against the splint. Indeed, in a clinical study, 37% of the patients were unable to wear a static splint for at least 8 hours a day (Andringa et al., 2013). Therefore, the authors pleaded for a dynamic orthosis capable of allowing the wrist to flex during high levels of spasticity while also providing a prolonged stretch by gentle mechanical loading.

1.2 Dynamic orthoses

Dynamic splints are well tolerated by patients with poststroke spasticity. In a 6-week program of conventional rehabilitation (2 h/week) combined with wearing a customized dynamic splint at home for about 8 h/day, stroke patients reported high levels of satisfaction, and none of them ceased wearing the splint (Yang et al., 2021). Preliminary studies have demonstrated that spasticity can be reduced by stretching only three fingers: the thumb, the index, and the middle finger. Built from ABS, using an FDM 3D printer, at an estimated cost of 80 USD, the personalized splint shown in Fig. 4.3 allows for unrestricted finger movement and tactile feedback, but it can also enforce finger extension due to a locking mechanism. Optimal fit is warranted by a patient-specific design based on a set of measurements (the lengths of the treated fingers and the widths of the forearm, wrist, and fingers).

Self-reported pain was similar in patients who wore the splint and those who only received conventional therapy, but splinting caused a statistically significant decrease in spasticity (Yang et al., 2021).

Additive manufacturing combined with the Internet of Things gave rise to smart splints capable of monitoring the evolution of the patient and the state of the splint itself via sensors placed in critical positions (De Agustín Del Burgo et al., 2020). In particular, pressure and temperature sensors were incorporated into the splint to detect swelling and inflammation. Using this strategy, changes in skin color and humidity can also be monitored in real time. The recorded data can be sent, via Bluetooth, to a smartphone and analyzed by a dedicated app, which can also issue alarms or exchange information with e-health servers.

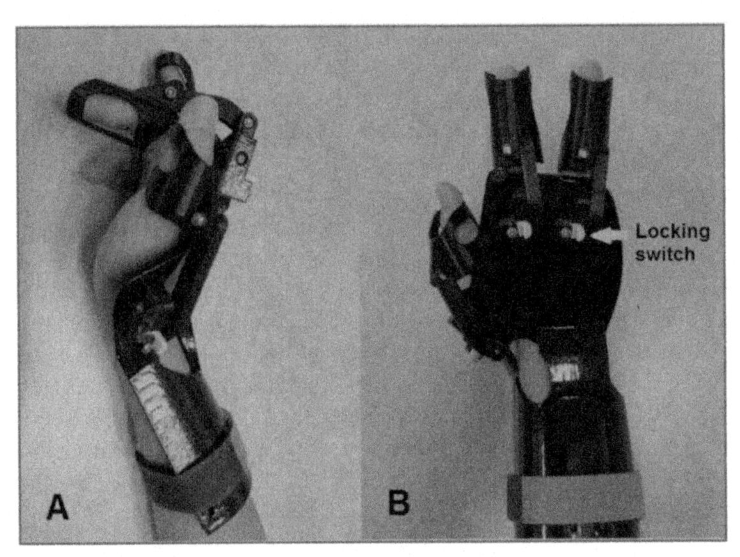

FIGURE 4.3 Personalized dynamic splints for treating poststroke spasticity.

Pictures of the 3D-printed splint in (A) lateral and (B) dorsal view.

From Yang, Y.-S., Tseng, C.-H., Fang, W.-C., Han, I.-W., & Huang, S.-C. (2021). Effectiveness of a new 3D-printed dynamic hand–wrist splint on hand motor function and spasticity in chronic stroke patients. Journal of Clinical Medicine, 10, *4549. https://www.mdpi.com/2077-0383/10/19/4549. Reprinted under the terms of the Creative Commons Attribution 4.0 International License (https://creativecommons.org/licenses/by/4.0/legalcode).*

In case of motor dysfunction, an active orthosis (a wearable assistive robot) can improve the patient's quality of life. For example, to facilitate daily tasks of patients with cervical spinal cord injury, Yoo and coworkers created a 3D-printed active orthosis controlled by surface electromyography (sEMG) signals—transient voltages generated between different regions of the skin as nearby muscles are activated (Yoo et al., 2019). The patient was examined to identify muscles left unaffected by the spinal cord injury, such as the upper trapezius or the ipsilateral biceps. Two electrodes were attached to the skin overlying the muscle belly, 2 cm apart, and the ground electrode was placed over the elbow of the dominant arm. The sEMG signal was amplified, filtered to retain components from the 10–500 Hz interval of frequencies, digitized, and converted into a root mean square (RMS) signal. The control unit operated the linear motor when the RMS signal exceeded a certain threshold, set high enough to prevent accidental actuation by unintended muscle contractions. Whenever the RMS signal exceeded the threshold, the glove switched from open to closed state or vice versa (Fig. 4.4).

Nylon threads (highlighted by yellow arrows in Fig. 4.4G) run along the palmar side of the fingers, through guiding holes included in the finger rings; they are further

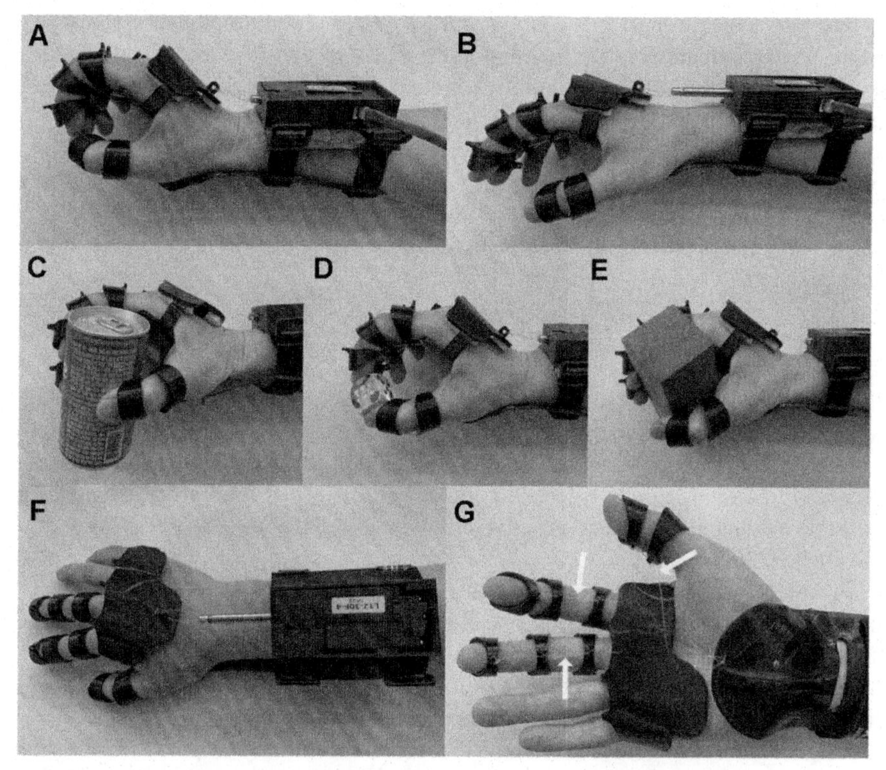

FIGURE 4.4 A 3D-printed hand orthosis intended to help paralyzed fingers grasp.

Photographs of the myoelectric orthosis: (A) lateral view in closed state; (B) lateral view in open state; (C–E) lateral views during functional tests; (F) dorsal view in open state; to extend the wrist, the linear motor pulls on the nylon threads fastened to the dorsal side of the hand part; and (G) palmar view in open state; yellow arrows (white arrows in print) point at nylon threads guided through holes and tied to the palmar forearm splint.

From Yoo, H.-J., Lee, S., Kim, J., Park, C., & Lee, B. (2019). Development of 3D-printed myoelectric hand orthosis for patients with spinal cord injury. Journal of NeuroEngineering and Rehabilitation, 16, *162. https://doi.org/10.1186/s12984-019-0633-6. Reprinted under the terms of the Creative Commons Attribution 4.0 International License (https://creativecommons.org/licenses/by/4.0/legalcode).*

guided through the palmar side of the hand part and tied to the palmar forearm splint. A linear motor, housed by the dorsal forearm splint, extends a mobile stainless steel rod toward the hand in the relaxed (open) state of the orthosis (Fig. 4.4, panels B and F). When the patient wants to grasp an object within reach, she/he contracts the wired muscle (hard enough to generate a supra-threshold RMS sEMG signal), so the controller activates the motor. While the rod is retracted, it pulls on the nylon threads tied to the dorsal hand part, extending the wrist. Thus, the fingers move away from the palmar forearm splint; consequently, the nylon threads become tensed and flex the interphalangeal and metacarpophalangeal joints (Fig. 4.4C–E).

The personalization of this robotic orthosis relied on anthropometric measurements, whereas manufacturing was based on FDM-printed PLA rings and support plates, nylon threads, a 12 V lithium-ion battery, three sEMG electrodes, the linear motor, and an Arduino Mega 2560 microcontroller board (Arduino, Torino, Italy), adding up to about 230 USD in material costs (Yoo et al., 2019). For comparison, the 2019 Red Dot Design Award winner Neomano Robotic Grasp Assist (manufactured by Neofect, Watertown, MA, USA) sells for 2000 USD (Neomano, 2019). The above example illustrates the power of enabling technologies to promote mass customization.

1.3 Prostheses

Prostheses are medical devices created to replace missing or diseased parts of the body. State-of-the-art limb prostheses are built to replace partially/fully amputated or congenitally absent extremities and to provide degrees of freedom and functionality akin to an intact limb. According to the FDA, external limb prostheses fall in the same medical device class as active orthoses (i.e., Class 2) (Classify Your Medical Device, 2020).

Prosthetic sockets are designed to closely match the shape of the patient's residual limb. They can incorporate multiple electrodes and electronic circuitry to acquire sEMG signals and process them to trigger a variety of powered joint movements in the prosthetic device. Additive manufacturing is increasingly used for building prostheses because it allows for personalization and cost reduction.

Targeting the specific needs of tens of thousands of military amputees, Liacouras and coworkers applied computed tomography (CT) and 3D printing to develop custom prosthetics (Liacouras et al., 2017). Using CT instead of 3D optical scanning was justified economically, by the clinical availability of CT scanners, and technically, because CT produces fewer imaging artifacts. Their team built devices for highly active people with well-defined physical activity options, such as playing ice hockey or working out in the gym. For instance, a partial forearm amputee desired to lift weights on his own during resistance training composed of both pushing and pulling exercises. The custom prosthesis designed for him incorporated a G-shaped hook protruding from a sturdy overlay mounted onto the prosthetic socket via carbon fiber weaving and lamination (Fig. 4.5).

Another interesting example concerns a 9-year-old girl with congenital left-hand deficiency, passionate about gymnastics. She attended beginner-level classes, and the coaches found her successful in all activities, except for hanging from a horizontal bar with her feet off the ground. She was referred to the University of Jamestown Physical Therapy Program (Fargo, North Dakota, USA) for receiving a prosthetic hand specifically built for gymnastics-related tasks. The experts evaluated the residual function of the affected hand and opted for a wrist-powered prosthesis. A suitable design, known as the Talon hand, was identified on the website of e-NABLE, a global community of 40,000 volunteers working under the motto "Give the World a Helping Hand" (Enabling the Future (e-NABLE), n.d.). Design files were

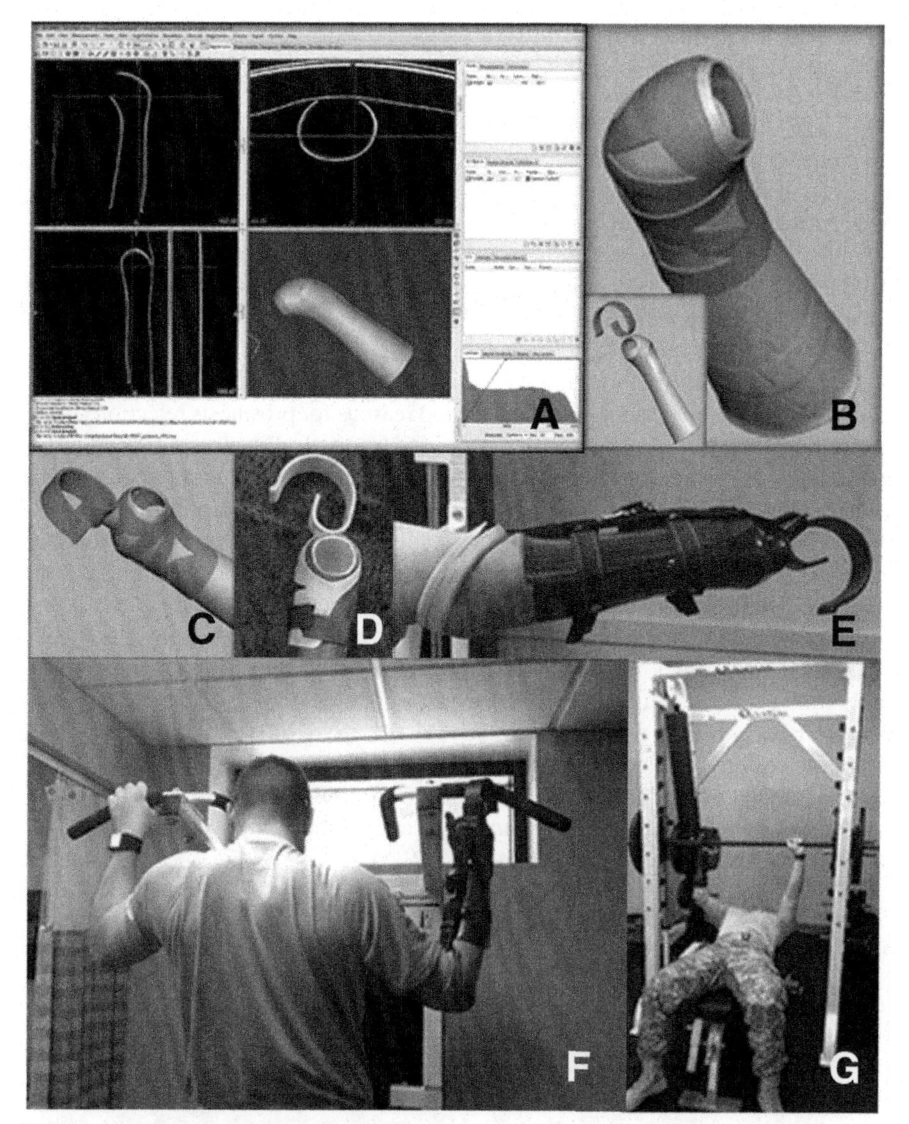

FIGURE 4.5 Design and fabrication of a forearm prosthesis customized for resistance training.

(A) Computed tomography (CT) scan of the preexisting prosthetic socket; (B) 3D reconstruction of the socket and design of a robust overlay (gray); the hook (green, dark gray in print), designed in CAD software, is shown in the insert; (C) hook and overlay merged in the same design file; (D) the merged structure's prototype (white) created by stereolithography for fit testing; if deemed satisfactory, the entire structure was 3D-printed from a titanium alloy by selective electron beam melting; (E) final product, obtained by weaving carbon fiber through the titanium structure and laminating it over the socket; (F) pulling exercise for back and biceps workout; a ratchet strap mounted around the patient's elbow prevents prosthesis detachment; and (G) pushing exercise for chest and triceps training.

From Liacouras, P. C., Sahajwalla, D., Beachler, M. D., Sleeman, T., Ho, V. B., & Lichtenberger, J. P. (2017). Using computed tomography and 3D printing to construct custom prosthetics attachments and devices. 3D Printing in Medicine, 3, *8. https://doi.org/10.1186/s41205-017-0016-1.*

downloaded, scaled to fit the girl's hand, and modified to withstand the forces encountered during gymnastics training. (Changes were needed to strengthen the joints between the digits and the palm, and the connection between the tension cords and digit tips.) The resulting device, 3D-printed from PLA, with a total cost of 35 USD, enabled the girl to participate in gymnastics training for 14 weeks, 1 hour weekly. The prosthetic hand worked as expected and boosted her confidence and satisfaction (Anderson & Schanandore, 2021).

Patients with less ambitious fitness goals would probably trade biomechanics for beauty. Indeed, poor aesthetics is one of the top reasons for prosthesis rejection (Biddiss & Chau, 2007). A natural appearance and aesthetically pleasing design are important for the general population. An anthropomorphic look hides the disability, whereas art enables the users to express themselves and, thereby, boosts their self-esteem. Positively perceived prostheses have a low abandon rate, promote a healthy and independent lifestyle, foster social life, and, importantly, help their users to develop a personal acceptance of their condition (Manero et al., 2019).

In the case of children, for instance, a prosthetic hook is an affordable option, but its abandon rate is high because of its look and the associated stigma. Commercial upper-limb prostheses of natural appearance cost between 4000 and 20,000 USD if they are body-powered (Zuniga et al., 2015), and up to 100,000 USD if they are externally powered (Calado et al., 2019). Replacing such devices as a child grows is financially daunting for most families. The Cyborg Beast (Fig. 4.6) was created to satisfy the need for affordable, customized, and aesthetically pleasing prosthetic hands for children (Zuniga et al., 2015).

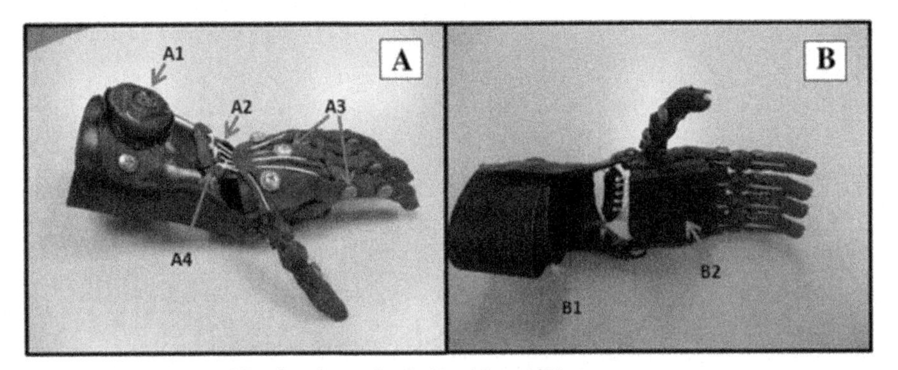

FIGURE 4.6 Low-cost, 3D-printed prosthetic hand for children.

Photographs of the (A) dorsal and (B) palmar view of the Cyborg Beast; (A1) tensioner dial, (A2) nylon cords, (A3) Chicago screws, (A4) tension balance system; (B1) Velcro strap for forearm; (B2) Velcro strap for hand.

From Zuniga, J., Katsavelis, D., Peck, J., Stollberg, J., Petrykowski, M., Carson, A., & Fernandez, C. (2015). Cyborg beast: a low-cost 3d-printed prosthetic hand for children with upper-limb differences. BMC Research Notes, 8, *10. https://doi.org/10.1186/s13104-015-0971-9. Reprinted under the terms of the Creative Commons Attribution 4.0 International License (https://creativecommons.org/licenses/by/4.0/legalcode).*

The dial located on the dorsal wrist splint serves for setting the pretension in the nylon cords that run along the palmar face of each finger. A roughly 20 degrees wrist flexion increments the tension in the nylon cords (by moving the prosthetic hand's dorsal side away from the tensioner), which causes a synchronous flexion of all the fingers. When the wrist flexion ceases, the elastic cords that run along the fingers' dorsal side (not shown in Fig. 4.6) extend them spontaneously (Zuniga et al., 2015).

Low cost, cool look, and ease of use are not the only strong points of the Cyborg Beast; it also allows for remote customization by analyzing a set of photographs of the patient's arms in the presence of a measuring tape. Three pictures need to be taken from above, showing the patient's upper limbs resting on a table in (i) lateral view with wrists extended, (ii) lateral view with wrists flexed, and (iii) dorsal view. The lateral views are used for angular measurements, whereas the dorsal one serves for linear measurements. A reliability study conducted on a sample of 11 children has shown that, on average, there were no differences between in-person anthropometric measures and those extracted from pictures. Nevertheless, a small bias was observed in range of motion assessments, but they did not affect the fitting and the function of the 3D-printed prosthetic hand (Zuniga et al., 2015). By contrast, the fitting procedure of most commercially available prosthetic hands is based on wrap casting of the affected limb with plaster bandages or on 3D scanning—beyond reach for children who live in poor countries and/or remote places.

The new generation device created by Zuniga and collaborators, named Cyborg Beast 2, was investigated in a quantitative study of patient satisfaction related to the use of remotely fitted prostheses. With a fresh look, but the same working principles, the new device features silicone fingertips, a thermoplastic socket, and a brace leverage system. The latter provides better finger flexion control by wrapping the patient's forearm. Essentially, the Velcro straps have been replaced by 3D-printed components. Two standardized self-report questionnaires, the Quebec User Evaluation of Satisfaction with assistive Technology (QUEST) and the Orthotics Prosthetics Users Survey (OPUS) indicated patient satisfaction and comfort levels similar to those reported in the literature for conventionally fitted prostheses (Zuniga et al., 2019).

The holy grail of upper limb prosthetics is an externally powered, anthropomorphic, versatile prosthetic hand, with 6 degrees of freedom, adaptive grip, and tactile feedback. Before the advent of 3D printing, the prices of externally powered high-end hand prostheses ranged from 25,000 to 100,000 USD (Calado et al., 2019; Ten Kate et al., 2017). Electric prosthetic hands were built during the last decade by 3D printing for less than 500 USD in material costs, albeit with a variable level of functionality and prototype appearance. Most of them are myoelectric devices, but voice and electroencephalography (EEG)—recording of electric signals produced by the brain at the level of the scalp—have also been investigated as a means for prosthesis motion control (Ten Kate et al., 2017; Wendo et al., 2022).

The open-source movement revolutionized the production and access to prosthetic devices. Just as in the case of the young gymnast, open-source designs distributed under the Creative Commons Attribution-ShareAlike 4.0 International (CC BY-SA 4.0) license can be downloaded and modified according to the recipient's needs. Then, their components can be fabricated by additive manufacturing and assembled according to the instructions provided by the developers. Fortunately, fabrication is not the concern of the end-user: the e-NABLE community takes care of scaling, customization, assembly, and fitting of the prosthesis at no charge (Enabling the Future (e-NABLE), n.d.). Fig. 4.7A represents Robotic Arm Version 2, an open-source myoelectric robotic hand (Robotic Arm Version 2, 2020). Its previous version, known as the "El Medallo" Bionic Hand, became popular as the world's first sEMG-controlled, open-source, trans-radial hand prosthesis ("El Medallo" Bionic Arm, 2019).

The myoelectric hand prosthesis created by Open Bionics grew from a research prototype, the Bionic Hand (Ten Kate et al., 2017), into a full-fledged commercial product, the Hero Arm, bundled with medical services and updates (Meet the Hero Arm – a Prosthetic Arm for Adults and Children, 2022). With a price range of 10,000–20, 000 USD, depending on the patient's age and condition, the Hero Arm is about twice cheaper than its nearest competitor.

Although it is only authorized for carrying weights of up to 8 kg, the Hero Arm is made of sturdy, wear-resistant nylon 12. It comes in two different versions: a three motor hand, in which the index and the middle fingers move together, and a four motor hand, in which each finger moves individually. It notifies the user of device status via the hand button's color, mechanical vibration, and a beeper. For motion control, the bionic arm includes two sEMG sensors to record voltages developed, while the user tenses the muscles responsible for similar functions in a healthy hand. The sEMG signals are analyzed and translated into finger movements. To close the Hero Arm's hand, the user needs to contract the muscles located on the palmar side of the forearm; to open it, the antagonist muscles need to be contracted.

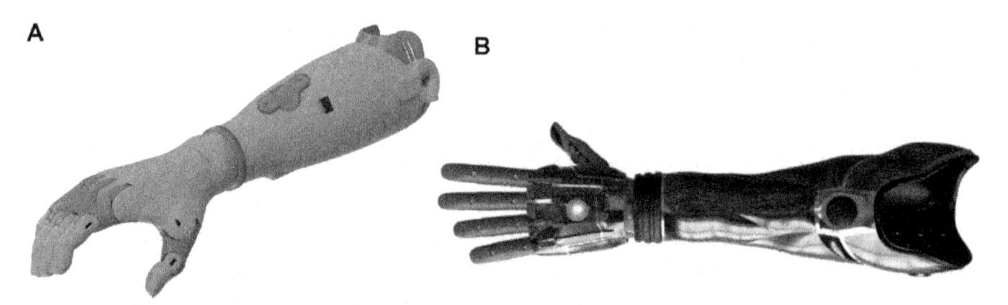

FIGURE 4.7 Examples of myoelectric robotic arms.

(A) The Robotic Arm Version 2, developed within the e-NABLE community and (B) the Hero Arm, commercialized by the British company Open Bionics.

Panel (A) image source: Robotic Arm Version 2. (2020). https://github.com/Humanos3D/RoboticArmV2/wiki/Robotic-Arm-Version-2; reprinted under the terms of the Creative Commons Attribution-ShareAlike 4.0 International (CC BY-SA 4.0) license (https://creativecommons.org/licenses/by-sa/4.0/legalcode); panel (B) image source: https://www.nicepng.com/ourpic/u2e6w7q8w7a9w7o0_open-bionics-hero-arm/, a free image library.

Movement speed is proportional to the extent of muscle activation—that is, gently tensed muscles elicit slow hand movement, suitable for manipulating delicate objects. The three-motor version can generate four different grip patterns (fist, hook, and two tripod grips meant for picking up small objects); the four-motor version provides two additional pinch patterns, in which the index finger meets the thumb, while the other fingers remain extended or flexed (Hero Arm User Manual 6.0, 2021).

The producer of the Hero Arm, Open Bionics, is a company founded in 2014 by Samantha Payne and Joel Gibbard in Bristol, UK. Under the motto "*Turning disabilities into superpowers,*" the company developed the world's first FDA-approved 3D-printed bionic arm, which combines multigrip functionality with memorable design (Our story - making 3D prosthetics beautiful - Open Bionics, 2021). Besides clinically proven performance, the Hero Arm is attractive because of its detachable decorative covers created by world-class artists from Disney and Eidos-Montreal. The Hero Arm hit the market in April 2018 and proved to be a successful product, with more than 1000 people on the waiting list. In an interview given to the Elite Business magazine, Payne explained part of her entrepreneurial motivation: "*I absolutely love receiving photos and videos from parents of our young users doing actions for the first time, of them cleaning their teeth with ease for the first time or brushing their hair or holding a fork or eating dinner or taking their dog for a walk. I get all of these photos and videos and it just totally makes my day.*" (Sandy & Johansson, 2018).

2. Assistive devices for daily living

In this section, the generic name *assistive devices* is used as a collective term for small, nonmedical devices specifically designed to help disabled people in daily activities. Different tasks deemed problematic by people with various impairments, of different degrees, explain the need for a vast variety of assistive devices. Over the years, thousands of CAD models have been constructed. To manufacture them, producers mainly used 3D printers, but 3D laser cutters and milling machines with computer numerical control have also been used.

2.1 Precise fit by 4D printing

An increasing body of evidence suggests that 4D printing can facilitate assistive device customization via postprinting shape morphing (Qamar et al., 2018; Sun et al., 2021). Sun and collaborators proposed to design assistive devices to approximately fit the objects intended to be manipulated and to achieve a precise fit by the shape transition triggered by a stimulus (Sun et al., 2021). For example, a single type of device can be fastened over different pieces of cutlery. The user inserts the utensil's handle into a specifically designed sleeve and places the system in hot water; the sleeve will tightly embrace the handle due to stress relaxation. During the FDM of PLA, thin filaments are extruded through a narrow nozzle heated to over

200°C. As the filament cools down in contact with the target surface, internal stress builds up. When the 3D-printed object is warmed up, to about 60°C, the polymer chains relax and the extruded filament shrinks along the printing path. The shape change of a complex 3D structure depends on the printing parameters, as well as on infill pattern and percentage—e.g., for a 0.1 mm layer height, a printout with a 20% honeycomb infill shrinks by roughly 25% during stress relaxation (Sun et al., 2021). Unfortunately, this method is unfeasible for fitting 3D-printed PLA devices onto the patient's limbs, but further progress in material science will certainly allow for accommodating the user's anatomy by shape morphing elicited by milder stimuli. Then, optimal comfort will be achieved despite small errors in measurement (or 3D imaging) and/or additive manufacturing.

2.2 The user as a co-designer

Since low-cost 3D printers became widely available, a shift in manufacturing paradigm took place, in which the end-user is not a mere consumer, but an active player in his own customized product's design. Hence, users became co-designers and co-producers. Despite their impairment, people who need assistive technology are increasingly valued as collaborators of professional designers. The literature abounds with examples of assistive devices created with the implication of the recipient. Ostuzzi and coworkers, for instance, investigated the idea of applying 3D printing for producing custom tools for elderly people with rheumatic diseases (RDs) (Ostuzzi et al., 2015). The stiff and painful joints of RD patients hamper their movement, limiting the spectrum of daily activities and undermining their self-confidence.

Additive manufacturing, if properly conducted, has the potential to satisfy the needs of RD patients. In the production paradigm proposed by Ostuzzi et al., the development team includes four players: (i) the end user, who knows his condition and sets device requirements, (ii) the designer, who proposes technical solutions to satisfy the requirements, (iii) the occupational therapist, who validates the solutions and evaluates potential risks posed by them, and (iv) the researcher, who integrates the developers by permanent communication and makes sure the team employs state-of-the-art techniques and materials (Ostuzzi et al., 2015). Furthermore, as co-designers, the patients get actively involved in conceiving and testing the product, so they perceive it as their own creation. The satisfaction associated with a personal accomplishment favors device acceptance. The psychological bond between the user and the assistive device is further consolidated by the iterative optimization process. As the product is tested and redesigned, the patient becomes familiar with its capabilities and grows fond of it. The emotional bond with the product is strengthened not just directly, as a result of the invested time and effort, but also indirectly, due to the self-expressive value of the product. These factors are important because "*emotions enrich a person's life and can increase one's general experience of well-being*" (Mugge et al., 2009).

In the +TUO project, 10 participants were recruited from a group of patients who volunteered to fill out a questionnaire about daily activities and assistive device needs. Dressing up, opening bottles and jars, cooking, and climbing stairs were deemed problematic by many responders (Ostuzzi et al., 2015).

The first meeting of all the players served for team building and informing the end users about additive manufacturing. Then, the team engaged in the co-design process—several cycles of conception, prototype fabrication, and testing. Finally, when a prototype was found satisfactory, the team moved to the next stage, called co-production, in which the final product was built using the same 3D printers as for prototype fabrication, but with special care for optimizing the printing process and the printout's appearance. Fig. 4.8A depicts assistive devices made for and with RD patients during the implementation of the +TUO project (Ostuzzi et al., 2015).

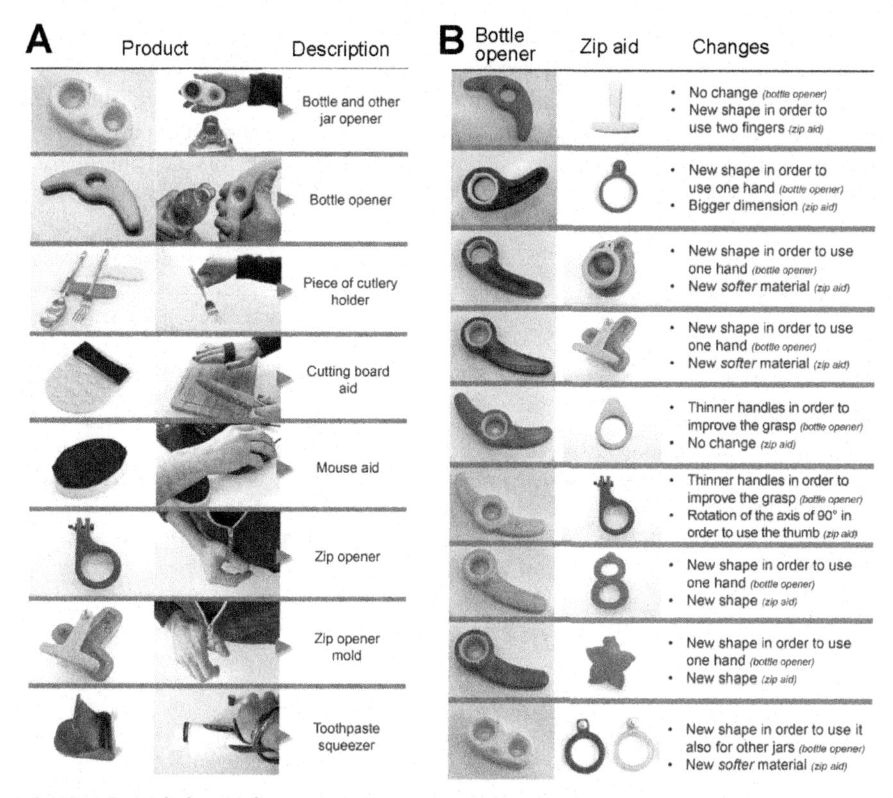

FIGURE 4.8 Assistive devices created for people with rheumatoid diseases.

(A) Examples of 3D-printed assistive devices produced by a team that included prospective users as co-designers; (B) different versions of bottle openers and zipper aids created during the iterative co-design process.

Reprinted with permission from Ostuzzi, F., Rognoli, V., Saldien, J., & Levi, M. (2015). TUO project: low cost 3D printers as helpful tool for small communities with rheumatic diseases. Rapid Prototyping Journal. *Emerald Publishing Limited all rights reserved.*

Thoroughly tested assistive devices helped the patients to eat and drink on their own, maintain their oral hygiene, cook, operate the zippers of their clothes and bags, and navigate their computers. The iterations performed during the co-design of two products (bottle opener and zip aid), shown in Fig. 4.8B, demonstrate the creativity of the development team.

The main strengths of the co-design and co-production approach included the emergence of unexpected solutions, low cost, as well as product usability, and accessibility. On the social side, the sense of community dismantled barriers and catalyzed creative thinking. Satisfied end-users reached out to other patients, raising new challenges because it is not clear how to scale the collaborative development to engage larger groups of geographically separated users. In this respect, the experience of open-source communities, such as e-NABLE (Enabling the Future (e-NABLE), n.d.) and Thingiverse (MakerBot Thingiverse, n.d.), might provide valuable insights.

Already in 2015, Thingiverse provided free access to hundreds of assistive device models created by over 100 designers (Buehler et al., 2015). Some of them were as simple as a mushroom-shaped knife blade attachment, a beverage can holder, or a wearable pill box, while others embodied multiple 3D-printed parts. As of May 2022, the search term "prosthesis" returned more than 200 hits on Thingiverse, including the Cyborg Beast, the Talon Hand 3.0, the Cryptic Pattern Leg Prosthesis, as well as simpler devices, such as a wrist brace or the Solo Finger Pen, to mention just a few, liked by thousands. Navigating Thingiverse is quite fun and it may become a life-changing experience for a disabled person!

References

Anderson, B., & Schanandore, J. V. (2021). Using a 3D-printed prosthetic to improve participation in a young gymnast. *Pediatric Physical Therapy, 33*. https://journals.lww.com/pedpt/Fulltext/2021/01000/Using_a_3D_Printed_Prosthetic_to_Improve.18.aspx.

Andringa, A., van de Port, I., & Meijer, J.-W. (2013). Long-term use of a static hand-wrist orthosis in chronic stroke patients: A pilot study. *Stroke Research and Treatment*. https://doi.org/10.1155/2013/546093, 546093–546093.

Assistive Technology Industry Association (ATIA). (n.d.). Retrieved May 20, 2022, from www.atia.org/about-atia/.

Biddiss, E., & Chau, T. (2007). Upper-limb prosthetics: Critical factors in device abandonment. *American Journal of Physical Medicine & Rehabilitation, 86*. https://journals.lww.com/ajpmr/Fulltext/2007/12000/Upper_Limb_Prosthetics__Critical_Factors_in_Device.4.aspx.

Buehler, E., Branham, S., Ali, A., Chang, J. J., Hofmann, M. K., Hurst, A., & Kane, S. K. (2015). Sharing is caring: Assistive technology designs on thingiverse. In *Proceedings of the 33rd annual ACM conference on human factors in computing systems* (pp. 525–534). Association for Computing Machinery. https://doi.org/10.1145/2702123.2702525

Calado, A., Soares, F., & Matos, D. (2019). A review on commercially available anthropomorphic myoelectric prosthetic hands, pattern-recognition-based microcontrollers and sEMG sensors used for prosthetic control. In *2019 IEEE international conference on autonomous robot systems and competitions (ICARSC)* (pp. 1–6). https://doi.org/10.1109/ICARSC.2019.8733629

Classify Your Medical Device. (2020). *United States Food and Drug administration*. www.fda.gov/medical-devices/overview-device-regulation/classify-your-medical-device.

Cleveland Clinic Health Library (n.d.). Retrieved May 23, 2022, from https://www.my.clevelandclinic.org/health.

Convention on the Rights of Persons with Disabilities and Optional Protocol. (May 3, 2008). United Nations. www.un.org/development/desa/disabilities/convention-on-the-rights-of-persons-with-disabilities.html#Fulltext.

De Agustín Del Burgo, J. M., Blaya Haro, F., D'Amato, R., & Juanes Méndez, J. A. (2020). Development of a smart splint to monitor different parameters during the treatment process. *Sensors, 20*, 4207. www.mdpi.com/1424-8220/20/15/4207.

Delda, R. N., Basuel, R. B., Hacla, R. P., Martinez, D. W., Cabibihan, J.-J., & Dizon, J. R. (2021). 3D printing polymeric materials for robots with embedded systems. *Technologies, 9*, 82. https://doi.org/10.3390/technologies9040082

Dollar, A. M., & Herr, H. (2008). Lower extremity exoskeletons and active orthoses: Challenges and state-of-the-art. *IEEE Transactions on Robotics, 24*, 144–158. https://doi.org/10.1109/TRO.2008.915453

"El Medallo" Bionic Arm. (2019). https://hub.e-nable.org/s/e-nable-devices/wiki/page/view?title=El+Medallo+Bionic+Arm.

Enabling the Future (e-NABLE). (n.d.). Retrieved May 20, 2022, from http://enablingthefuture.org/.

Faustini, M. C., Neptune, R. R., Crawford, R. H., & Stanhope, S. J. (2008). Manufacture of passive dynamic ankle-foot orthoses using selective laser sintering. *IEEE Transactions on Biomedical Engineering, 55*, 784–790. https://doi.org/10.1109/tbme.2007.912638

Halanski, M., & Noonan, K. J. (2008). Cast and splint immobilization: Complications. *JAAOS-Journal of the American Academy of Orthopaedic Surgeons, 16*, 30–40. https://journals.lww.com/jaaos/Fulltext/2008/01000/Cast_and_Splint_Immobilization__Complications.5.aspx.

Hero Arm User Manual 6.0. (June 2, 2021). *Open bionics*. https://openbionics.com/wp-content/uploads/2021/06/d100161_06-00-EXTERNAL-USE_Hero-Arm-User-Manual.pdf.

ISO 8549-1:2020(en) Prosthetics and orthotics—Vocabulary—Part 1: General terms for external limb prostheses and external orthoses. (2020). International Organization for Standardization. www.iso.org/obp/ui/#iso:std:iso:8549:-1:ed-2:v1:en.

Liacouras, P. C., Sahajwalla, D., Beachler, M. D., Sleeman, T., Ho, V. B., & Lichtenberger, J. P. (2017). Using computed tomography and 3D printing to construct custom prosthetics attachments and devices. *3D Printing in Medicine, 3*, 8. https://doi.org/10.1186/s41205-017-0016-1

Li, J., & Tanaka, H. (2018). Rapid customization system for 3D-printed splint using programmable modeling technique—a practical approach. *3D Printing in Medicine, 4*, 5. https://doi.org/10.1186/s41205-018-0027-6

Lunsford, C., Grindle, G., Salatin, B., & Dicianno, B. E. (2016). Innovations with 3-dimensional printing in physical medicine and rehabilitation: A review of the literature. *PM&R, 8*, 1201–1212. https://doi.org/10.1016/j.pmrj.2016.07.003

MakerBot Thingiverse. (n.d.). Retrieved May 25, 2022, from www.makerbot.com/thingiverse/.

Manero, A., Smith, P., Sparkman, J., Dombrowski, M., Courbin, D., Kester, A., Womack, I., & Chi, A. (2019). Implementation of 3D printing technology in the field of prosthetics: Past, present, and future. *International Journal of Environmental Research and Public Health, 16*. https://doi.org/10.3390/ijerph16091641

Meet the Hero Arm – a prosthetic arm for adults and children. (2022). *Open bionics*. https://openbionics.com/hero-arm/.

Mugge, R., Schoormans, J. P. L., & Schifferstein, H. N. J. (2009). Emotional bonding with personalised products. *Journal of Engineering Design, 20*, 467–476. https://doi.org/10.1080/09544820802698550

Neomano. (2019). *Neofect*. Watertown, MA, USA www.neofect.com/us/neomano.

Ostuzzi, F., Rognoli, V., Saldien, J., & Levi, M. (2015). + TUO project: Low cost 3D printers as helpful tool for small communities with rheumatic diseases. *Rapid Prototyping Journal, 21*(5), 491–505. https://doi.org/10.1108/RPJ-09-2014-0111

Our story - making 3D prosthetics beautiful - Open Bionics. (2021). *Open bionics*. https://openbionics.com/en/our-story/.

Popescu, D., Zapciu, A., Tarba, C., & Laptoiu, D. (2020). Fast production of customized three-dimensional-printed hand splints. *Rapid Prototyping Journal, 26*, 134–144. https://doi.org/10.1108/RPJ-01-2019-0009

Qamar, I. P. S., Groh, R., Holman, D., & Roudaut, A. (2018). HCI meets material science: A literature review of morphing materials for the design of shape-changing interfaces. In *Proceedings of the 2018 CHI conference on human factors in computing systems* (p. Paper 374). Association for Computing Machinery. https://doi.org/10.1145/3173574.3173948

Robotic Arm Version 2. (2020). https://github.com/Humanos3D/RoboticArmV2/wiki/Robotic-Arm-Version-2.

Sandy, E., & Johansson, E. (October 9, 2018). *Open bionics is reimagining disabilities and the Dalai Lama is totally onboard. Elite Business*. http://elitebusinessmagazine.co.uk/interviews/item/open-bionics-is-reimagining-disabilities-and-the-dalai-lama-is-totally-onboard.

Sun, L., Yang, Y., Chen, Y., Li, J., Luo, D., Liu, H., Yao, L., Tao, Y., & Wang, G. (2021). ShrinCage: 4D printing accessories that self-adapt. In *Proceedings of the 2021 CHI conference on human factors in computing systems* (p. 433). Association for Computing Machinery. https://doi.org/10.1145/3411764.3445220

Ten Kate, J., Smit, G., & Breedveld, P. (2017). 3D-printed upper limb prostheses: A review. *Disability and Rehabilitation: Assistive Technology, 12*, 300–314. https://doi.org/10.1080/17483107.2016.1253117

Wendo, K., Barbier, O., Bollen, X., Schubert, T., Lejeune, T., Raucent, B., & Olszewski, R. (2022). Open-source 3D printing in the prosthetic field—the case of upper limb prostheses: A review. *Machines*, 10. https://doi.org/10.3390/machines10060413

WHO standards for prosthetics and orthotics. (2017). Geneva: World Health Organization. https://apps.who.int/iris/handle/10665/259209.

WIPO Technology Trends 2021: Assistive Technology. (2021). *World intellectual property organization (WIPO)*. www.wipo.int/edocs/pubdocs/en/wipo_pub_1055_2021.pdf.

Yang, Y.-S., Tseng, C.-H., Fang, W.-C., Han, I.-W., & Huang, S.-C. (2021). Effectiveness of a new 3D-printed dynamic hand–wrist splint on hand motor function and spasticity in chronic stroke patients. *Journal of Clinical Medicine, 10*, 4549. www.mdpi.com/2077-0383/10/19/4549.

Yoo, H.-J., Lee, S., Kim, J., Park, C., & Lee, B. (2019). Development of 3D-printed myoelectric hand orthosis for patients with spinal cord injury. *Journal of NeuroEngineering and Rehabilitation, 16*, 162. https://doi.org/10.1186/s12984-019-0633-6

Zuniga, J., Katsavelis, D., Peck, J., Stollberg, J., Petrykowski, M., Carson, A., & Fernandez, C. (2015). Cyborg beast: A low-cost 3d-printed prosthetic hand for children with upper-limb differences. *BMC Research Notes, 8*, 10. https://doi.org/10.1186/s13104-015-0971-9

Zuniga, J. M., Young, K. J., Peck, J. L., Srivastava, R., Pierce, J. E., Dudley, D. R., Salazar, D. A., & Bergmann, J. (2019). Remote fitting procedures for upper limb 3d printed prostheses. *Expert Review of Medical Devices, 16*, 257–266. https://doi.org/10.1080/17434440.2019.1572506

CHAPTER

3D Bioprinting techniques

5

Bioprinting involves the positioning of live cells and biomaterials (if needed) to produce tissue substitutes. A bioprinter is a numerically controlled device capable of material deposition according to a digital model inferred from medical imaging or created from scratch using 3D modeling software. The definition and a brief history of 3D bioprinting are presented in Chapter 1.

Given the ambitious goal of replicating native tissue morphology and function, several bioprinting technologies have been proposed. Although none of them is up to the task, the progress made so far is a reason to look forward with optimism. A decade ago, an inspiring review article stated that "the majority of published work to date is at a relatively low level of technological readiness and has used a very limited range of materials." Derby, (2012). In contrast, "The bioprinting roadmap"—a recent article published by important players in the field—considers that the set of bioink formulations "has expanded greatly in the last decade, with numerous materials—primarily natural and synthetic hydrogels—being applied or developed to meet the stringent demands of bioprinting." (Sun et al., 2020). The opinion expressed in this recent article is that each bioprinting method available to date has limitations, but "the advances of each method seem to complement the shortcomings of other methods." Therefore, the authors conjecture that "future technologies will encompass and utilize multiple modalities into single platforms along with the integration of novel processes, such as cell aggregate bioprinting techniques, in order to fabricate scalable, structurally stable, and perfusable tissue constructs with enhanced biomimicry and desired functionality" (Sun et al., 2020).

This chapter describes the most common bioprinting methods. According to their physical principles, current bioprinting technologies are classified into three broad categories: (i) extrusion-based bioprinting, (ii) droplet-based bioprinting, and (iii) light-based bioprinting. These will be discussed in the first three sections, whereas the last section will treat tissue spheroid bioprinting techniques.

1. Extrusion-based bioprinting

Extrusion-based bioprinting (EBB) is rooted in experimental investigations of computer-guided pneumatic dispensing of polymer solutions through a nozzle of less than 1 mm in inner diameter (Landers & Mülhaupt, 2000). Single- and dual-cartridge systems were built, with independent heating of cartridge and nozzle,

Towards 4D Bioprinting. https://doi.org/10.1016/B978-0-12-818653-4.00004-8

and dispensing was performed in liquid media to ensure solidification as a result of precipitation, curing, or polyelectrolyte formation. Matching the density of the medium with that of the extruded material conferred stability to the dispensed structure due to buoyancy. This technique was initially called 3D plotting, but it was soon renamed 3D bioplotting because the authors were aware that their instrument is suitable to create tissue engineering scaffolds as well as to handle hydrogels, proteins, and even cells (Pfister et al., 2004). Indeed, the possibility of 3D plotting "biocomponents such as proteins, growth factors, cells" was mentioned already in their first work on this topic (Landers & Mülhaupt, 2000).

The advent of EBB started with the development of multinozzle systems capable of extruding cell-laden hydrogels (Khalil et al., 2005), cell spheroids (Jakab et al., 2006), and multicellular cylinders (Norotte et al., 2009). The commercial availability of affordable EBB instruments along with dedicated bioinks has facilitated the widespread use of EBB. A variety of tissue constructs were produced by EBB (e.g., skin, cartilage, bone, liver, heart), as well as tissue-on-chip devices (Ozbolat & Hospodiuk, 2016).

1.1 Physical principle

An extrusion-based bioprinter includes one or more print heads (cartridges) outfitted with micro-needle nozzles. The print head exerts pressure on the enclosed bioink to dispense it while the nozzle moves under computer control. Bioink extrusion can be driven pneumatically (by air pressure), by exerting a force on a piston, or by a screw mechanism Fig. 5.1.

Pneumatic micro-extrusion is the most common option. It is affordable and cell-friendly, but it might pose problems when the pressure needs to be adjusted during printing. Due to the large volume of compressed air found in the tubing, there is a delay between the pressure change and the corresponding change in bioink flow. A piston-based print head offers better control over the bioink flow through the nozzle. Screw-based print heads are especially suited for highly viscous bioinks, or biomaterial inks such as molten polycaprolactone (PCL), and offer the best spatial control. Nevertheless, when it is used for bioink delivery, a screw-based print head might affect cell viability by exerting excessive pressure on the bioink. Therefore, off-the-shelf screw-based mechanisms are not recommended for bioprinting; one needs to design the screw mechanism by taking into account the needs of cells (Malda et al., 2013). Actually, both piston- and screw-based systems are prone to generate a huge pressure gradient in the nozzle, and the associated shear stress inflicts damage on cell membranes. Multimaterial bioprinters can take advantage of different types of print heads; for example, a screw-based head can be used for hot-melt extrusion, whereas pneumatic or piston-based print cartridges can be used for cell-laden hydrogel dispensing (Ozbolat & Hospodiuk, 2016).

The drawbacks of pneumatic micro-extrusion have been mitigated by building mechanical or magnetic valves into the print heads. A mechanical valve-based nozzle is depicted in Fig. 5.2A. In the absence of pressurized air, the spring blocks

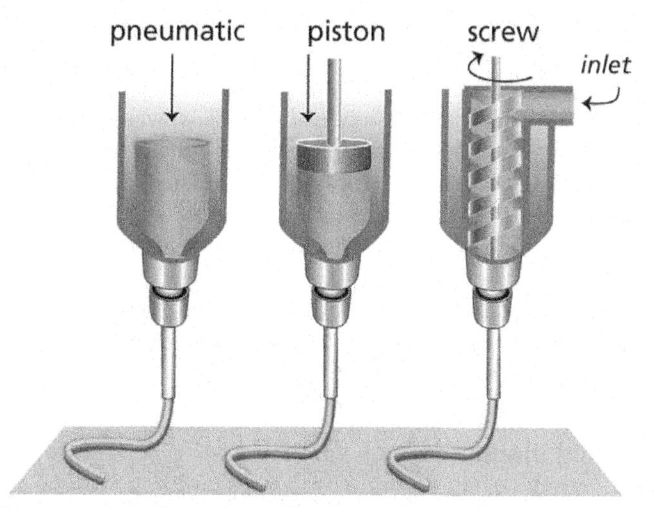

FIGURE 5.1 Physical mechanisms of extrusion-based bioprinting (EBB).

Pneumatic microextrusion (*left*) relies on air pressure. In a piston-based configuration (*middle*), extrusion is accomplished by a force exerted onto the piston via a mechanical device driven by an electric motor. In a screw-based system (*right*), an electric motor rotates the screw that advances the bioink.

Reproduced with permission from John Wiley and Sons from Malda, J., Visser, J., Melchels, F.P., Jüngst, T., Hennink, W.E., Dhert, W.J.A., Groll, J. & Hutmacher, D.W. (2013). 25th anniversary article: Engineering hydrogels for biofabrication. Advanced Materials, 25, *5011–5028. https://doi.org/10.1002/adma.201302042.*

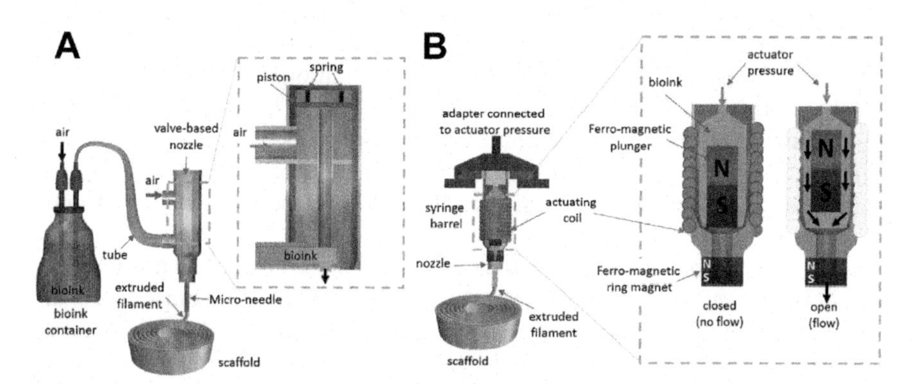

FIGURE 5.2 Micro-extrusion bioprinting modulated by valves.

(A) Pneumatic print head outfitted with a mechanical valve and (B) Solenoid-gated pneumatic print head.

Reprinted from Ozbolat, I.T. & Hospodiuk, M. (2016). Current advances and future perspectives in extrusion-based bioprinting. Biomaterials, 76, *321–343. https://doi.org/10.1016/j.biomaterials.2015.10.076. Copyright (2016), with permission from Elsevier.*

the central tube and, therefore, no force acts on the bioink filament. When air pressure is applied, the spring becomes compressed and lets air into the vertical tube. The bioink filament advances under the action of the pressure applied from the left and from the top, remaining with the sole option to move downwards through the nozzle. As soon as the pressure is canceled, the spring pushes the piston against the vertical tube, and the extrusion ceases instantaneously. This system allows for an accurate dosage of hydrogel strands. The magnetic valve device shown in Fig. 5.2B, also known as the solenoid micro-extrusion system, modulates pneumatic extrusion via a magnetic stopper. The pressure is permanently on, but the ferromagnetic plunger is attracted toward the magnet that embraces the nozzle, blocking it. When an electric current flows through the solenoid, the generated magnetic field lifts the plunger, releasing the hydrogel from the cartridge.

1.2 Bioinks and bioprinters

The printability of a bioink depends on its rheological properties. In the most common case of a hydrogel-based bioink, the extruded material is a hydrogel precursor (a polymer solution) mixed with live cells. Once the structure is printed, it should have, or be endowed with, mechanical stability. To this end, the hydrogel precursor is converted into a gel via crosslinking accomplished by a temperature change, light exposure, the addition of a chemical, or a combination thereof. The rheological properties of a hydrogel precursor are mainly determined by the polymer concentration and molecular weight. Constructs made of highly viscous materials do not collapse upon deposition (more precisely, they do it slowly, providing time for gelation); nevertheless, such materials require a high pressure for extrusion (potentially affecting cell viability) and represent a restrictive environment for cell migration, proliferation, differentiation, and ECM deposition. A cell-friendly hydrogel contains low concentrations of high-molecular-weight polymers, which results in far from ideal rheological properties. A compromise is needed to assure both shape fidelity and cell viability (Malda et al., 2013). Therefore, the study of bioink formulations has been and remains a vivid research field (Tarassoli et al., 2021). Here, we discuss the strengths and weaknesses of widely used bioinks and bioprinters, whereas the problem of printability will be examined more closely in Chapter 6.

Besides the harsh conditions experienced by cells in the course of extrusion, cell viability is affected also by lengthy printing. The extruded bioink filament is not an ideal medium for cells: temperature and humidity are suboptimal, while gas exchange and nutrient transport are hampered. To shorten the fabrication time for a given bioink, one can raise the applied pressure or use a nozzle with a larger orifice, which, in turn, results in lower resolution. The success of EBB hinges on numerous factors, as illustrated in Fig. 5.3 (Malda et al., 2013).

EBB technology evolved in a remarkable synergy with bioink development Fig. 5.4 (Ozbolat & Hospodiuk, 2016). Even the simplest EBB system, shown in Fig. 5.4A, is highly versatile: depending on nozzle diameter, applied pressure, and

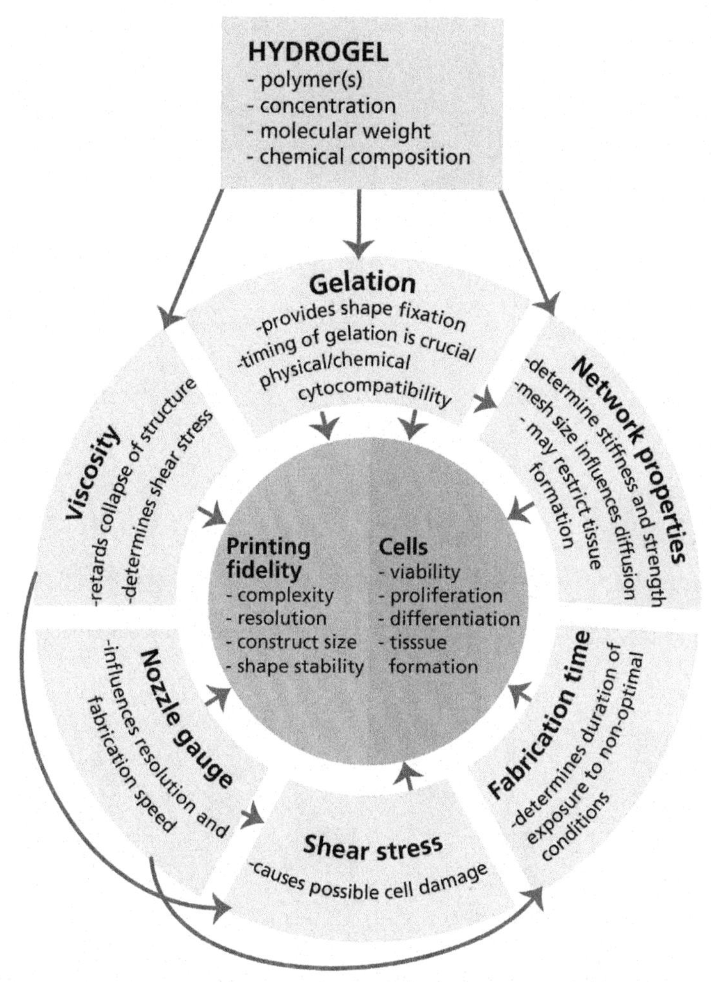

FIGURE 5.3 The impact of material properties and printing parameters on the outcome of extrusion-based bioprinting (EBB).

The composition of the hydrogel precursor solution determines its rheological and gelation properties; these properties, as well as the printing parameters (e.g., nozzle gauge, fabrication time), affect the geometry, viability, and functionality of the bioprinted construct.

Reprinted with permission from John Wiley and Sons from Malda, J., Visser, J., Melchels, F.P., Jüngst, T., Hennink, W.E., Dhert, W.J.A., Groll, J. & Hutmacher, D.W. (2013). 25th anniversary article: Engineering hydrogels for biofabrication. Advanced Materials, 25, *5011–5028. https://doi.org/10.1002/adma.201302042.*

print head velocity, EBB enables the delivery of a variety of bioinks whose viscosities range from 30 mPa·s to over 6×10^7 mPa·s.

Certain EBB techniques, shown in Fig. 5.4B1–B5, have been specifically tailored for building tissue constructs made of cells dispersed in alginate (Ozbolat

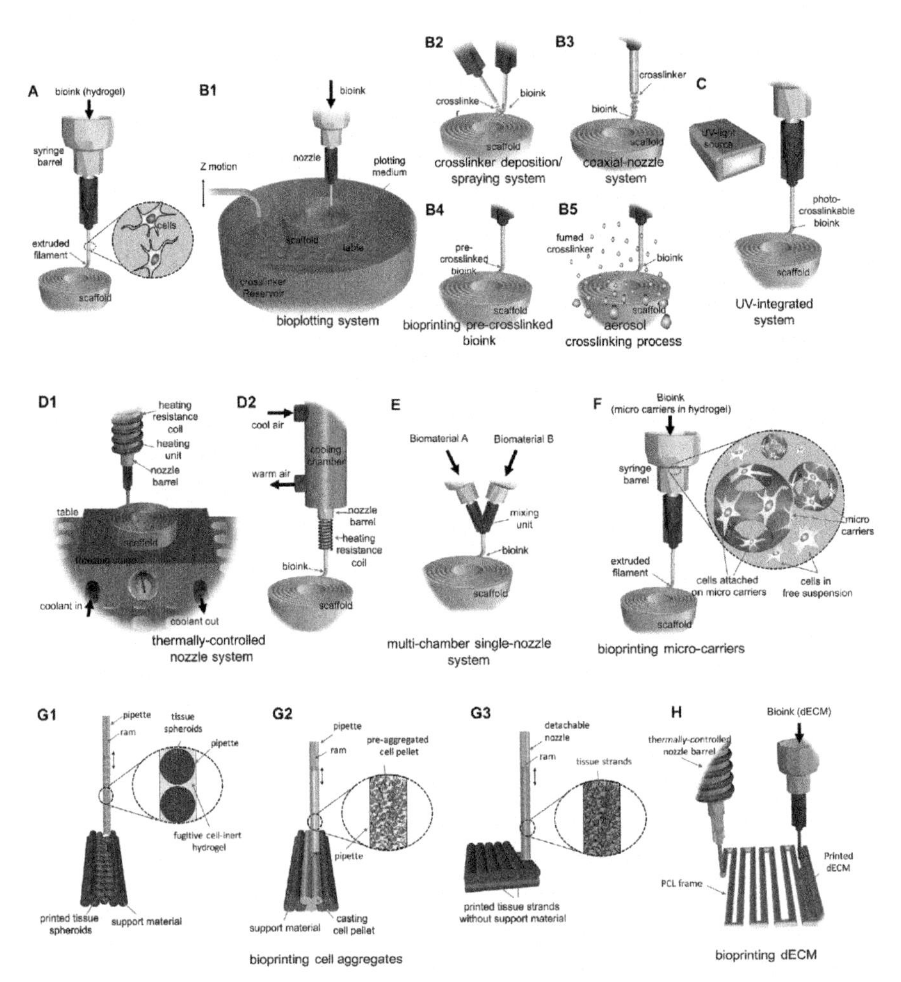

FIGURE 5.4 Extrusion-based bioprinting (EBB) instruments adapted for specific bioinks.

(A) Direct extrusion of cell-laden hydrogel bioinks; (B1) the original concept of bioplotting in which bioink is dispensed into a crosslinker solution; (B2) dual nozzle system for bioink delivery in tandem with crosslinker deposition or spraying; (B3) coaxial extrusion system in which the bioink filament is bathed by a simultaneously delivered crosslinker solution; (B4) extrusion of pre-crosslinked bioink; (B5) bioink extrusion into a crosslinker aerosol—much finer droplets than the ones obtained by spraying; (C) microextrusion combined with UV-curing; (D1) extrusion through a heated nozzle onto a cooled print bed; (D2) preservation of the bioink in a cooled fluid chamber and extrusion through a heated nozzle; (E) extrusion of multiple bioinks through a single nozzle—enables simultaneous, as well as sequential deposition of two or more sorts of bioink; (F) delivery of composite bioinks composed of cell-laden microcarriers dispersed in a hydrogel; (G1) extrusion of tissue spheroid chains along with support structures made of a sacrificial hydrogel that lacks cell adhesion motifs; (G2) extrusion of cylindrical bioink filaments into a

& Hospodiuk, 2016). Alginate is a generic name for a family of polysaccharides composed of variable ratios of beta-D-mannuronic acid (M) and alpha-L-guluronic acid (G) of natural origin—extracted from various genera of brown seaweed or from two genera of bacteria (*Pseudomonas* and *Azotobacter*) (Hay et al., 2013). Alginate undergoes instantaneous gelation in contact with aqueous solutions that contain Ca^{2+} ions. This property is exploited in different flavors of EBB instruments. One of them is the original bioplotting setup (Landers & Mülhaupt, 2000), suitable for alginate bioink extrusion into a crosslinker pool (plotting medium) Fig. 5.4B1. The construct remains immersed in the plotting medium until its completion, and buoyancy provides mechanical support provided that the bioink has a slightly higher density than the plotting medium (Ozbolat & Hospodiuk, 2016). In another approach, a secondary nozzle is included to deliver the crosslinker simultaneously with the bioink Fig. 5.4B2. The secondary nozzle revolves around the primary nozzle and can eject the crosslinker solution either as a continuous jet or as droplets. A third option is to use a coaxial nozzle to extrude the bioink through the core nozzle, while the crosslinker is supplied via the slightly longer outer nozzle, wrapping the alginate filament Fig. 5.4B3. The fourth technique dedicated to alginate bioprinting consists in extruding mildly precrosslinked alginate, thereby assuring structural integrity at the cost of exposing the cells to higher shear stress; the process is completed by bathing the bioprinted structure in a concentrated crosslinker solution Fig. 5.4B4. In the fifth method, crosslinker fumes generated by an ultrasonic humidifier envelope the bioprinted alginate filaments, inducing gelation gradually Fig. 5.4B5. Unlike spraying, ultrasonic fuming generates minuscule droplets, which envelope the entire construct uniformly and induce crosslinking between adjacent layers; this approach endows the construct with structural integrity and mechanical stability—from these points of view, microextrusion combined with ultrasonic fuming is only surpassed by coaxial printing and precrosslinked alginate printing (Ozbolat & Hospodiuk, 2016).

Gelatin is yet another hydrogel that inspired the development of specific bioprinters. Gelatin is a protein derived from collagen by controlled hydrolysis. It forms a thermoreversible hydrogel that is biocompatible, nonimmunogenic, and biodegradable. Nevertheless, it is only stable at low temperatures and melts right below 37°C. Among many physicochemical methods devised to improve the printability and stability of gelatin-based hydrogels, the most successful was a chemical modification with methacrylamide side groups. The result is a photocrosslinkable

fugitive non-adhesive hydrogel support structure; (G3) extrusion of tissue strands in a self-supportive 3D arrangement; (H) hybrid EBB involving the deposition of thermoplastic (e.g., PCL) filaments to support bioactive hydrogel (e.g., decellularized ECM) threads during their gelation.

Reprinted from Ozbolat, I.T. & Hospodiuk, M. (2016). Current advances and future perspectives in extrusion-based bioprinting. Biomaterials, 76, *321–343. https://doi.org/10.1016/j.biomaterials.2015.10.076.*

hydrogel, gelatin methacryloyl (GelMA), which can be printed by a microextrusion printer outfitted with a UV light source (Fig. 5.4C).

Hyaluronic acid (HA) is a nonsulfated glycosaminoglycan present in most connective tissues. Since it is the major constituent of cartilage ECM, it is used in rehabilitation medicine to lubricate painful arthritic knee joints, but it was found useful also in other clinical areas, including dermatology, plastic surgery, and ophthalmology. Chondrocytes suspended in HA hydrogel had high viability, making it an appealing candidate for EBB. Nevertheless, chemical modifications are needed to control the degradation rate and enhance the mechanical properties. For example, functionalization with methacrylate groups resulted in a UV-curable HA-MA hydrogel with excellent biocompatibility and shape fidelity. Jeon et al., on the other hand, prepared hydrogels with enhanced mechanical properties and reduced the degradation rate by crosslinking HA with poly(ethylene glycol)-diamine (Jeon et al., 2007).

Poly(ethylene glycol) (PEG), also known as poly(ethylene oxide) (PEO), is a hydrophilic, biocompatible polyether compound synthesized via a ring-opening polymerization of ethylene oxide. PEG is widely used in the pharmaceutical, cosmetic, and food industries. The FDA approved PEG-based hydrogels for internal use (e.g., as wound dressings). Also, photocrosslinkable PEG-diacrylate (PEG-DA) hydrogels were used in the 3D bioprinting of aortic valve scaffolds seeded with porcine aortic valve interstitial cells. During 3 weeks in culture, the cells remained anchored to the scaffolds, preserved their viability, but did not spread and did not divide. Hence, PEG-based hydrogels need to be functionalized with cell adhesion motifs in order to promote cell proliferation and migration. HA-MA, HA-PEG-diamine, and PEG-DA hydrogels can be printed using the UV-integrated setup of Fig. 5.4C (Ozbolat & Hospodiuk, 2016).

Certain hydrogels require precise control of the extruder's temperature. For example, agarose can be printed by using a heated print head and a cooled print bed (Fig. 5.4D1). Agarose is a polysaccharide extracted from seaweed, available as a powder that dissolves in hot water and becomes a gel when it cools. Agarose gels and melts at different temperatures: when hot agarose solution is cooled down, it undergoes the sol-gel transition at the gelling temperature; when the gel is warmed up, it melts at a higher temperature. For bioprinting, it is advantageous to induce chemical modifications in natural agarose. The hydroxyethylation of the agarose molecule lowers both transition temperatures, such that the physiological temperature falls between the gelling temperature (30°C) and the melting temperature (65°C). Cells can be dispersed in the liquid polymer solution, they become embedded in the hydrogel upon cooling, and remain so as the system is warmed up to 37°C. The interest in the bioprinting of agarose hydrogels is justified by its cytocompatibility, ease of handling, and by its cell-adhesion inert nature. Adipose-derived adult stem cells encapsulated in agarose gel had undergone chondrogenic differentiation and remained viable for weeks. Also, agarose gel has been used as a nonadhesive substrate for cell aggregate formation (Ozbolat & Hospodiuk, 2016).

Other hydrogels, such as collagen type I, matrigel, and pluronic, need to be kept below room temperature to preserve them in sol state in the printer cartridge. Their gelation is induced by heating toward physiological temperature, and the fine-tuning of the nozzle temperature provides control over their gelation kinetics and printability. Such hydrogels motivated the development of EBB instruments with cooled cartridges and heated nozzles Fig. 5.4D2.

Collagen is the name of a family of extracellular proteins composed of three polypeptide chains that include at least one repeating sequence of amino acids. The three chains wrap around each other in a right-handed, rope-like, and triple-helical arrangement. Collagen hydrogels are extensively used in tissue engineering and regenerative medicine (Pawelec et al., 2016). Collagen type I is the prevalent protein in native ECM, contributing to its structural and mechanical properties. Pathological fibrosis observed in a variety of conditions (such as myocardial infarction, scar formation, or cancer invasion) is associated with changes in collagen microarchitecture, which, in turn, modulates the behavior of myofibroblasts. Seo et al. prepared collagen type I hydrogels of distinct microarchitecture via thermal regulation of the gelation speed (Seo et al., 2020). Importantly, they observed that adipose-derived stem cells cultured in collagen type I scaffolds with thicker fibers and larger pores had a more pronounced differentiation into myofibroblasts and secreted larger amounts of proangiogenic factors—vascular endothelial growth factor (VEGF) and interleukin-8 (IL-8). These findings demonstrate the importance of precise temperature control in the EBB of collagen type 1 hydrogels.

Matrigel is the commercial name of a soluble extract of basement membrane proteins derived from an ECM-rich mouse tumor, the Engelbreth–Holm–Swarm (EHS) sarcoma (Kleinman & Martin, 2005). It mainly contains laminin, collagen type IV, perlecan, and nidogen, as well as proteases (such as matrix metalloproteinase-2 (MMP-2) and MMP-9), and various growth factors. Therefore, it is extremely bioactive, supporting the proliferation of a wide variety of cell types, and promoting the differentiation of stem cells. A Matrigel solution is prepared in concentrations ranging from 10 to 15 mg/mL; it is stored at −20°C, and, before use, it is slowly thawed on ice inside a refrigerator. It is liquid at 4°C. Upon warming, it gels between 24 and 37°C within half an hour and does not redissolve when cooled (Kleinman & Martin, 2005). In EBB, the extrusion of Matrigel should be completed before it becomes fully crosslinked; hence, the cooled print head is essential, and, for shape fidelity, a heated print bed is recommended to accelerate the crosslinking of the deposited hydrogel (Hospodiuk et al., 2017).

Pluronic is the trade name of a nonionic triblock copolymer composed of poly(-ethylene glycol)–poly(propylene glycol)–poly(ethylene glycol) (PEG–PPG–PEG). Poly(ethylene glycol) is a hydrophilic polymer, whereas poly(propylene glycol) is hydrophobic, so pluronic is amphiphilic. At a concentration of over 14% by mass, pluronic F-127 forms micelles at a temperature of about 12°C, whereas around 30°C it turns into a physically crosslinked hydrogel due to micelle entanglement (Lippens et al., 2011). Subsequent cooling leads to a gel-sol transition below 10°C. Pluronic F-127 has been approved by the FDA as a material suitable for in vivo applications in humans. It is injected into the body as a liquid and forms a gel at the destination, serving as a versatile drug delivery system. Nevertheless, it erodes fast in contact with body fluids, as well as in cell culture media (Lippens et al., 2011). The fast degradation of the Pluronic F-127 hydrogel is a drawback in certain applications, but in bioprinting it enabled an elegant solution to a long-standing problem of tissue engineering: the fabrication of thick, perfusable tissue constructs (Kolesky et al., 2014, 2016). Pluronic F-127 augmented with thrombin was deposited wherever a lumen was desired and cast into a cell-laden hydrogel (fibrinogen and gelatin). Released by the Pluronic filaments, thrombin triggered the formation of fibrin gel. Finally, the construct was cooled to liquefy and evacuate the Pluronic, leaving behind channels suitable for being connected to a perfusion bioreactor. In this context, Pluronic F-127 hydrogel was used as a sacrificial (fugitive) biomaterial ink.

Fibrin is appreciated in tissue engineering due to its facile gelation and cell-friendly microstructure. An aqueous solution of fibrinogen, thrombin, and Ca^{2+} ions forms a fibrin gel: the proteolytic enzyme thrombin cleaves the fibrinogen into fibrin monomers, which, upon polymerization, turn into a fibrin network. Calcium ions contribute to this process by accelerating the polymerization of fibrin monomers (Brass et al., 1978). Fibrin exposes multiple cell adhesion motifs, including the amino acid sequence arginine-glycine-aspartic acid (RGD), which has a strong affinity for integrins—transmembrane glycoproteins involved in cell–ECM interaction. The fibrin gel has a nonlinear elasticity, allowing for ample deformations without rupturing (Hospodiuk et al., 2017). Importantly, Nakatsu et al. observed the formation of new blood vessels (angiogenesis) when human umbilical vein endothelial cells (HUVECs) were co-cultured with fibroblasts in fibrin gel (Nakatsu et al., 2003). Although the polymerization conditions allow for a certain level of control over the mechanical properties and degradation kinetics of fibrin, it is far from satisfactory from these points of view. Moreover, its fast gelation can lead to nozzle clogging during EBB. To prevent it, one option is to blend fibrinogen and thrombin on ice, keep the mixture in a cooled print head to hamper gelation, and print it using a heated nozzle to control the rate of crosslinking (Fig. 5.4D2). Also, the multi-chamber, single-nozzle system shown in Fig. 5.4E enables the EBB of fibrin by mixing the two components right before extrusion (Ozbolat & Hospodiuk, 2016).

EBB is feasible also for building composite structures, such as tissue constructs made of a hydrogel mixed with cell-laden microcarriers (Fig. 5.4). The material properties, geometry, and porosity of the microcarriers have been optimized for culturing various cell types of the osteoarticular system, but the approach is promising for other cell types, as well. Studies conducted to date indicate that microcarriers favor cell aggregation and proliferation. Challenges remain, however, when it comes to adjusting the microcarrier concentration to enable them to interact. When they are close enough to one another, they tend to aggregate and block the extruder. Also, the degradation products of the microcarrier material might raise concerns (Ozbolat & Hospodiuk, 2016).

EBB can also be used to fabricate tissue constructs from multicellular building blocks, such as cell aggregates, multicellular cylinders, and tissue strands (Fig. 5.4G1–G3). Cell aggregates can be prepared in many ways (Breslin & O'Driscoll, 2013), and some of them provide high throughput and reproducibility; they might comprise several cell types, including stromal and vascular cells (De Moor et al., 2018). Therefore, cell aggregates have long been contemplated as building blocks of engineered tissues (Mironov et al., 2009). A typical cell aggregate is several hundred microns in diameter and is composed of 10^3-10^5 cells. Larger aggregates can be prepared from cell types that can cope with hypoxia (e.g., chondrocytes) because diffusion within dense clusters of cells is ineffective as a means of gas and nutrient transport over distances larger than about 200 μm. Printing cell aggregates has several advantages, including fabrication speed, high cell density, and biomimetic cell-cell interactions that favor the stability of cell phenotypes (Ozbolat & Hospodiuk, 2016). On the downside, the preparation of cell aggregates needed for engineered tissues of clinical relevance is laborious and requires up to 10^8 cells. Loading the cells into the nozzle (a glass micropipette) is challenging, and it is difficult to ensure that the new nozzle continues the printing where the old one left off. Also, cell aggregates need to be used within 10 days of their preparation; otherwise, maturation strips them of their ability to fuse. Better fabrication speed and precision can be attained by printing cylindrical bioink (Norotte et al., 2009). In this approach, cells are resuspended and centrifuged; the resulting cell pellet is sucked into a glass micropipette (of 300–500 μm in diameter), incubated at 37°C and 5% CO_2 for 15 min, and extruded in a grooved nonadhesive mold. Overnight incubation makes them sturdy enough to be printed via the method illustrated in Fig. 5.4G2. The EBB of spherical or cylindrical bioinks involves the use of supportive structures that are inert to cell adhesion (e.g., agarose or alginate) (Fig. 5.4G1 and G2); these provide temporary support until the bioprinted construct develops structural integrity via tissue fusion and need to be removed afterward (Norotte et al., 2009). To overcome these limitations, Akkouch et al. took advantage of alginate microtubes created by coaxial extrusion; they aspirated the cell pellet into a custom-made syringe and injected it into alginate microtubes (Akkouch et al., 2015). The tubes were sealed with surgery clips, and the resulting multicellular sausage was incubated for about a week to develop tight bonds between the constituent cells. Finally, the alginate tube was decrosslinked within 5 min by adding sodium citrate to the cell culture medium, leaving behind a sturdy tissue strand suitable for EBB (Fig. 5.4G3). Fig. 5.5 illustrates a tissue engineering workflow based on the scaffold-free bioprinting of tissue strands (Yu et al., 2016).

In their proof-of-principle study, Yu et al. prepared cartilage strands and proved that the biofabrication procedure did not harm the cells—they remained viable, proliferated, and secreted sulfated glycosaminoglycans. Cartilage strands were printed next to each other, in a sheet-like arrangement. Adjacent strands started to fuse within 12 h from printing and completed the process in about 1 week.

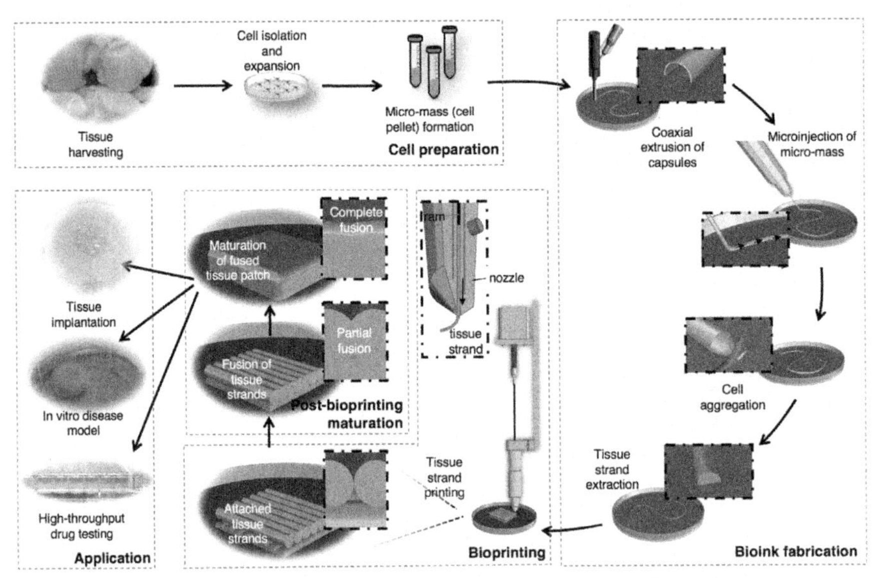

FIGURE 5.5 Biofabrication and extrusion-based bioprinting (EBB) of tissue strands.

Primary cells are harvested and cultured in large numbers; then, they are resuspended and centrifuged to obtain a cell pellet (*top left*). Alginate capsules (microtubes) are fabricated by coaxial extrusion, the cell pellet is aspirated into a microinjection device and injected into the capsules, cells are cultured for about a week to enable cell aggregation; then, the alginate capsule is dissolved by adding sodium citrate and the tissue strand is retrieved (*right*). Finally, tissue strands are loaded in a print head of specific design and extruded in a scaffold-free arrangement (*bottom middle*); subsequent tissue fusion and maturation give rise to a biomimetic tissue patch for various applications (*bottom left*).

Reprinted from Yu, Y., Moncal, K.K., Li, J., Peng, W., Rivero, I., Martin, J.A. & Ozbolat, I.T. (2016). Three-dimensional bioprinting using self-assembling scalable scaffold-free "tissue strands" as a new bioink. Scientific Reports, 6, *28714. https://doi.org/10.1038/srep28714. Reprinted under the terms of the Creative Commons Attribution 4.0 International License (https://creativecommons.org/licenses/by/4.0/legalcode).*

This section discussed several extrusion-based 3D bioprinting techniques, but it is far from comprehensive. Microextrusion is an essential component in complex bioprinting techniques discussed in the next section.

1.3 Freeform bioprinting

The stability of the printed structure is an old problem of bioprinting, as old as the field itself (Mironov et al., 2003). To sustain the fragile structure during and right after bioprinting, one option is to deliver a biocompatible material, layer-by-layer, in tandem with the bioink. The support material prevents the collapse of the printout under its own weight until it becomes structurally integrated and resilient due to chemical crosslinking or cellular processes. Printing one layer at a time, however, severely limits the print complexity. Another alternative is to adopt an embedded 3D printing approach pioneered by Wu et al. in their seminal work on the fabrication of biomimetic microvascular networks with hierarchical 3D architecture (Wu et al., 2011). They printed fugitive ink (23% wt Pluronic F-127) filaments into a photocurable physical hydrogel reservoir—25% wt Pluronic F-127 diacrylate (F127-DA). F127-DA was obtained by replacing the terminal hydroxyl group from each of the two PEG segments of Pluronic F-127 with an acrylate group. The reservoir was covered with a less concentrated, 20% wt F127-DA solution meant to fill the void left behind by the moving extrusion needle. Once the printing was completed, the F127-DA solution was transformed into a chemically crosslinked hydrogel via photopolymerization, and the fugitive ink was liquefied by cooling the construct to 0°C and removed by gentle suction, leading to perfusable microvascular channels. This approach, called omnidirectional printing, enabled seamless control over microchannel paths and calibers, and inspired freeform bioprinting techniques.

Embedded 3D bioprinting has been developed independently by three research groups (Bhattacharjee et al., 2015; Highley et al., 2015; Hinton et al., 2015). Unlike (Wu et al., 2011), in embedded bioprinting the fugitive component is the reservoir—a yield-stress support bath with the role of holding the bioink in place until cured and is liquefied afterward to release the printed construct. Using this technique, known as Freeform Reversible Embedding of Suspended Hydrogels (FRESH), a wide range of bioink materials can be printed in a variety of support baths, including gelatin, agarose, and alginate-based yield-stress materials (Shiwarski et al., 2021). The most reliable support bath materials consist of a slurry of jammed hydrogel microparticles in a minimal amount of aqueous solution (e.g., phosphate buffered saline (PBS)). To prepare a support bath, one can mix gel microparticles (of tens of μm in diameter) with an aqueous buffer suitable for hosting the cells dispersed in the bioink. Then, the mixture is centrifuged to compact the microparticles, and the excess buffer is removed. Such a material is viscoplastic: it behaves as a solid until a high enough shear stress (the so-called yield stress), is applied, at which point it becomes fluid-like. The smooth transition from solid to fluid state enables the free passage of the extruder's needle. As the shear stress ceases, the granular material resolidifies swiftly, trapping the extruded ink in place (Shiwarski et al., 2021).

Fig. 5.6 represents the working principle of FRESH printing.

The liquid component of the support bath might also contain a crosslinking agent, such as $CaCl_2$ desirable when an alginate bioink is extruded. For bioinks based on acidified collagen or decellularized ECM, a pH buffer (e.g., HEPES) can be added to the support bath to neutralize the pH of the bioink and trigger its gelation. The aqueous phase of the support bath can be supplemented with bioactive compounds (e.g., growth factors) for enhancing cell survival. FRESH printing is more cell-friendly than bioprinting performed in the air because the reservoir prevents bioink dehydration (Shiwarski et al., 2021).

FRESH printing was used to print biomimetic structures of anatomic size and complexity. Examples include a tenfold magnified alginate replica of the heart of a chick embryo (Hinton et al., 2015), functional components of the human heart built

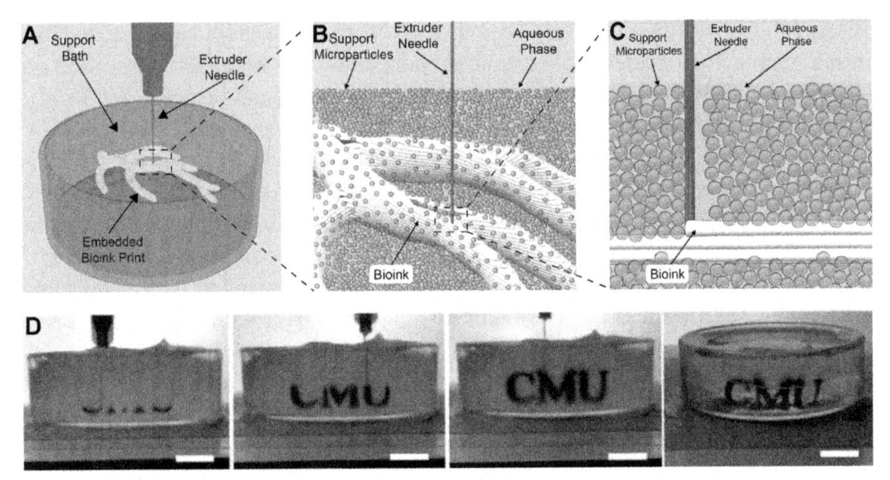

FIGURE 5.6 The principle of FRESH printing.

(A–C) Schematic representation of the FRESH printing approach: (A) the overall picture of the print bed and extrusion nozzle; (B) zoom-in view of the extruder needle and the bioprinted construct suspended in a support bath of jammed gelatin microparticles; (C) close-up view of the extruder needle advancing within the yield-stress support bath and delivering bioink strands. (D) successive snapshots of an actual FRESH printing process, wherein CMU, the acronym Carnegie Mellon University, is printed by depositing black dyed alginate ink in a gelatin slurry support bath (*light-gray*); the *bottom-right* panel shows the printed structure in the transparent, melted gelatin obtained by warming the system to 37°C.

Adapted with permission from Shiwarski, D.J., Hudson, A.R., Tashman, J.W. & Feinberg, A.W. (2021). Emergence of FRESH 3D printing as a platform for advanced tissue biofabrication. APL Bioengineering, 5, *010904. https://doi.org/10.1063/5.0032777;* (panels *A–C) and from Hinton, T.J., Jallerat, Q., Palchesko, R.N., Park, J.H., Grodzicki, M.S., Shue, H.-. J., Ramadan, M.H., Hudson, A.R. & Feinberg, A.W. (2015). Three-dimensional printing of complex biological structures by freeform reversible embedding of suspended hydrogels.* Science Advances, 1. *https://doi.org/10.1126/sciadv.1500758;* (panel *D). Reprinted under the terms of the Creative Commons Attribution 4.0 International License (https://creativecommons.org/licenses/by/4.0/legalcode).*

from cell-laden collagen hydrogel (Lee et al., 2019), vascularized cardiac patches, and a downscaled tissue construct with the architecture of a human heart (Noor et al., 2019).

Among the most widely used suspension media are jammed microgels of polyacrylic acid. Carbopol is the trade name of a family of high-molecular-weight polymers of acrylic acid crosslinked with allyl ethers of polyalcohols. One variant, called Carbopol 980-NF, contains no additional functional groups, whereas others, such as Carbopol ETD 2020 and Carbopol Ultrez 20, are copolymers of polyacrylic acid and a C10–C30 alkyl acrylate functional group (O'Bryan et al., 2018). In their groundbreaking work, Bhattacharjee et al. investigated extrusion bioprinting in Carbopol granular gel composed of hydrogel particles, of about 7 μm in diameter, immersed in a minimal amount of interstitial liquid. They used glass micropipette extrusion nozzles with internal diameters of 100–200 μm to print delicate structures from photopolymerizable poly(vinyl alcohol) (PVA) but also from uncrosslinkable media, achieving feature sizes of the order of the nozzle diameter (Bhattacharjee et al., 2015). As the nozzle advanced within the reservoir, the granular medium was locally fluidized under low shear stresses (<200 Pa), but it regained its solid consistency right after the nozzle's passage. Particle image velocimetry demonstrated that the liquefied trail was comparable in size with the nozzle diameter, and the extruded material favored the resolidification of the granular gel, becoming firmly entrapped in it. The embedded constructs remained stable for months even in the absence of photopolymerization. UV-cured PVA structures of biologically inspired architectures (octopus, jellyfish, branched vascular tree) could be easily released from the support bath by gentle stirring in an aqueous solution, which gradually dispersed the microgel particles and left the constructs afloat (Bhattacharjee et al., 2015). The same research group conducted comprehensive tests of the feasibility of EBB in a Carbopol microgel bath prepared by using a cell type-specific cell culture medium as a solvent. Eleven types of live cells were printed in the support bath, and their viability was evaluated along with their proliferative capacity. Moreover, arrays of cell aggregates were printed in the reservoir; they remained in place and increased in volume as a result of cell division (Bhattacharjee et al., 2016). EBB of carefully mobilized epithelial cell pellets resulted in multicellular threads of a controllable thickness (via the rate of extrusion and the speed of nozzle movement) with the lower limit corresponding to a chain of individual cells (Bhattacharjee et al., 2015). Nevertheless, multicellular threads and sheets fabricated this way displayed spontaneous shrinking and buckling due to cell traction forces. Such undesirable phenomena could be mitigated by fine-tuning the rheological properties of the support medium (Morley et al., 2019).

Highley et al. used hyaluronic acid (HA) modified with adamantane (Ad-HA) or β-cyclodextrin (CD-HA), in various proportions, for both the extruded bioink and support bath. For extrusion, they modified 25% of the HA repeat units to create two hydrogel types, Ad_{25}-HA and CD_{25}-HA, and mixed them in equal proportions to prepare a colloidal hydrogel with a total polymer concentration of 5% wt/v. For the support gel, the percentage of the modified repeat units was 40%, and the resulting Ad_{40}-HA and CD_{40}-HA hydrogels were mixed in equal proportions at a total concentration of 4% wt/v. The increased extent of modification ensured higher stability while preserving the shear-thinning and self-healing behavior given by noncovalent, reversible bonds responsible for supramolecular assembly through guest–host complexes. Hence, this approach was named direct writing of guest–host hydrogels (GHost writing) (Highley et al., 2015). These authors endowed the support gel with a secondary, covalent crosslinking mechanism by introducing methacrylate groups in about 20% of the repeating unit of HA before further modifications with either Ad or CD. The resulting materials formed hydrogel networks through supramolecular bonding, enabling suspended extrusion, as well as covalent crosslinking achieved by UV exposure, assuring mechanical stability. These features allowed for the fabrication of perfusable channels using the methodology devised by Wu et al. (2011).

Support media can be prepared also from hydrogel precursors of natural origin. For example, Brassard et al. prepared a support bath from an ice-cold solution of native bovine dermis collagen type one (5 mg/mL) supplemented with 10% v/v DMEM (10×), 8% v/v Advanced DMEM/F12%, and 2% v/v sodium bicarbonate solution of 0.5 M (Brassard et al., 2021). This collagen solution was used for printing vascular structures, whereas for stem cell constructs, the collagen solution was mixed with an equal amount of ice-cold Matrigel. Brassard et al. combined bioprinting in such bioactive support baths with organoid technology to produce centimeter-sized tissue constructs. The name proposed for their approach, bioprinting-assisted tissue emergence (BATE), emphasizes that 3D bioprinting merely directs multicellular self-assembly, which ultimately leads to tissue-specific microarchitecture and function. Fig. 5.7 illustrates the principle of BATE along with a few representative results.

The BATE setup, shown in Fig. 5.7A, includes an extrusion system outfitted with a pulled glass micropipette. The extruder is coupled to an inverted microscope with a manually controlled stage, providing visual feedback; this enables the user to adjust the extrusion rate and nozzle speed on the fly to optimize the print quality. Multiple cell types can be printed sequentially, even in discrete locations. The user notices when she/he needs to stop printing because the nozzle starts to drag the gelling support bath. Then, the entire structure needs to be incubated for about 10 min to finalize the crosslinking of the embedding gel, and the system should be covered with cell culture medium augmented with a cocktail of growth factors and/or differentiation factors. Within a few days of incubation, the deposited cells self-assemble into biomimetic, functional tissue constructs Fig. 5.7B and C. To release them, one can degrade the support matrix by adding collagenase type IV to the cell culture medium (5 mg/mL); after about 12 min, the printouts can be washed with PBS and placed in the appropriate cell culture medium (Brassard et al., 2021).

The above examples demonstrate that suspended EBB can take a variety of forms and can lead to diverse applications. In a powerful biofabrication approach called sacrificial writing into functional tissue (SWIFT), Skylar-Scott et al. prepared a support bath from organ building blocks (embryoid bodies, organoids, or tissue spheroids) made of patient-derived induced pluripotent stem cells (iPSCs). They cultured iPSCs, transferred them into nonadherent microwell arrays to form many ($\sim 10^5$) embryoid bodies, mixed them with a solution of collagen type one and Matrigel kept between 0 and 4°C and compacted them by centrifugation (Skylar-Scott et al., 2019). Then, they printed sacrificial gelatin ink into this support bath and melted it by warming the entire system to 37°C, thereby obtaining a dense mass of tissue building blocks endowed with a set of perfusable channels akin to a vascular tree. Channel diameters ranged between 0.4 and 1 mm; thinner channels could not be printed because the granules of the support bath (the embryoid bodies) were about 0.2 mm in diameter. Finally, the channels were coated with endothelial cells by perfusing them with a suspension of 10^7 HUVECs/mL. SWIFT was used to build a functional, perfusable cardiac tissue comprising more than 2×10^8 cells/mL (Skylar-Scott et al., 2019).

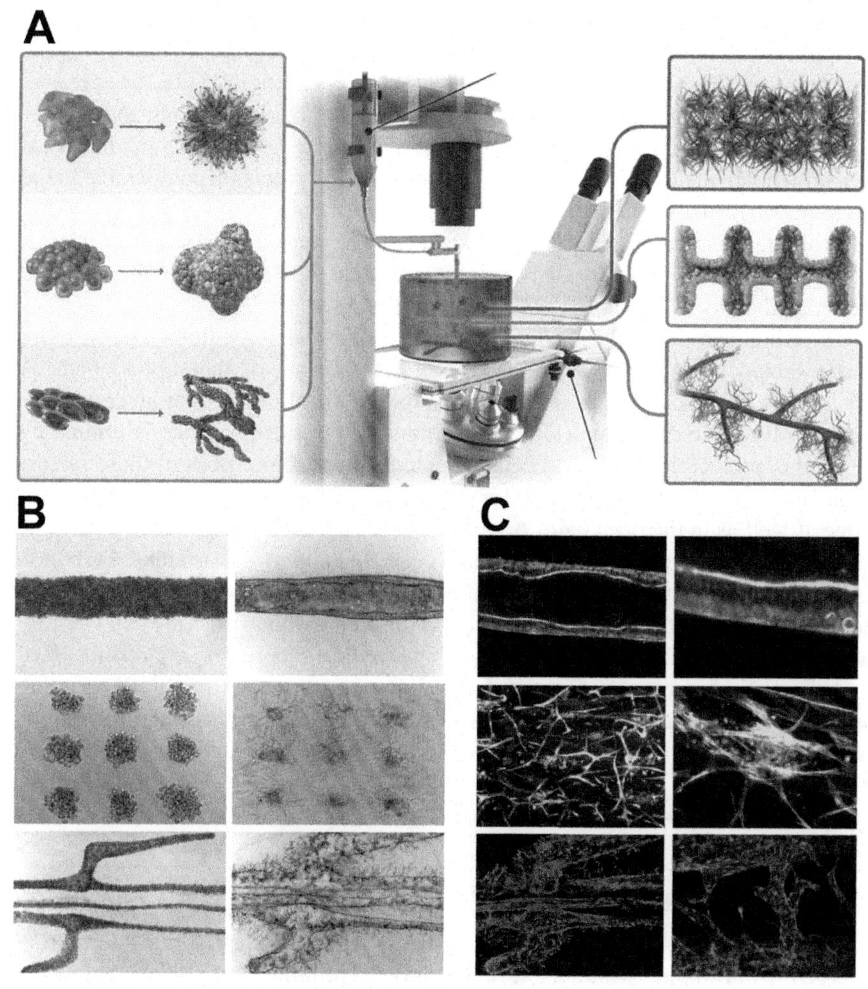

FIGURE 5.7 Bioprinting-assisted tissue emergence (BATE).

(A) Schematic representations of the working principle and potential applications of BATE; (B) Bright-field microscopy snapshots of typical printouts before and after a few days of self-organization of the constituent cells (*left* and *right*, respectively)—human intestinal stem cells (hISCs), human mesenchymal stem cells (hMSCs), and human umbilical vein endothelial cells (HUVECs); scale bars, 500 μm; (C) Confocal microscopy images of self-organized constructs shown in low and high magnification (*left* and *right*, respectively); cell nuclei are blue, F-actin is green, whereas CD31, an endothelial cell adhesion protein, is pink; scale bars, 250 μm (*left*) and 75 μm (*right*). For interpretation of the references to color in this figure legend, please refer online version of this title.

Reprinted by permission from Springer Nature from Brassard, J.A., Nikolaev, M., Hübscher, T., Hofer, M. & Lutolf, M.P. (2021). Recapitulating macro-scale tissue self-organization through organoid bioprinting. Nature Materials, 20, *22–29. https://doi.org/10.1038/s41563-020-00803-5.*

EBB in a support bath is an extremely versatile bioprinting technique with potential applications in encapsulation technologies, drug delivery, soft robotics, flexible electronics, and tissue engineering. In the context of tissue engineering, the strengths of this technology include (i) the possibility of omnidirectional printing and discontinuous printing in arbitrary locations, (ii) the ability to deliver fine bioink filaments (of a few cell diameters in thickness), with a positioning precision comparable to the cell size, (iii) the feasibility of dispensing low viscosity bioinks with slow crosslinking or biological maturation kinetics (iv) the ability to prevent the collapse of a large, fragile tissue construct, and (v) the capacity to prevent the dehydration of the printed construct, thereby allowing for prolonged printing time.

There are also limitations to bioprinting in a suspension medium. The extruder nozzle should move slowly to prevent excessive stirring of the support bath. A fast-moving print head might distort or disrupt previously deposited layers. While using small nozzle diameters (for high resolution), the extrusion rate is limited by the need to protect the dispensed cells from high shear stress. Both of these factors lengthen the printing time, pleading for physiological conditions in the printer's cartridge as well as in the print bath. A gelatin-based support bath, however, cannot be heated to 37°C because it would melt and cease to support the printout. Carbopol, agarose, or alginate microgel support baths are not affected by physiological temperatures but removing the printout from them is difficult without affecting cell viability (Shiwarski et al., 2021). Challenges posed by freeform bioprinting include the need for dedicated software tools. Current slicing software, inherited from the 3D printing field, decomposes the 3D model into a set of planar slices; the resulting print path is inherently 2D except for short transitions from one plane to another (McCormack et al., 2020).

2. Droplet-based bioprinting

The historical perspective provided in Chapter 1 shows clearly that bioprinting emerged from inkjet printing—a conventional printing technology based on dispensing tiny volumes (tens of picoliters) of ink on a sheet of paper. Since the pioneering work of Robert J. Klebe on printing fibronectin-patterned substrates for cell adhesion (Klebe, 1988), and the seminal paper of Thomas Boland's team on printing live cells (Wilson & Boland, 2003), a variety of droplet-based bioprinting (DBB) methods emerged (Fig. 5.8), which share the basic idea of dropwise dispensing of low viscosity bioinks. The currently known DBB techniques can be classified into three categories inkjet bioprinting (Fig. 5.8 panels A1-A5), acoustic bioprinting (Fig. 5.8B), and micro-valve bioprinting (Fig. 5.8C). Inkjet technologies include continuous inkjet bioprinting, drop-on-demand bioprinting, and electrohydrodynamic jet bioprinting (Gudapati et al., 2016).

In continuous inkjet bioprinting (Fig. 5.8A1), the bioink is pneumatically ejected through a narrow orifice to form a fine jet, which breaks up into individual droplets prior to reaching the target surface. It is quite common for a liquid exiting a nozzle at

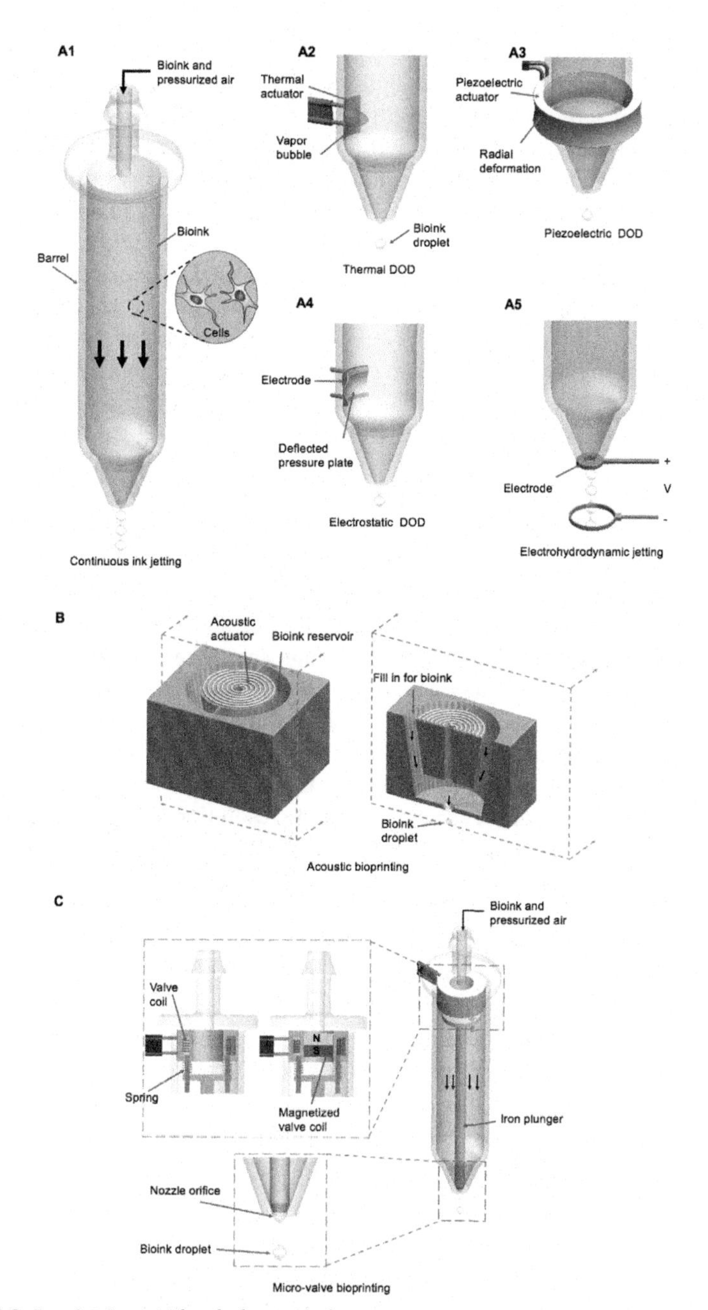

FIGURE 5.8 Droplet-based bioprinting techniques.

(A1–A5) Inkjet-based bioprinting methods include continuous inkjet bioprinting (A1), thermal inkjet bioprinting (A2), piezoelectric inkjet bioprinting (A3), electrostatic drop-on-demand bioprinting (A4), and electrohydrodynamic jetting (A5); (B) Acoustic bioprinting, low flow rates to transition from a smooth cylindrical jet to a chain of droplets. This phenomenon, known as capillary instability or Rayleigh-Plateau instability, stems from disturbances acting on the jet and the tendency of the liquid to minimize its surface area. Consider a jet of initial radius R_0. When periodic perturbations lead to a modulation of the jet radius, resulting in a regular arrangement of pinched and bulging portions, the pressure exerted by the surface layer (the Laplace pressure) is higher in the pinched region, creating a pressure gradient and, consequently, a liquid flow toward the bulging regions. Eventually, the necks rupture, and the bulged segments turn into droplets. Joseph A. F. Plateau demonstrated that the instability occurs when $kR_0 < 1$, where $k = 2\pi/\lambda$ is the wavenumber; here, λ is the wavelength of the perturbation (the minimum distance between two similarly bulged portions). In other words, the jet becomes unstable when the wavelength of the perturbation is larger than the perimeter of the initial jet. Lord Rayleigh analyzed the dynamics of this phenomenon and proved that the most amplified perturbation has the wavelength $\lambda_m \approx 2\pi R_0/0.7$ (Eggers & Villermaux, 2008).

Drop-on-demand inkjet bioprinting relies on ejector systems that issue individual droplets whenever needed, providing better control over the printing process than continuous inkjet printers. The bioink is loaded in the fluid chamber of the print head and is retained by the surface tension force acting at the nozzle opening. A sudden pressure pulse expels a liquid column through an orifice of tens of micrometers in diameter; the column gradually becomes sperm shaped, turning into a leading droplet with a thin tail, which can also break up into tiny satellite drops. Due to the higher friction force experienced by the leading droplet, the satellite drops catch up and merge with the leading droplet. Such a merger is desirable because the presence of satellite drops at impact can affect printing resolution. To favor drop merging in flight, it is recommended to maintain a stand-off distance of a few millimeters between the print head and the target surface (Derby, 2010). Droplet ejection may be accomplished by thermal, piezoelectric, or electrostatic mechanisms.

In a thermal inkjet printer (Fig. 5.8A2), an electric heater generates a minuscule vapor bubble, which forces a fluid droplet out of the nozzle before it collapses by cooling. Despite the harsh jetting procedure, cell viability and proliferative capacity are well preserved during thermal inkjet printing, presumably because droplet generation is a brief event affecting a small portion of the bioink. At typical cell densities (10^6 cells/mL), there is one cell in 1000 pL of bioink. The droplet volume, on the other hand, is of the order of 100 pL (Xu et al., 2008), so there is a 1/10 probability for a cell to reside in the heated region. Hence, about 10% of the cells are likely to be affected by the hot vapor bubble. This rough estimate is in good

based on an acoustic actuator that expels droplets from a bioink reservoir; and (C) microvalve bioprinting, based on an electromagnetically actuated plunger.

Reprinted from Gudapati, H., Dey, M. & Ozbolat, I. (2016). A comprehensive review on droplet-based bioprinting: Past, present and future. Biomaterials, *102. https://doi.org/10.1016/j.biomaterials.2016.06.012, with permission from Elsevier.*

agreement with the experimentally observed cell viability in tissue constructs built by thermal inkjet bioprinting, 75%–90% (Xu et al., 2005; Xu et al., 2006).

In a piezoelectric inkjet printer (Fig. 5.8A3) a piezoelectric transducer placed in the fluid chamber expands under the action of an electric signal, generating a pressure pulse that expels a droplet. The shape and amplitude of the actuating pulse can be adjusted to control drop size and velocity. Nevertheless, while designing the actuating pulse, special care needs to be taken to avoid acoustic resonances (Derby, 2010). A number of studies attempted to optimize the printing process by varying the properties of the voltage pulse used for transducer actuation, such as amplitude, frequency, rise and fall times, dwell time, and echo time. Although high temperatures are not involved in this technique, the high mechanical stress exerted during droplet ejection is potentially detrimental to cells. In cell-laden alginate constructs fabricated using piezoelectric inkjet printing, cell viability was found between 70% and 95% (Gudapati et al., 2016).

Electrostatic bioprinting (Fig. 5.8A4) is similar to piezoelectric bioprinting in that droplet generation is an athermal process based on a brief deformation of the bioink chamber. Part of the chamber wall is a deformable pressure plate. When voltage is applied between the pressure plate and a rigid conductor plate located next to it, the pressure plate suffers elastic deformation, slightly incrementing the chamber's volume. As soon as the voltage is removed, the pressure plate regains its original shape, exerting a sudden pressure pulse upon the bioink, which results in droplet ejection. Electrostatic bioprinting of alginate loaded with human cervical adenocarcinoma (HeLa) cells (6×10^6 cells/mL) resulted in 70% cell viability (Nishiyama et al., 2008). In these experiments, the adverse factors experienced by cells include high shear stress during ejection, as well as osmotic stress caused by the relatively high osmolar concentration of the substrate—an aqueous solution of 10% $CaCl_2$ (crosslinker) and 15% polyvinyl alcohol (viscosity enhancer).

The inkjet bioprinters discussed so far have one weak point in common: they generate a high-pressure pulse to force out a small volume of bioink through a tiny orifice (of the order of 10 μm in diameter), exposing the cells to high mechanical stress. Electrohydrodynamic jet bioprinting alleviates this stress by pulling instead of pressing bioink out of the fluid chamber's nozzle (Fig. 5.8A5) (Gudapati et al., 2016). The bioink chamber has a gold-coated glass nozzle with an orifice similar to or even smaller in diameter than other inkjet print heads. The nozzle is placed at a stand-off distance of about 100 μm from the substrate placed on a computer-controlled stage with 5 degrees of freedom, enabling translations along the x, y, z axes and rotations around two axes for tilting the print bed. Mild, steady pressure is applied onto the bioink to form a spherical meniscus at the nozzle opening. When a voltage is applied between the nozzle and a grounded electrode placed under the substrate, ions from the bioink accumulate at the surface of the meniscus, making it conical with a rounded, highly curved tip. At sufficiently high voltage, the electrostatic stress matches the maximum capillary stress, at which point an electrically charged droplet is extracted from the meniscus. Consequently, the meniscus volume and electric charge decrease, the electrostatic stress drops below the capillary tension, the ejection halts, and the meniscus recovers its spherical shape (Park et al., 2007). The jetting mode depends on bioink composition as well as on the applied voltage (0.1–20 kV). While the electric field is kept on, periodic dripping is obtained at relatively low voltages, a stream of droplets is generated at intermediate voltages, whereas a continuous bioink jet is obtained at high voltages. The advantages of this technique include high cell viability (of over 90%), the ability to generate droplets smaller than the nozzle's orifice, and the capability to deliver bioinks of over 2 Pa s in viscosity. Besides numerous advantages, electrohydrodynamic jet bioprinting also has limitations: it is expensive, requires custom-made instruments assembled from commercially available components, exposes the user to electric shock hazard, and does not allow for on-demand delivery of individual droplets (Gudapati et al., 2016).

Acoustic bioprinters eject droplets by producing focused acoustic waves in an open pool of bioink. Their print heads are microfluidic chambers, with openings of 50–300 μm in diameter (Demirci, 2006), in which the bioink is retained by surface tension forces (Fig. 5.8B) (Gudapati et al., 2016). The actuator is a piezoelectric substrate (such as lithium niobate, lithium tantalate, or quartz) plated with two sets of interdigitated gold rings—one set is connected to the ground pad, and the other is connected to the signal pad. Via the inverse piezoelectric effect, a periodic electrical signal is converted into acoustic waves of circular geometry, which form an acoustic focal point at the bioink–air interface at the center of the ejector opening. A droplet is ejected when the force exerted by the acoustic radiation at the focal point exceeds the surface tension force (Demirci, 2006). The droplet size is independent of the diameter of the ejector window; instead, it depends on the actuation frequency (the larger the frequency, the smaller the droplet), and on the speed of sound in the bioink (for a given frequency, the larger the speed of sound, the larger the droplet). Droplet speed can be controlled by adjusting the stand-off distance and the number of acoustic cycles involved in droplet generation. In experiments aimed at incorporating five different cell types into agarose gel droplets, droplet diameters were about 40 μm and cell viabilities exceeded 89.8% (Demirci & Montesano, 2007).

In a micro-valve bioprinter, air pressure is applied onto the bioink, while the print nozzle is occluded by a ferromagnetic plunger. When a voltage pulse is applied to the solenoid coil wrapped around the bioink chamber (Fig. 5.8C), the plunger is lifted by the magnetic field of the electric current passing through the coil. Depending on the applied pressure and voltage pulse profile, such a printer ejects a continuous jet or individual droplets. Droplet volume and cell viability depend on bioink composition, cell concentration, and printing parameters, such as nozzle geometry and applied pressure. Typical droplet diameters obtained by this technique range from 100 to 600 μm, providing a smaller resolution than inkjet bioprinting, but cell viability, proliferative capacity, and differentiation are favored by the lower mechanical stress. Cell viability of over 90%, moderate cost, and the ability to handle highly viscous bioinks are among the strengths of microvalve bioprinting (Gudapati et al., 2016).

3. Light-based bioprinting

Light is employed in current bioprinting technologies for either photopolymerization or bioink transfer (Sun et al., 2020). Photopolymerization of a biocompatible resin is accomplished, one voxel at a time, by a laser beam in stereolithography, or one layer at a time, by light projected onto the surface of the resin in digital light processing, also known as dynamic optical projection stereolithography, or digital micromirror device projection stereolithography. Computer-controlled bioink transfer by laser light is achieved either by radiation forces or by thermal effects, by vaporizing a small amount of material to propel bioink from a donor surface onto a nearby receiving substrate, in a set of techniques collectively known as laser-assisted bioprinting.

3.1 Bioprinting via photopolymerization

Hydrogels are networks of interconnected hydrophilic polymer fibers soaked by an aqueous solution (see also Chapter 2). In the absence of connections among the fibers, the polymer solution is liquid. One might picture such a system as boiled, wet spaghetti noodles nicely arranged in a bowl, next to each other. If we mix them vigorously, we entangle the noodles, endowing the system with elastic solid-like properties, which could be reinforced by adding cheese cubes to glue some noodles together. Similarly, creating crosslinks in an aqueous polymer solution results in a hydrogel. The crosslinking method has a deep impact on the mechanical and biochemical properties of the hydrogel. Crosslinking can be achieved by physical or chemical means. Physical crosslinks include entanglements created by in-situ polymerization, electrostatic interactions mediated by ions, hydrogen bonds, hydrophobic interactions, and crystallization. Chemical crosslinks are (usually strong and permanent) chemical bonds created by photopolymerization, enzymatic reactions, click reactions, and other, less common methods (Hu et al., 2019).

Light-activated crosslinking is one of the most widely used methods of hydrogel preparation. It is fast, does not require extreme conditions, and provides control over the hydrogel's properties via the parameters of light exposure—such as duration, light intensity, and location. The synthesis of hydrogels relying on photo-initiated polymerization is largely based on the incorporation of (meth)acrylate groups. Their double-bonded carbon groups promote photo-initiated free-radical polymerization in the presence of a photoinitiator, such as Irgacure2959, Irgacure1173, Irgacure 819, Irgacure 651, lucirin TPO, etc. Photoinitiator molecules are cleaved by photons of a specific wavelength, releasing radicals that initiate chain polymerization (Hu et al., 2019). Fig. 5.9 represents the generic structure of biocompatible hydrogels that include methacryloyl groups (Yao et al., 2018).

The strategy outlined in Fig. 5.9 proved to be effective for a variety of natural and synthetic polymer backbones. Gelatin methacryloyl (GelMA)—also known as gelatin methacrylate, methacrylated gelatin, or gelatin methacrylamide—is among the most appreciated hydrogels in biofabrication. Its methacrylamide groups and

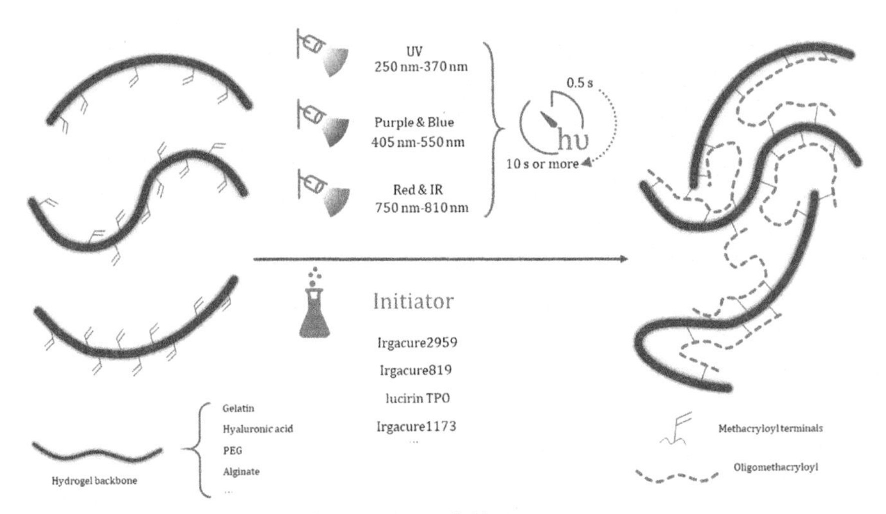

FIGURE 5.9 Hydrogels prepared by photocrosslinking.

Schematic drawing of hydrogel formation by the photo-initiated chain polymerization of methacryloyl groups.

From Yao, H., Wang, J. & Mi, S. (2018). Photo processing for biomedical hydrogels design and functionality: A review. Polymers, 10, *11. https://www.mdpi.com/2073-4360/10/1/11. Reprinted under the terms of the Creative Commons Attribution 4.0 International License (https://creativecommons.org/licenses/by/4.0/legalcode).*

methacrylate groups do not interfere with the Arg-Gly-Asp (RGD) sequences from gelatin, so it enables cell adhesion; moreover, it can be degraded by proteolytic enzymes (matrix metalloproteinases (MMPs)), and, therefore, it does not impede cell migration, proliferation, and differentiation (Yue et al., 2015). Poly (ethylene glycol) (PEG)-based hydrogels, on the other hand, are synthetic, biologically inert materials, which can be tailored to become cell-friendly. Along the way, they unveil physicochemical factors that are essential for cell function. For example, Dhariwala et al. incorporated Chinese hamster ovary cells into a hydrogel obtained by the photopolymerization of a 3:2 mixture of PEG and PEG-dimethacrylate (PEGDMA) in the presence of Irgacure 2959. Cell viability dropped with increasing photoinitiator concentration, but it was acceptable ($\approx$65%) at the lowest concentration needed for photopolymerization (Dhariwala et al., 2004). To promote cell adhesion, Arcaute et al. incorporated Arg−Gly−Asp−Ser (RGDS) peptide sequences into the PEGDMA hydrogel (Arcaute et al., 2006). Chan et al. explored 20% poly(ethylene glycol) diacrylate (PEGDA) hydrogels with molecular weights (M_w) ranging from 700 to 10^4 Da with and without RGDS motifs. A live/dead assay indicated that, regardless of M_w, most of the encapsulated cells survived the photopolymerization process, but they died by day 2 in hydrogels with $M_w = 700$ Da, presumably because the small (3.1 nm) pores of the hydrogel hindered gas and nutrient diffusion. For $M_w = 3400$ and larger, cell viability was satisfactory (>65% after 1 week in culture). In PEGDA hydrogels of $M_w = 3400$ functionalized with 5 mM RGDS, cell numbers increased for 1 week, remained stable for another week, and started to decrease afterward (Chan et al., 2010). Further investigations will likely lead to the development of light-curable hydrogels specifically adapted to match the mechanical properties of the target tissue without affecting the incorporated cells (Raman & Bashir, 2015; Yao et al., 2018).

3.1.1 Stereolithography

Stereolithography (SL) was the groundbreaking invention that opened the era of 3D printing (see Chapter 1). It was not the first, however, to be applied in bioprinting, because it was challenging to assure that cells are not affected by the resin and/or the photocuring process. Indeed, cytotoxicity was the main concern addressed by the pioneering study conducted by Thomas Boland's research team (Dhariwala et al., 2004), and 16 years later, the most notable limitation of photopolymerization-based bioprinting is the relatively small set of biocompatible and photocurable resins. Another downside of techniques involving light-curing is low cell density ($<10^7$ cells/mL), which hampers the interactions between the encapsulated cells (Sun et al., 2020). Nevertheless, remarkable progress has been made in SL (Raman & Bashir, 2015), and its numerous advantages, including high resolution (5–100 μm) and the stimulus-responsiveness of certain bioprinted structures (Miao et al., 2018), will certainly motivate further research.

In a stereolithography apparatus (SLA), a vat is filled with a liquid photopolymer resin, and the build platform is positioned right below the free surface of the liquid. A scanning laser beam selectively illuminates the surface of the liquid, triggering photochemical reactions that harden the resin. Then, the build platform is lowered to flood the hardened polymer sheet with a thin layer of liquid (the same result is achieved faster by dragging a resin-covered blade over the cured structure). This sets the stage for the next round of light-curing, and the entire procedure is repeated until the digital model is materialized as a 3D object. Such an instrument, in which the laser is placed above the vat, is called a top-down SLA. Another option is to place the laser below the vat (whose bottom should be transparent) and start with the build platform placed slightly above the bottom. As the first layer is hardened under the action of the laser beam, the build platform is lifted by the height of one layer and liquid resin flows in from the periphery of the hardened structure. In this bottom-up approach, the construct is built upside-down and, unlike in the top-down SLA, its height is not capped by the depth of the resin bath.

When the vat contains live cells dispersed in a biocompatible photosensitive prepolymer solution, SLA becomes a form of bioprinting. Fig. 5.10A depicts the working principle of a top-down SLA adapted for bioprinting (Chan et al., 2010).

The top-down approach is expensive because it requires a large volume of prepolymer solution. Biomaterial costs add up with cell culture expenses, and part of the bioink is inevitably discarded in the bioprinting process. Also, a top-down SLA is limited to handling just one type of bioink. Furthermore, in experiments performed on NIH/3T3 murine cells suspended in PEGDA hydrogel, the bioprinted structure created by the top-down approach suffered from inhomogeneous cell

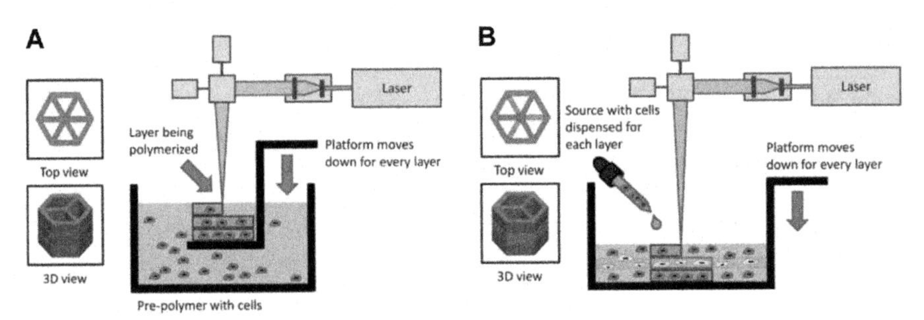

FIGURE 5.10 Bioprinting by stereolithography (SLA).

Schematics of two bioprinting methods based on SLA: (A) the conventional top-down approach, and (B) the stepwise bioink addition approach.

Republished with permission of The Royal Society of Chemistry, from Chan, V., Zorlutuna, P., Jeong, J.H., Kong, H. & Bashir, R. (2010). Three-dimensional photopatterning of hydrogels using stereolithography for long-term cell encapsulation. Lab on a Chip, 10, *2062–2070. https://doi.org/10.1039/C004285D; permission conveyed through Copyright Clearance Center, Inc.*

distribution. The cell density was highest at the bottom of the construct and decreased toward the top because the cells gradually settled down in the PEGDA solution under the action of gravity. To mitigate these disadvantages, Chan et al. modified the top-down approach by lowering the entire vat and supplementing the bioink after every photocuring step, as shown in Fig. 5.10B. The new methodology asks for smaller amounts of bioink, avoids cell sedimentation, and enables the user to build tissue constructs featuring multiple layers made of different types of cells and/or materials. The authors called this method a "bottoms-up" approach (Chan et al., 2010). This name might be puzzling at first sight, given that the laser is kept in the upright position. Actually, the "top-down" and "bottom-up" qualifiers in SL do not refer to the direction of light propagation; they tell us how the construct is built—from the top of the vat toward its bottom (*top-down*) or in the opposite direction (*bottom-up*). The plural in the name proposed by Chan et al. suggests that SL proceeds from the bottoms of distinct layers.

3.1.2 Digital light processing

Digital light processing (DLP) is another 3D printing technique based on photocuring. Unlike SL, which employs an optical scanning module to focalize a laser beam on selected points of the resin surface, DLP takes advantage of projection optics and a digital mask to send patterned LED light onto the resin surface, thereby hardening an entire layer at once. The digital mask may be a digital micromirror device (DMD) (Fig. 5.11A) or a liquid-crystal display (LCD).

Zhu et al. employed DLP to build prevascularized tissue constructs made of cells dispersed in GelMA (Zhu et al., 2017). The photoinitiator used in this study was lithium phenyl-2,4,6 trimethylbenzoylphosphinate (LAP). The right side of Fig. 5.11A depicts the digital model of the construct, along with representative digital masks generated by the slicing software. The first mask (*left*) was used to print

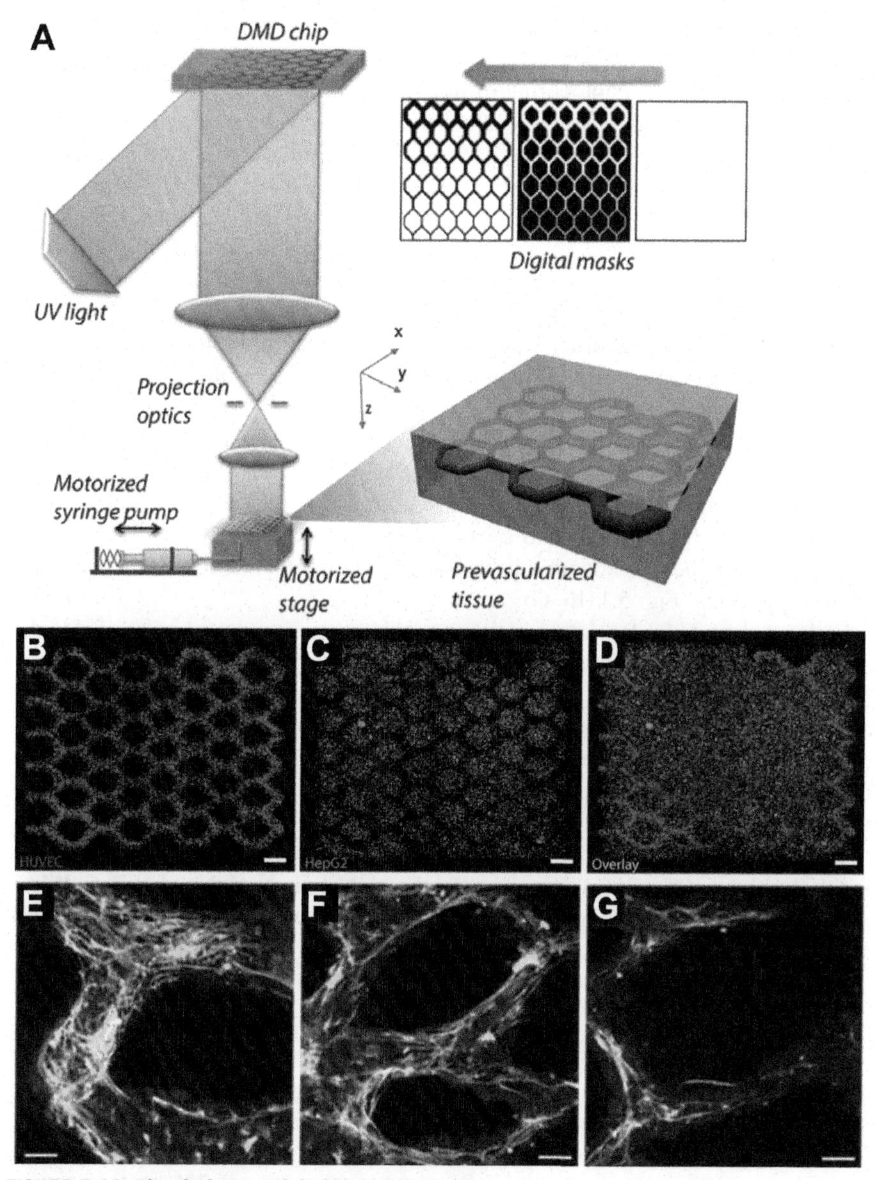

FIGURE 5.11 Bioprinting by digital light processing.

(A) Schematic representation of the digital light processing (DLP) printing process: a digital micromirror device (DMD) comprising an array of individually controlled micromirrors reflects light onto the projection system that illuminates the surface of the prepolymer solution according to a digital pattern (*top right*); once the first layer is photocured, more prepolymer solution is injected into the vat to cover the hardened structure with a fresh layer of solution; (B–D) fluorescence microscopy images of a bioprinted tissue construct the hexagonal islands embraced by the vascular channels, whereas the third mask (*right*) served to print a uniform layer (e.g., the top of the construct). Both of these masks were used for the photocuring of bioink A, composed of 5% (w/v) GelMA, 0.15% (w/v) LAP, and 2×10^7 HepG2 cells/mL. The middle mask was employed for curing bioink B, made of 2.5% (w/v) GelMA, 1% (w/v) glycidal methacrylate-hyaluronic acid (GM-HA), 0.15% (w/v) LAP, 2×10^7 HUVECs/mL, and 4×10^5 10T1/2 cells/mL. First, the islands were printed. Then, the unpolymerized fraction of bioink A was aspirated, and the construct was flushed with Dulbecco's phosphate-buffered saline. Subsequently, bioink B was injected into the vat and polymerized to create the vascular channels. The remaining bioink B was removed, and the construct was flushed again. Finally, bioink A was loaded into the vat and cured to form a homogeneous cover. The resulting structure was about 0.6 mm thick.

Cells were evenly distributed in DLP-printed tissue constructs (Fig. 5.11B–D) and remained viable in a proportion of about 85% (Zhu et al., 2017). Viability did not decline during 1 week of static in vitro culture. On the contrary, during this time, HUVECs self-organized into a vascular network, and 10T1/2 cells acquired a pericyte phenotype—α-smooth muscle actin is a common marker for the identification of pericytes (Fig. 5.11E–G).

Early attempts at DLP printing in tissue engineering opened the way toward high resolution and high throughput fabrication of GelMA scaffolds (Gauvin et al., 2012). The mechanical properties of millimeter-sized scaffolds, with woodpile or hexagonal architecture, could be controlled by tuning the polymer concentration and strut geometry. Individual layers were about 0.1 mm thick, whereas the in-plane resolution was even better ($\approx$50 μm). At low strains, of up to 40%, the compressive moduli of the scaffolds were of about 50 kPa regardless of their microarchitecture; at high strains (70%–90%), the woodpile architecture provided higher compressive moduli (by about 10%–25%). For example, the high strain modulus of 10% GelMA scaffold was roughly 560 kPa for the hexagonal structure and about 700 kPa for the woodpile structure.

outfitted with interconnected vascular channels of uniform width (scale bars, 250 μm): panel (B) shows human umbilical vein endothelial cells (HUVECs) (*red*, dark gray in print), panel (C) shows human hepatocyte carcinoma (HepG2) cells (*green*, dark gray in print), whereas panel (D) shows both cell populations; (E–G) immunofluorescence confocal microscopy images of the endothelial network formed during 1 week of in vitro culture; here, HUVECs are green due to CD31 staining, whereas mesenchymal cells (10T1/2 cells) are purple as a result of α-smooth muscle actin staining (scale bars, 100 μm).

Reprinted from Zhu, W., Qu, X., Zhu, J., Ma, X., Patel, S., Liu, J., Wang, P., Lai, C.S.E., Gou, M., Xu, Y., Zhang, K. & Chen, S. (2017). Direct 3D bioprinting of prevascularized tissue constructs with complex microarchitecture. Biomaterials, 124, *106–115. https://doi.org/10.1016/j.biomaterials.2017.01.042, with permission from Elsevier.*

Fig. 5.11 illustrates clearly that conventional DLP printers are anything but handy when it comes to printing multiple materials. To address this limitation, Bhusal et al. built a multimaterial DLP bioprinter (Bhusal et al., 2022). It includes a light source that generates a uniform beam of UV radiation, of 380 nm wavelength, assuring a light intensity of about 0.7 W/cm^2 at the focal plane of the projection lens placed about 2 cm below the transparent bottom of the bioink reservoir. Along its way, the UV beam encounters a DMD chip, whose micromirrors reflect selected portions of the beam toward the projection system. Several types of bioink are placed in a UV-grade Petri dish divided into radial compartments; one compartment is reserved for the washing bath. In this bottom-up system, the build platform, attached to a *z*-stage, plunges into the bioink. The central piece of the multimaterial bioprinter is a rotating table Fig. 5.12.

To assure a firm attachment of GelMA and PEGDA hydrogels, the glass surface of the build platform is coated with 3-(trimethoxysilyl)-propyl methacrylate. The hardened layer, however, might also stick to the bottom of the bioink chamber unless it is spin-coated with poly(dimethylsiloxane) (PDMS). As soon as one bioink is cured, the *z*-stage lifts the build platform. The table rotates, placing the washing bath below the build platform, and the *z*-stage dips the

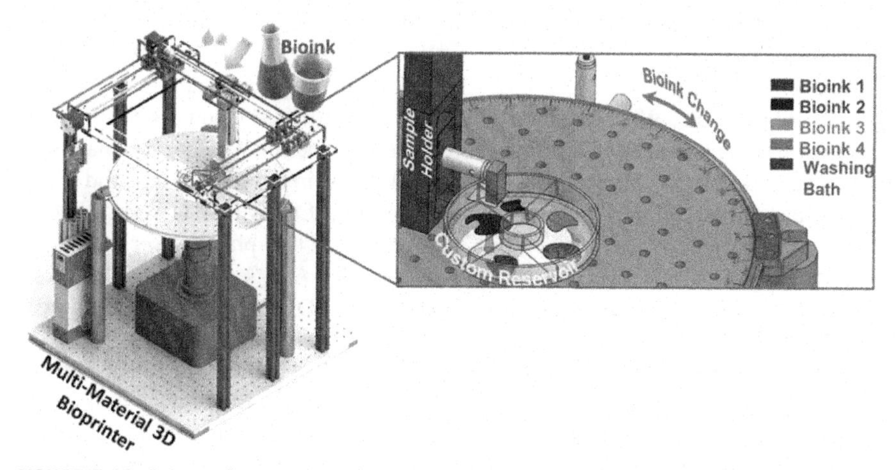

FIGURE 5.12 Schematic representation of a multimaterial digital light processing (DLP) bioprinter.

Patterned UV light is focused on the bottom of the bioink reservoir. Due to a specific coating, the hardened construct remains attached to the build platform—a glass plate fastened onto a *z*-stage. The concerted action of the *z*-stage and the rotating table enables the fast transfer of the construct from one bioink into the washing bath and into another bioink.

Republished with permission of IOP Publishing, from Bhusal, A., Dogan, E., Nguyen, H.A., Labutina, O., Nieto, D., Khademhosseini, A. & Miri, A.K. (2022). Multi-material digital light processing bioprinting of hydrogel-based microfluidic chips. Biofabrication, 14, *014103. https://doi.org/10.1088/1758-5090/ac2d78; permission conveyed through Copyright Clearance Center, Inc.*

construct into the bath to rinse the residual uncured bioink out of it. Then, the build platform is lifted again, and the table puts the second bioink under the construct. The cure-rinse-transfer sequence is repeated until all the components of a given layer are in place. Then, the print bed is immersed again into the first bioink, leaving one layer thickness (50–200 μm) between the construct and the vat's bottom, and the procedure is repeated.

3.1.3 Two-photon polymerization

Compared to SL or DLP, an impressive boost in resolution has been achieved by two-photon polymerization (TPP). The two-photon absorption (TPA) process has been described theoretically by Maria Goeppert Meyer in 1931 and discovered experimentally in 1961 by W. Kaiser and C.G.B Garrett from Bell Laboratories. They observed the generation of blue light by fluorescence as a result of illuminating a $CaF_2:Eu^{2+}$ crystal with red light emitted by a ruby laser—see (Lee, Moon, & West, 2008) and references therein. The so-called degenerate TPA process consists of the simultaneous absorption of two photons of the same energy, whereas in the nondegenerate TPA process, two photons of different energies are absorbed. Most applications known today rely on the degenerate TPA process. In 3D nano/microfabrication, a two-photon absorbing chromophore emits, by fluorescence, photons that excite a photoinitiator; consequently, the latter undergoes a chemical reaction that yields the radicals capable to initiate polymerization. TPA can be induced by Ti:sapphire femtosecond pulse lasers with 780 nm wavelength and pulse durations of 100 fs or less. When the laser beam is closely focused using a 100× objective lens with immersion oil to ensure a high numerical aperture (≈ 1.4), the material response (proportional to the photon density) is highest in the focal point, and TPP provides a spatial resolution below the diffraction limit (Lee, Moon, & West, 2008). In a representative, pioneering study, Lee et al. employed two-photon laser scanning photolithography to link cell adhesive ligands (Arg-Gly-Asp-Ser-Lys (RGDSK) sequences) to PEG hydrogels according to a predefined 3D micropattern (S.-H. Lee, Moon, & West, 2008). The laser beam generated by a Ti:sapphire laser was focused on selected regions of the PEG hydrogel using a 10× plan-apochromat objective lens. The photocured layer thickness was 3 μm, and the in-plane resolution was comparably good. Human dermal fibroblasts actively migrated within the micropatterned regions without leaving them (Lee, Moon, & West, 2008). A similar laser was used to build PEGDA scaffolds by TPP (Ovsianikov et al., 2010). Laser pulses of 120 fs duration were delivered at a repetition frequency of 80 MHz. Focused using a 20 × microscope objective, these pulses triggered the photopolymerization of the monomer solution as a result of TPA. Since TPA requires very high light intensity, polymerization occurs strictly within the focal volume, providing a spatial resolution of the order of 100 nm. The scaffolds had a precise microarchitecture, consisting of superimposed layers of PEGDA hydrogel rings in a close-packed hexagonal arrangement. Their vertical pore design made them especially suitable for cell seeding via laser-induced forward transfer (LIFT) (see next section) (Ovsianikov et al., 2010).

3.2 Laser-assisted bioprinting

3.2.1 Laser-guided direct writing

Laser-guided direct writing (LGDW) takes advantage of radiation pressure to deposit cells on solid or soft substrates with micrometer precision (Nahmias et al., 2005; Odde & Renn, 2000).

Light has long been used to manipulate transparent microparticles suspended in transparent media (Ashkin, 1970). When a laser beam encounters the surface of a particle of different refractive index, part of it is reflected and the rest is refracted. Consider a cell partially illuminated by a laser beam whose cross-section is larger than the cell. The cell has a higher refractive index than the medium; therefore, in the course of refraction, photons are scattered away from the axis of the laser beam. By momentum conservation, the cell receives an equal and opposite momentum—i.e., the cell is "sucked" into the core of the laser beam, where the light intensity is highest. A small fraction of the laser light is reflected on the cell surface, so the cell is also pushed forward by the photons that bounce back from its membrane.

An LGDW system includes a diode laser that generates a beam of 830 nm wavelength, 200 mW power, and an elliptical cross-section. The beam passes through an anamorphic prism pair (which makes its cross-section circular), traverses a low numerical aperture lens of 60 mm focal length (which makes it weakly focused), and is reflected by a gold-coated mirror toward a cell suspension that covers a mobile receiving substrate. A camera pointing perpendicularly to the beam axis provides a sideways view of the cells captured and propelled by the laser beam (Odde & Renn, 1999). Fig. 5.13A represents schematically the working principle of LGDW.

To prevent optical damage to cells, the wavelength of the guiding laser is chosen in the near-infrared domain (700—900 nm). This interval lies beyond the wavelengths absorbed by most proteins and below the infrared absorption bands of water. Therefore, it came as no surprise that cell function and viability remained unaltered in LGDW experiments performed on neurons, endothelial cells, multipotent adult progenitor cells, and bacteria. Also, laser tweezers working in this spectral domain have been used for decades to manipulate a variety of cell types with no detectable harm (Nahmias & Odde, 2006).

The most prominent limitation of LGDW is its low throughput: a single laser beam is able to deliver a few hundred cells per hour. Also, its working distance is relatively short (up to 0.3 mm) due to the convective motion of the culture medium, which overcomes optical confinement forces. Nevertheless, coupling the laser into a hollow optical fiber (Fig. 5.13B) allows for guiding the cells inside the fiber along several millimeters without affecting them. Cells do not adhere to the fiber wall because the light intensity is highest at the center and zero at the wall. This guiding strategy has several advantages that reach beyond mere range extension. Within fibers, cells can be transported along curved paths and placed where needed by precisely positioning the fiber's tip. Since the cell source compartment is separated from the target compartment, the printed pattern cannot be spoiled by spontaneous cell

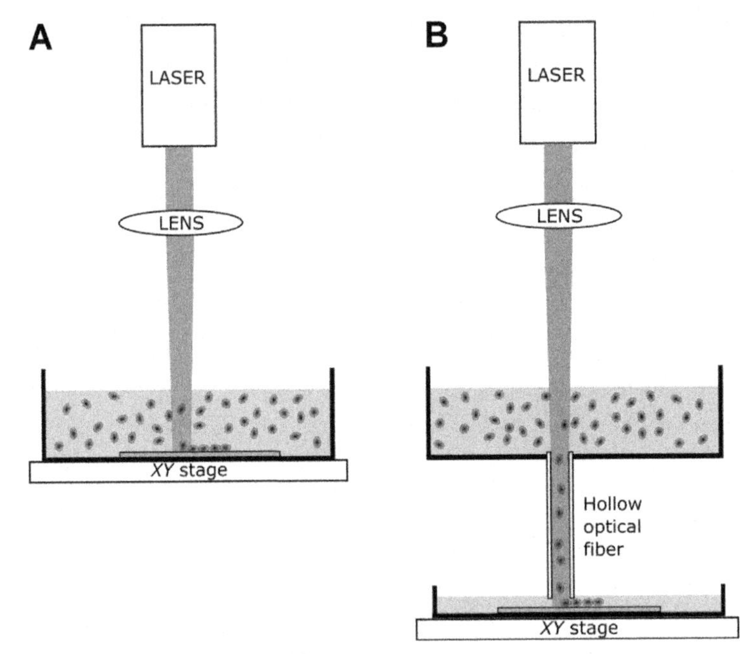

FIGURE 5.13 Laser-guided direct writing.

(A) Schematics of laser-guided direct writing (LGDW); (B) long-range LGDW by guiding cells along the lumen of hollow optical fibers.

adhesion. Finally, several fibers coupled to suspensions of different cell types can be used to build heterotypic tissue constructs, thereby boosting printing rate and complexity (Odde & Renn, 1999).

The capability of LGDW to dispense multiple cell types with single-cell resolution makes this technique especially suitable for building small but complex tissue constructs such as stem cell niches or model tissues with incorporated microvasculature (e.g., liver, pancreas, tumor microenvironment). Nahmias et al., for example, printed contiguous lines of HUVECs on Matrigel and observed that endothelial cells elongated and formed hollow structures. Then, they covered the vascular pattern with a suspension of rat hepatocytes (4×10^5 cells/mL). Within one day, the hepatocytes aggregated around the microvessels, giving rise to 3D tissue constructs akin to liver sinusoids (Nahmias et al., 2005).

3.2.2 Laser-induced forward transfer

Laser-induced forward transfer relies on the thermal effects of pulsed laser beams. Bohandy et al. vacuum-deposited a thin (0.41 μm thick) copper film on a transparent support (a fused silica plate), and bombarded it through the support with a high-energy laser pulse focused onto the film (Bohandy et al., 1986). The metal was ablated from the support and propelled onto the silicon substrate placed near the film (Fig. 5.14A). The term laser-induced forward transfer was coined by Fogarassy and

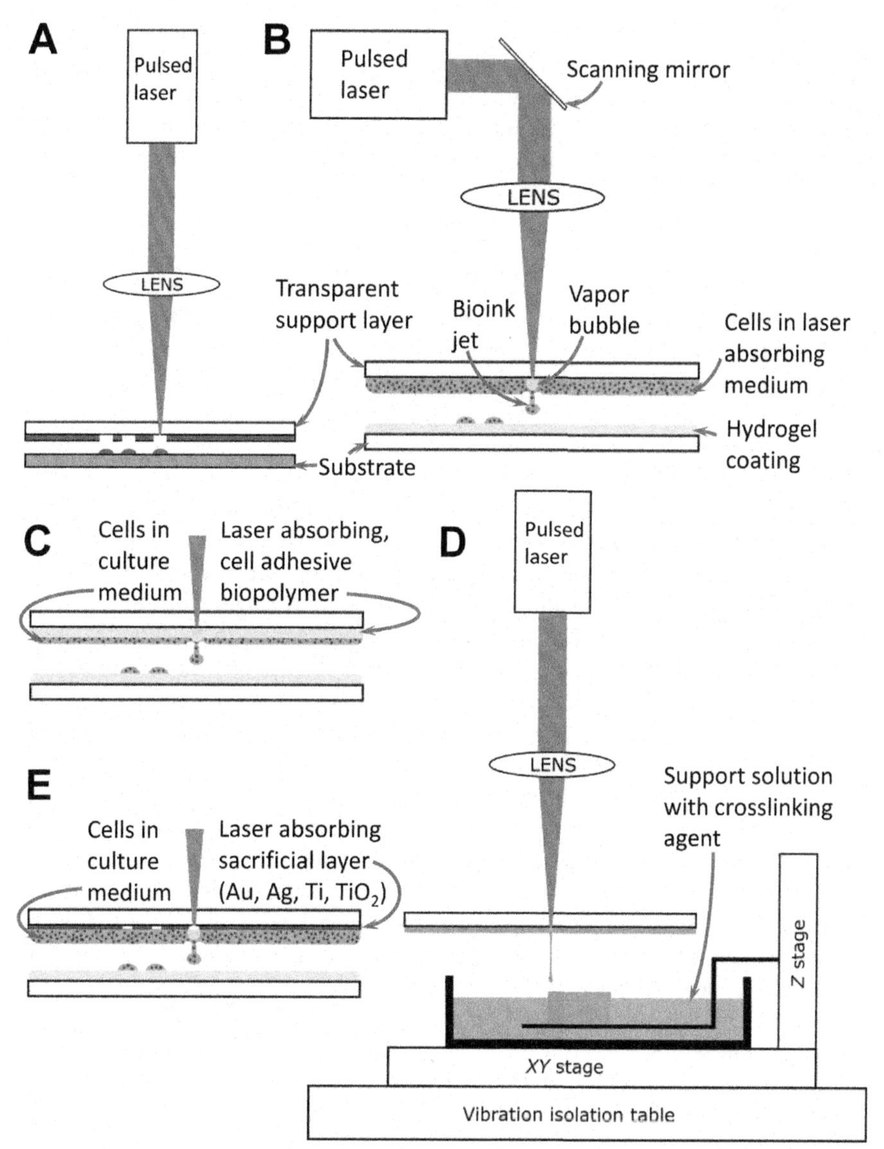

FIGURE 5.14 Schematics of bioprinting techniques based on laser-induced forward transfer (LIFT).

(A) LIFT of thin sheets of laser-absorbing material precoated on a transparent support plate; (B) matrix-assisted pulsed-laser evaporation direct write (MAPLE DW) of cells suspended in a laser-absorbing culture medium or prepolymer solution; (C) MAPLE DW using a ribbon coated with a laser-absorbing hydrogel layer seeded with cells; (D) high-throughput MAPLE DW of cell-laden alginate constructs; the print bed is immersed in a crosslinker pool, which also provides mechanical support via buoyancy; and (E) absorbing-film-assisted laser-induced forward transfer (AFA-LIFT).

Fuchs in their investigation of pulsed laser deposition and patterning of superconducting thin films (Fogarassy et al., 1989). Besides the ArF excimer laser used by Bohandy et al. (193 nm wavelength), Fogarassy and Fuchs also employed a neodymium-doped yttrium aluminum garnet (Nd:YAG) laser, delivering near-infrared (1064 nm wavelength) light pulses. Successful LIFT was observed when the UV laser delivered 20 ns pulses with energy density (fluence) in the interval 0.18–0.3 J/cm^2, as well as when the Nd:YAG laser generated 5 ns pulses with 0.5–1 J/cm^2 fluence.

The LIFT technique has been modified in various ways to enable the patterned deposition of biomolecules and live cells (Guillemot et al., 2010). Known as matrix-assisted pulsed-laser evaporation direct write (MAPLE DW), the modified approach involves coating the transparent support with cells dispersed in a biocompatible matrix (Wu et al., 2001). The support may be a quartz plate if a UV laser is used. The bioink-laden transparent plate is called a ribbon to reinforce the analogy between a MAPLE DW printer and a classical typewriter. The simplest ribbon configuration is the one shown in Fig. 5.14B, in which the bare support is coated with a cell suspension—cells uniformly dispersed in a hydrogel or in a prepolymer solution (Xiong et al., 2015). Another option is to precoat the ribbon with cell-free hydrogel, immerse it in a cell culture medium, seed the hydrogel with cells, and let them grow to confluence (Fig. 5.14C). This elaborate approach, adopted in early MAPLE DW experiments (Ringeisen et al., 2004; Wu et al., 2001), ensures a thermal and mechanical buffer layer between the vapor bubble and nearby cells.

The ArF excimer laser is a good choice for MAPLE DW because water absorbs UV light of 193 nm wavelength. Under the action of a laser pulse, part of the biolayer is suddenly volatilized; the resulting vapor bubble expands for a while (≈ 1 μs) and collapses afterward, a process which, under the right conditions, leads to jetting the overlying bioink toward the nearby receiving substrate, such as a glass slide covered with a thin hydrogel layer to attenuate the impact on landing. Indeed, the transferred embryonal carcinoma cells died upon landing on a hard surface, about half of them survived when the target surface was covered with a 20 μm thick Matrigel layer, and essentially all of them remained viable when the cushion layer on the receiving substrate was at least 40 μm thick (Ringeisen et al., 2006). In certain high-throughput setups, the print bed is covered by a large 3D construct, which is progressively immersed in a liquid medium containing a crosslinking agent—e.g., Ca^{2+} if the ribbon is loaded with cells suspended in sodium alginate (Fig. 5.14D) (Xiong et al., 2015).

In 2004, Hopp et al. proposed to alleviate the need for a laser-absorbing matrix in LIFT-based bioprinting. To prepare the print ribbon, they first coated the transparent holder with a 50 nm thick silver layer deposited by vacuum evaporation; then, they covered the silver sheet with cell suspension (Fig. 5.14E). In this approach, called absorbing-film-assisted laser-induced forward transfer (AFA-LIFT), cells are still suspended in a culture medium or a hydrogel, but these media do not have the role of absorbing the laser light; the interlayer is meant to do that (Fig. 5.14E). The thickness of the interlayer should be slightly higher than the laser penetration depth to protect the cells from the powerful laser pulse and to avoid the contamination of the bioink (Hopp et al., 2004). Since then, a variety of laser-absorbing materials have been used with AFA-LIFT, including Au, Ti, TiO2, and triazene (a UV-absorbing volatile polymer) (Guillemot et al., 2010). For AFA-LIFT, certain research groups preferred the name biological laser printing (BioLP) (Ringeisen et al., 2006), laser bioprinting (LaBP) (Koch et al., 2013), whereas others called it simply LIFT (Ovsianikov et al., 2010).

In all of the above techniques, the target substrate is placed below the ribbon, at a distance ranging from a few hundred microns to a few millimeters, and is mounted on a high-speed *XYZ* stage. Fresh bioink can be exposed for printing by repositioning the ribbon via a motorized *XY* stage (Xiong et al., 2015), or by reorienting the laser beam via a scanning mirror (Fig. 5.14B) (Guillemot et al., 2010); their combination is also feasible.

Bioink transfer elicited by a laser pulse depends on several process parameters (e.g., bioink rheology and layer thickness, ribbon-target distance, and laser fluence), well-characterized both theoretically (Guillemot et al., 2010; Schiele et al., 2010) and experimentally (Duocastella et al., 2011; Kryou et al., 2021). For a given bioink layer, the laser fluence should exceed a threshold value for successful jetting. Further increasing the laser fluence results in larger transferred droplet volumes. Nevertheless, over a certain laser fluence the jet breaks up into tiny (<1 μm) droplets spattered over a large area of the receiving substrate. This undesirable behavior is called the plume regime. Jet formation involves a series of events triggered by the laser pulse. The cavitation bubble mainly extends toward the target surface because its lateral expansion is hampered by the bioink present on the ribbon. Hence, a large pressure builds up in the bioink that embraces the bubble, pushing bioink streams along the bubble wall, toward its tip, where they meet and form a jet that spans the gap between the ribbon and the receiving substrate (Fig. 5.15).

In the case of large gaps, the jet can break up into droplets due to the Rayleigh-Plateau instability. The bubble collapses within a few microseconds, but, eventually, the entire bioink volume mobilized during the bubble's expansion is transferred into the droplet formed on the target surface (Duocastella et al., 2011). A similar scenario has been observed also for bioinks composed of cells suspended in Dulbecco's Modified Eagle's Medium, with or without 10% glycerol (needed to tune viscosity), at cell concentrations of up to 7.5×10^7 cells/mL (Kryou et al., 2021).

One of the main advantages of LIFT-based bioprinting is its nozzle-free nature; therefore, it is unaffected by clogging issues. LIFT enables the deposition of cells and/or biomolecules in picoliter-sized droplets at micrometer resolution. It is able to dispense cells contiguously, one by one, at high printing speed, of the order of thousands of droplets per second. Despite the thermally actuated deposition of cell-laden liquid or gel media, heat shock protein expression was statistically similar in printed and control cells. Also, printed cell viability was above 95% when the receiving substrate was a hydrogel-coated glass plate (Guillemot et al., 2010). In the high-throughput printing procedure depicted in Fig. 5.14D, the postprinting viability of NIH 3T3 mouse fibroblasts incorporated in 2% (w/v) alginate

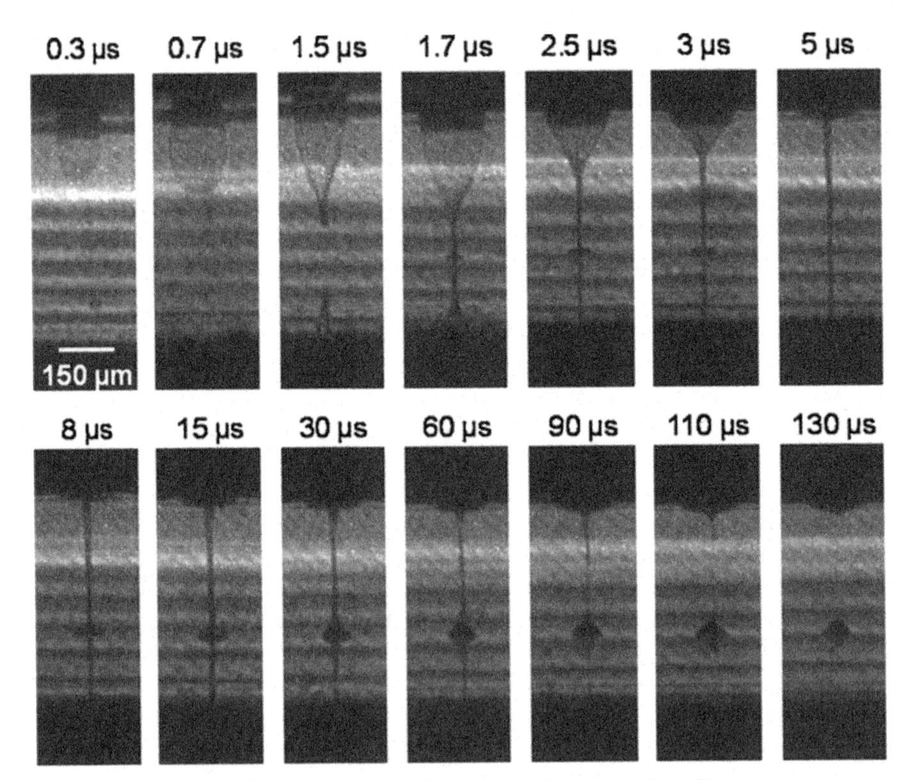

FIGURE 5.15 Jet formation during laser-induced forward transfer of liquids.

Stop-action movie of the laser-induced forward transfer (LIFT) process recorded with a fast-intensified CCD camera. The time delay between the laser pulse and the moment of image acquisition is listed above each frame.

Reprinted from Duocastella, M., Patrascioiu, A., Dinca, V., Fernández-Pradas, J.M., Morenza, J.L. & Serra, P. (2011). Study of liquid deposition during laser printing of liquids. Applied Surface Science, 257, *5255–5258. https://doi.org/10.1016/j.apsusc.2010.10.148, with permission from Elsevier.*

(10^7 cells/mL) was about 64% for straight tubes and 68% for Y-shaped bifurcated tubes. The higher viability in the latter case was attributed to the more effective attenuation of the impact forces exerted during bioink landing on the overhang portions of the crosslinked hydrogel construct (Xiong et al., 2015).

The downsides of LIFT bioprinting include high equipment costs, laborious preparation of the print ribbon, and poor scalability. Also, potential photonic damage of the cells needs to be tested carefully, especially in MAPLE DW systems, in which only a fraction of the laser pulse is absorbed at the interface between the transparent support and the adjacent bioink; the rest of it traverses the overlying bioink layer. During bioink ejection, the rapid expansion of the vapor bubble brings about enormous accelerations, of the order of 10^5–10^9 times the gravitational acceleration, depending on laser fluence (Schiele et al., 2010). The created jet advances toward the target surface at a velocity of 20–150 m/s (Guillemot et al., 2010). Hence, both the acceleration involved in jetting and the deceleration experienced in the course of landing can potentially inflict mechanical damage. These aspects have been investigated for diverse cell types and experimental setups, proving the feasibility of LIFT bioprinting, but they need to be tested in further studies that seek to render these techniques translational.

3.2.3 Laser induced side transfer

Laser induced side transfer (LIST) bioprinting is achieved by bioink droplet ejection from a glass microcapillary due to a laser pulse focused on the capillary lumen (Ebrahimi Orimi et al., 2020). Fig. 5.16 depicts the working principle of this technique (*panel* A) along with side-view photographs of droplet ejection (*panel* B).

LIST was demonstrated on a low-viscosity bioink prepared from HUVECs suspended in culture medium (10^6 cells/mL) supplemented with 13.2 μM fibrinogen, 7.7 μM aprotinin, and 10 mM Allura red AC—a food dye needed to enhance the absorption of the green laser light of 532 nm wavelength. Laser pulses of 6 ns duration were generated by a Q-switched pulsed Nd:YAG laser. A syringe pump replenished the bioink loaded in a 5 cm long glass capillary of 0.3 mm by 0.3 mm square internal cross-section and 0.15 mm thick walls. The laser beam was focused by a 4× plan-achromatic microscope objective lens at a point located on the axis of the capillary, 0.5 mm above its opening. The cavitation threshold of the bioink was reached at a laser pulse energy of 90 μJ, giving rise to a transient vapor bubble. Encountering less resistance from below, the bubble expanded downwards, ejecting a micro-jet of bioink (Fig. 5.16B). The jet front moved with a speed of 3.2 m/s across the 0.5 mm gap between the capillary tip and the receiving substrate—a 1 mm-thick fibrin gel layer deposited on a microscope cover glass (Ebrahimi Orimi et al., 2020).

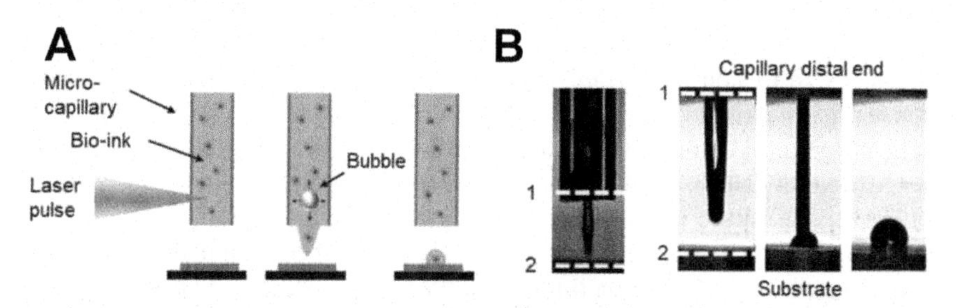

FIGURE 5.16 Laser induced side transfer (LIST) bioprinting.

(A) Schematic drawing of the stages of LIST bioprinting; (B) representative snapshots of bioink droplet ejection during LIST bioprinting; dashed lines mark the tip of the capillary (1) and the receiving substrate (2).

From Ebrahimi Orimi, H., Hosseini Kolkooh, S.S., Hooker, E., Narayanswamy, S., Larrivée, B. & Boutopoulos, C. (2020). Drop-on-demand cell bioprinting via laser induced side transfer (LIST). Scientific Reports, 10, *9730. https://doi.org/10.1038/s41598-020-66565-x. Reprinted under the terms of the Creative Commons Attribution 4.0 International License (https://creativecommons.org/licenses/by/4.0/legalcode).*

Printed droplets were about 165 μm in diameter, indicating that LIST bioprinting falls behind LIFT in what concerns print resolution, albeit further refinements (e.g., using thinner capillaries) might bring them on the same footing. For laser pulse energies in the 90–120 μJ interval, cell viability was found between 93.1% and 96.5% right after LIST bioprinting, with smaller viability at high energy. Moreover, HUVECs proliferated after printing and formed adherens junctions. In another experiment performed on dorsal root ganglion neurons, postprinting cell viability was high (86%) and gene expression was unaltered. Viable bioprinted neurons were comparable to their control counterparts in their ability to release the calcitonin gene-related peptide to communicate with surrounding cells. Nevertheless, neurite outgrowth was significantly suppressed by LIST bioprinting, especially at a high laser pulse energy, of 120 μJ (Roversi et al., 2021). Hence, LIST and LIFT seem to have similar impacts on cell viability and function.

Envisioned developments of LIST bioprinting include the use of higher pulse repetition rates, of up to the theoretical limit of 2.5 kHz computed from the duration of one droplet transfer (≈400 μs). Further investigations will be needed to evaluate the biological impact of a high repetition rate (e.g., excessive heating of the bioink, as observed in thermal inkjet printing at high droplet ejection rates). Also, the Nd:YAG laser could be operated at its most common wavelength, 1064 nm, which is much better absorbed by water than its second harmonic. Finally, multimaterial bioprinting can be readily implemented by using multiple capillaries and a scanning mirror to point the laser at them (Ebrahimi Orimi et al., 2020).

4. Spheroid-based bioprinting

Tissue spheroids (cell aggregates) can also be used as bioink droplets (Mironov et al., 2003). Shorter printing time is just one of the advantages of printing cell aggregates instead of individual cells. Indeed, a tissue spheroid small enough for perfusion-free (diffusion-based) delivery of oxygen and nutrients contains thousands or even tens of thousands of cells (depending on cell type). A typical tissue construct of clinical relevance comprises millions to billions of cells, suggesting that a bioprinter capable of dispensing one spheroid per second could build a properly sized engineered tissue within a few hours. Cell aggregates composed of tissue-specific cell types in the right proportions provide a biomimetic 3D environment for the constituent cells, which favor cell–cell interactions. In addition, the cell density of tissue spheroids is of the order of 10^8–10^9 cells/mL, which is at least one order of magnitude higher than the maximum cell density achieved in hydrogel-based bioinks (Hospodiuk et al., 2017). Therefore, spheroid-based bioprinting is seen as a promising avenue toward reproducible organ-on-a-chip devices, model tissues for drug testing, human disease models, organoid engineering, tissue engineering, and biofabrication. Also, it might revolutionize experimental systems built for fundamental research on synthetic biology, angiogenesis, and organogenesis (Ayan, Heo, et al., 2020).

Apart from technical challenges related to the fast and precise handling of delicate submillimeter tissue droplets, spheroid-based bioprinting still faces serious limitations. An inherent one is the inability of certain cell types to adhere to each other; such cell types do not form aggregates on their own, but some of them might be coaxed to do so by supplementing their culture medium with ECM components. Another problem with cell aggregate bioprinting lies in the relatively large size of the bioink droplets; that is, the minimum feature size achievable is of the order of the spheroid diameter, which results in limited resolution. Also, this type of bioprinting displays most strikingly a general feature of bioprinting, that the final outcome of the process is not what the printer builds; bioprinting merely sets the stage for the cells. The self-organizing capability of cells is a double-edged sword: on one hand, it raises the complexity of bioprinting, making it more difficult to plan for a certain outcome, but, on the other hand, it brings about tissue-specific features of single-cell resolution. One notable example in this respect is the network of capillary vessels created by endothelial cells incorporated in cell aggregates (De Moor et al., 2018, 2021). Spheroid formation is primarily driven by cell—cell interactions. Nevertheless, subsequent culturing is essential for ECM secretion, which enhances the mechanical properties of the spheroids. Postprinting fusion and tissue construct maturation depend on preprinting spheroid culture conditions (Ayan, Celik, et al., 2020). Spheroid size is constrained by the diffusion limit of oxygen, of about 200 μm in vivo. Hence, except for cell types that are adapted to hypoxia (e.g., chondrocytes), spheroids should be less than 0.4 mm in diameter; otherwise, cells in their core are deprived of oxygen and nutrients. Spheroid-based bioprinting is challenging in the case of tissue constructs that also include epithelial cells. Common spheroid preparation techniques yield a random mixture of cells within the bulk of spheroids, which is far from the physiological milieu of epithelial cells. Endothelial cells exceptionally can thrive within a cell mass because they assemble into cords on their own, which further evolve into capillaries. Other epithelial cell types might be seeded on the surface of tissue spheroids or on hydrogel beads. Finally, additional complexity arises because of cell sorting in heterotypic tissue spheroids: the most cohesive cells segregate in their cores and the other cell types form concentric layers in the decreasing order of cohesivity (Moldovan et al., 2017). We dive into these problems more deeply in Chapters 7 and 8.

Early attempts at spheroid-based bioprinting (Jakab et al., 2006) relied on microextrusion printers to deliver strings of contiguous cell aggregates loaded into glass micropipettes (Fig. 5.4G1). Although the positioning precision was satisfactory, tissue spheroid biofabrication and print cartridge loading were tedious, whereas changing the print cartridge also involved careful calibration to make sure that spheroid deposition resumes at the right spot. Progress made since then resulted in high-throughput production of monodisperse, heterotypic tissue spheroids, which can be incorporated into biocompatible hydrogels and printed using conventional EBB. Despite the limited cell density of such spheroid-laden hydrogels and poor control over the placement of individual spheroids, this approach preserves cell viability and function. For example, De Moor et al. (2021) used EBB to print tissue constructs made of 10% (w/v) GelMA loaded with spheroids composed of human umbilical vein endothelial cells (HUVEC), human foreskin fibroblasts (HFF), and adipose tissue-derived mesenchymal stem cells (ADSC). Cell viability was above 91% after 6 days in culture. HUVECs sprouted out of spheroids, and their protein expression was consistent with active blood vessel formation. When implanted in fertilized eggs, on chick chorioallantoic membranes, the self-assembled capillary-like network displayed anastomosis with the host vasculature (De Moor et al., 2021).

Gutzweiler et al. used piezoelectric drop-on-demand bioprinting to deliver individual HUVEC spheroids, of about 100 μm in diameter, encapsulated in fibrin gel droplets. When such droplets were printed on top of a fibrin gel sheet 500 μm apart and covered with an additional fibrin layer, capillary sprouting formed connections between adjacent spheroids within 3 days of incubation (Gutzweiler et al., 2017). In spite of highly precise droplet positioning, this approach has limited capabilities for building closely packed spheroid structures because the ejected droplet is larger than the spheroid and the ejection does not assure a central placement of the spheroid. Furthermore, shooting droplets contiguously might affect previously dispensed spheroids.

Although hydrogel-based bioprinting strategies might well lead to important discoveries and useful tissue constructs, this section is focused on bioprinting techniques specifically designed for individual spheroid dispensing.

4.1 Fluidics-based tissue spheroid singularization and bioprinting

Tissue spheroids placed in a cell culture medium next to each other are poised to develop adhesive bonds mediated by cell adhesion molecules. In time, the spheroids fuse because cells relocate to maximize the number and strength of bonds formed with their neighbors (see Chapter 7 for details). Spheroid bioprinter cartridges need to prevent their self-assembly and enable their delivery one by one. To this end, Mekhileri et al. built a tissue spheroid singularization and injection head. Fig. 5.17 presents the working principle of the spheroid singularization module (Mekhileri et al., 2018).

The singularization system is meant to collect spheroids into a reservoir hopper and guide one of them into the injection nozzle. It consists of a polycarbonate block outfitted with six valves to manipulate fluid flow such as to aspirate one spheroid, separate it from its peers, and drive it toward the print head. The first step of singularization, called agitation (Fig. 5.17A1), aims to clear the way toward the capture port, which is connected to both a vacuum source and a pressure source via dedicated valves (denoted in Fig. 5.17A by CV and CP, respectively). The second step consists in locking the leading spheroid to the capture port by a precisely controlled suction pressure (Fig. 5.17A2). The third step consists of flushing away lagging spheroids toward the reservoir (Fig. 5.17A3). Finally, the leading spheroid is released (by ceasing aspiration) and flushed toward the injection nozzle (Fig. 5.17A4).

The dimensions of the singularization unit are established in units of the spheroid size, and, therefore, the system works well for a monodisperse set of tissue spheroids

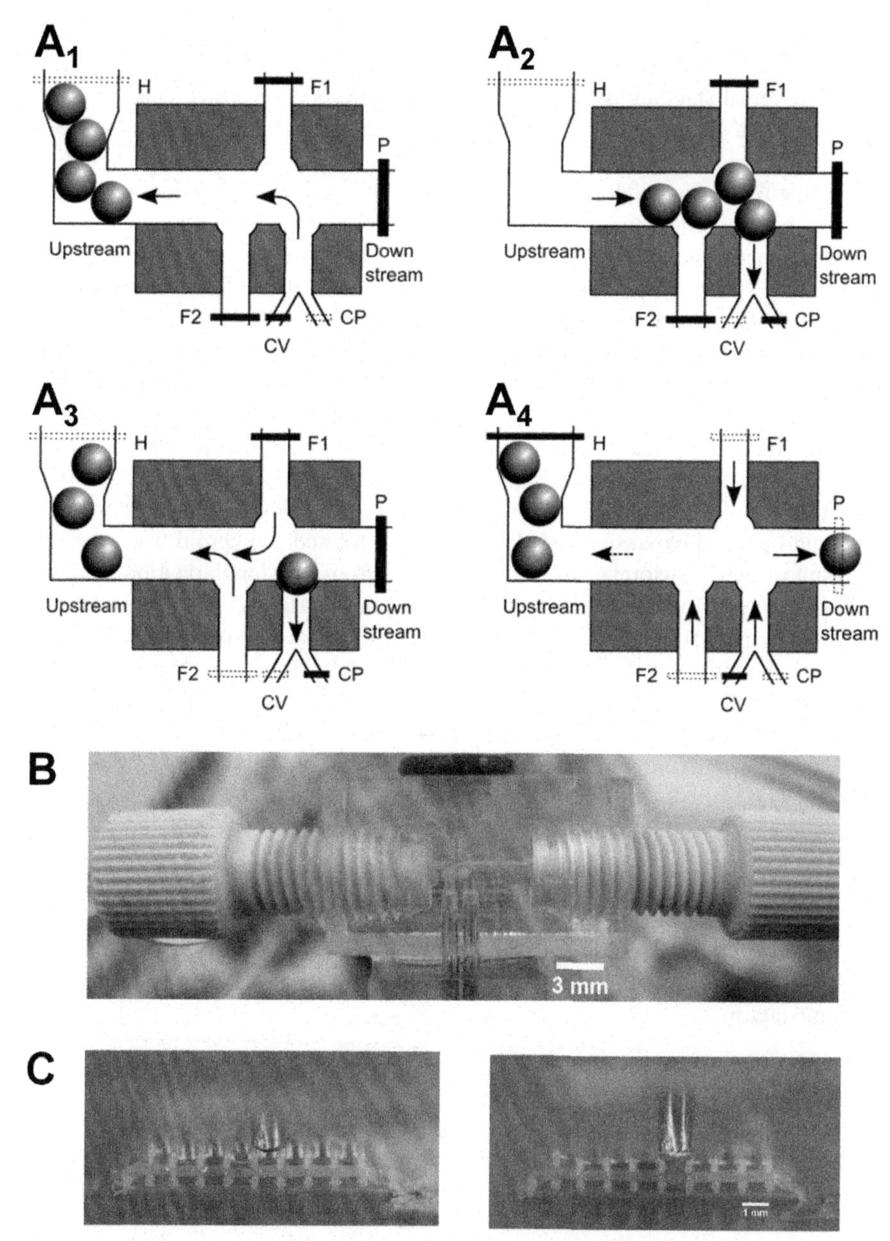

FIGURE 5.17 The working principle of fluidics-based tissue spheroid singularization.

(A_1–A_4) Schematic drawings of the stages of tissue spheroid singularization: (A_1) agitation; (A_2) capture of the leading spheroid by gentle aspiration; (A_3) separation of the leading spheroid; (A_4) release and expulsion of the leading spheroid downstream; Abbreviations: *H*, hopper valve; *P*, pinch valve; *F1* and *F2*, flush valves 1 and 2; *CV*, capture port vacuum of a given size. The diameter of the main fluidic chamber (horizontal tube in the center of Fig. 5.17B) is about 30% larger than the spheroid size to prevent blockage due to the transversal stacking of spheroids. The pressure ports (vertical channels) are 30% smaller in diameter than the spheroids to keep them from entering these channels.

The singularization module shown in Fig. 5.17B is part of a hybrid biofabrication platform that combines tissue spheroid bioassembly and 3D plotting of thermoplastic polymer structures. This platform was built by mounting a custom-made tool for tissue spheroid bioprinting on a commercial computer-aided tissue engineering instrument equipped with a temperature-controlled hydrogel extruder and a hot melt dispense head—BioScaffolder (SYS ENG, Germany) (Mekhileri et al., 2018).

Micro-tissue bioprinting is illustrated in Fig. 5.17C. In their hybrid bioprinting study, Mekhileri et al. 3D plotted PEGT:PBT scaffolds with 1 mm fiber spacing to fit microtissues of about the same size. To insert a tissue spheroid into a selected pore, the injection nozzle was first centered over the pore relying on the G-code used to plot the scaffold. Then, the injection nozzle was opened by retracting its stopper—a solenoid-operated expansion rod (Fig. 5.17C—*left*), and a spheroid was expelled as a result of fluid flow orchestrated by the valves from the singularization module. Finally, the expansion rod was lowered to close the nozzle and press the spheroid down into the pore (Fig. 5.17C—*right*). Experiments conducted on spheroids of human nasal chondrocytes demonstrated that neither singularization nor injection did affect cell viability: it was $80.4 \pm 4.8\%$ in control (not printed) spheroids, $83.5 \pm 3.4\%$ in singularized spheroids, $78.4 \pm 2.4\%$ in injected spheroids, and $83.6 \pm 5.5\%$ in spheroids subjected to both singularization and injection (Mekhileri et al., 2018).

4.2 The Kenzan method of scaffold-free bioprinting

Kenzan is the name of the hedgehog-like set of spikes emerging from a support plate used in Ikebana, the Japanese art of flower arrangement, to keep flower stalks in place. Kenzan literally means sword mountain (in Japanese, ken = sword and zan = mountain).

valve; and *CP*, capture port pressure valve; notations: solid rectangle—closed valve, dotted rectangle—open valve, solid arrow—direction of flow, dotted arrow—direction of applied pressure; (B) photograph of the spheroid singularization module; (C) The spheroid injection nozzle before (*left*) and after (*right*) the controlled insertion of a spheroid into an empty pore of a 3D plotted scaffold made of poly(ethylene glycol)-terephthalate (PEGT) and poly(butylene terephthalate) (PBT) block copolymers.

Adapted from Mekhileri, N.V., Lim, K.S., Brown, G.C.J., Mutreja, I., Schon, B.S., Hooper, G.J. & Woodfield, T.B.F. (2018). Automated 3D bioassembly of micro-tissues for biofabrication of hybrid tissue engineered constructs. Biofabrication, 10, *024103. http://stacks.iop.org/1758-5090/10/i=2/a=024103.*

In scaffold-free bioprinting, a microneedle array is used as temporary support for tissue spheroids (Itoh et al., 2015). In the medical-grade Kenzan devised by Itoh et al., 81 mutually parallel, 0.17 mm thick stainless steel microneedles are placed 0.4 mm from each other (center-to-center distance). Spheroids of about 0.4–0.6 mm in diameter are formed or placed in a nonadherent 96-well plate. They are picked up by a robotic arm equipped with a suction nozzle (of 0.45 mm outer diameter and 0.23 mm inner diameter) and inserted into the needles according to an arrangement created using the bio-3D designer software shipped with the Regenova Bio-3D printer (Cyfuse Biomedical K.K., Japan) (Fig. 5.18A). Eventually, the desired tissue construct forms via the fusion of contiguous tissue spheroids. As soon as the printout develops structural integrity, it is removed from the needles by gently lifting a sieve inserted into the Kenzan before printing. Actually, there are two removable plastic plates lying at the bottom of the needle array. One of them serves for construct retrieval, whereas the other one guides the reinsertion of the first. Within a few days of incubation, the cells relocate and seal the narrow channels left in the construct where the needles were before (Itoh et al., 2015; Moldovan et al., 2017).

The construct shown in Fig. 5.18D was removed from the Kenzan on day 4 post bioprinting. Subsequently, it was cultured for additional 4 days on a perfusion bioreactor, being pulled over a gauge 22 plastic catheter that featured an array of lateral holes. During this time, the holes left behind by the needles healed completely, as shown in the preimplantation histological sections (Fig. 5.18E and F). While implanted in a rat abdominal aorta, the bioprinted tube suffered ample remodeling, which resulted in an enhanced endothelial lining of the luminal surface and a roughly twofold increase in the lumen area (Fig. 5.18G) (Itoh et al., 2015).

The Regenova printer is housed by a sterile hood and includes a plate management and transport unit and a printing assembly composed of a Kenzan holder stand, a spheroid manipulation unit, and a pair of cameras (Fig. 5.19A). The Kenzan holder is a lidless container filled with sterile phosphate-buffered saline solution. The upper camera captures the tips of the microneedles, whereas the lower camera visualizes the spheroids lying in the plate wells as well as the orifice of the spheroid suction nozzle. The images acquired by the two cameras are processed by image analysis software to assess spheroid shape and size prior to printing, to check the positions of the spikes in the Kenzan, and to evaluate the state of the suction nozzle. The nozzle only picks up spheroids deemed suitable for printing. Once a spheroid is captured, the nozzle positions itself over the selected needle and dives into the Kenzan holder, inserting the spheroid into the needle; finally, it releases the spheroid by applying a gentle puff. The entire process is repeated according to the digital model until the construct is finalized, typically in less than one hour. Hence, the Kenzan method is highly effective for building centimeter-sized tissue replacements (Murata et al., 2020).

Various Kenzans have been developed for bio-3D printing (Fig. 5.19B), with microneedle configurations tailored according to the geometry of the target tissue. For example, the 9×9 Kenzan was used for tendon, liver, and urinary bladder

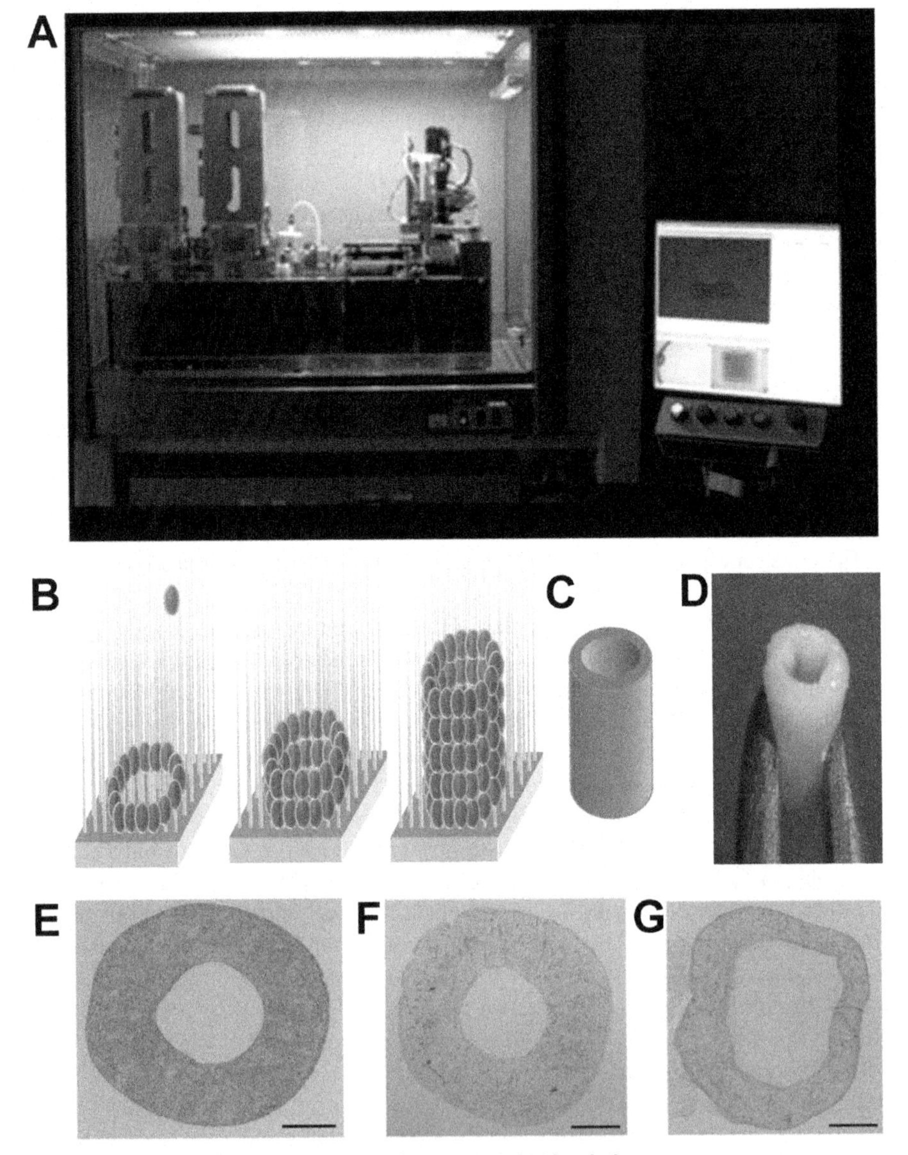

FIGURE 5.18 The Kenzan technique of tissue spheroid bioprinting.

(A) Photograph of the Regenova Bio-3D printer; (B) scheme of the Kenzan at three consecutive stages of tubular tissue bioprinting; (C) scheme of the tube expected to form after the removal of the construct from the microneedle array; (D) picture of the 7 mm long tubular construct; (E, F) histological sections of the bioprinted construct before implantation: (E) hematoxylin and eosin (H and E) staining, (F) immunohistochemical staining of von Willebrand factor showing the distribution of endothelial cells; (G) von constructs; the 9 × 9 circular Kenzan served for building rat blood vessel replacements, as well as tubular structures made for tissue repair in rats (peripheral nerves, trachea, esophagus); the 13 × 13 circular Kenzan was used for minipig osteochondral reconstruction; the 26 × 26 Kenzan served for rat heart patch and liver tissue bioprinting; finally, the 34 × 34 circular Kenzan was used for minipig blood vessel constructs, dog peripheral nerve reconstruction, and rat diaphragm tissue repair. A vast variety of in vivo studies demonstrated the usefulness of bio-3D printed structures in tissue repair. Also, liver tissue constructs created by the Kenzan method were found suitable for drug screening (Murata et al., 2020).

4.3 Aspiration-assisted bioprinting

Aspiration-assisted bioprinting (AAB) is specifically designed for the precise placement of tissue spheroids in a hydrogel support medium. It can be performed in a layer-by-layer fashion (Ayan, Heo, et al., 2020) or as freeform bioprinting in a suspension bath (Ayan, Celik, et al., 2020).

The bioink employed in this technique consists of multicellular spheroids prepared beforehand and placed in a cell culture medium. They are picked up, one by one, by exerting a gentle suction through a borosilicate glass micropipette attached to the print head. To this end, tapered micropipettes with an orifice diameter of about 80 μm were fabricated using a micropipette puller. A cell aggregate is captured and kept attached to the pipette tip owing to a pressure difference between the interior of the pipette and the surroundings. The suction pressure should exceed a critical value to assure a lifting force capable of overcoming the hydrodynamic drag and interfacial tension forces encountered while extracting the spheroid from the cell culture medium. Exaggerated suction, however, might damage the tissue spheroid.

To assure a controlled aspiration pressure, the pipette is connected to a pneumatic line monitored by a pressure sensor. The required pressure is generated using a compressor and a pressure regulator, as well as a vacuum pump and a vacuum regulator; both the pressure and vacuum lines are gated by solenoid valves that open only when their coils are powered. The electric signals needed to control the solenoid microvalves are generated as a result of the computational analysis of input signals received from the pressure sensor and the traveler camera that monitors the neighborhood of the pipette tip. Computer vision algorithms analyze the recorded images

Willebrand factor staining of a representative cross-section of a construct implanted for 5 days in a rat abdominal aorta; in panels E–G, scale bars = 0.5 mm.

Adapted from Itoh, M., Nakayama, K., Noguchi, R., Kamohara, K., Furukawa, K., Uchihashi, K., Toda, S., Oyama, J., Node, K. & Morita, S. (2015). Scaffold-free tubular tissues created by a bio-3D printer undergo remodeling and endothelialization when implanted in rat aortae. PLoS One, 10, *e0136681. https://doi.org/10.1371/journal.pone.0136681. Reprinted under the terms of the Creative Commons Attribution 4.0 International License (https://creativecommons.org/licenses/by/4.0/legalcode).*

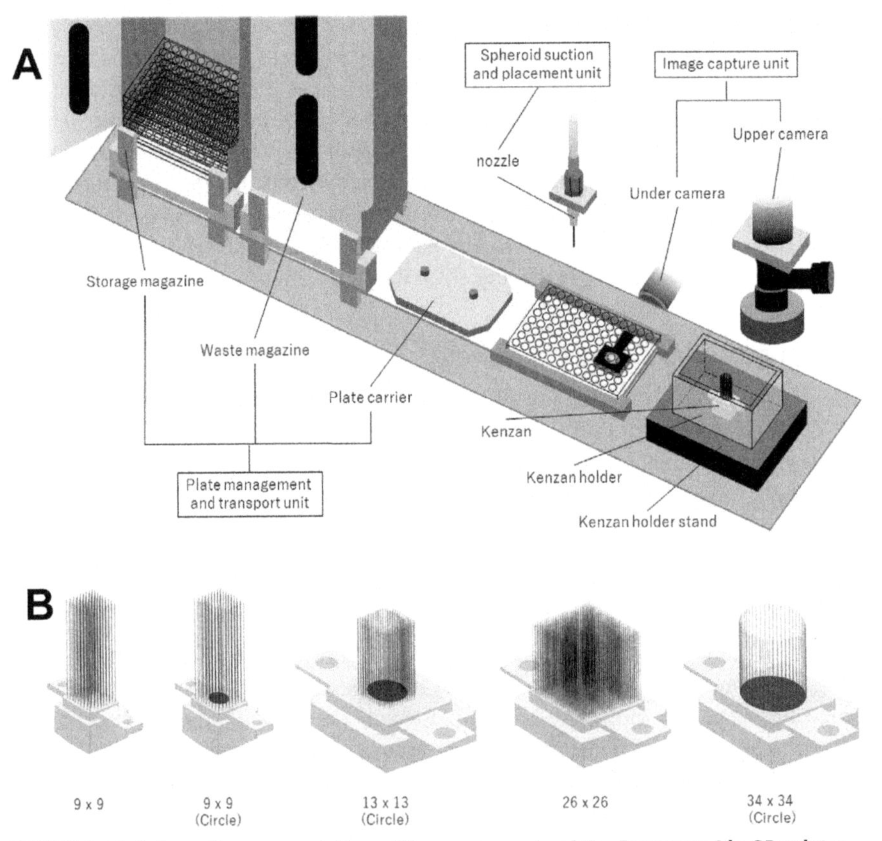

FIGURE 5.19 Schematic representation of the components of the Regenova bio 3D printer.

(A) The main parts of the bio 3D printer and (B) types of microneedle arrays (Kenzans) available for bio-3D printing.

From Murata, D., Arai, K. & Nakayama, K. (2020). Scaffold-free bio-3D printing using spheroids as "bio-inks" for tissue (re-)construction and drug response tests. Advanced Healthcare Materials, 9, *1901831. https://doi.org/10.1002/adhm.201901831. Reprinted under the terms of the Creative Commons Attribution 4.0 International License (https://creativecommons.org/licenses/by/4.0/legalcode).*

to identify spheroids within reach and assess their diameters. Based on spheroid size, the software determines the minimum back pressure needed to handle it. The medium flow caused by aspiration generates drag forces on a nearby spheroid oriented toward the micropipette tip. Thus, the spheroid blocks the pipette orifice and remains attached to it. The captured spheroid is inspected again before being used for bioprinting: it needs to have a proper shape and size, with no signs of mechanical damage. This second assessment also helps to determine the optimal offset distance between the pipette tip and the substrate to prevent damage to the spheroid during deposition.

During AAB, spheroids are partly immersed into a partially crosslinked gel substrate, such as 1% (w/v) alginate, and the aspiration pressure is removed. The spheroid quickly recovers its original shape because it was delivered within about 30 s—a short time compared to the viscoelastic relaxation time of multicellular spheroids (ranging from 3 to 18 min, depending on cell type) (Ayan, Heo, et al., 2020). The force due to tissue surface tension still acts between the spheroid and the pipette tip, but it is overwhelmed by an order of magnitude larger force of adherence exerted between the spheroid and the gel. As the tip is lifted, the spheroid remains nested in the gel substrate. The process is repeated to create a bioink pattern with a placement precision of a few cell diameters. Once a bioink layer is completed, the spheroids are covered with alginate, which will serve as a substrate for the next layer.

In aspiration-assisted freeform bioprinting (AAfB) complex multicellular constructs are built by the AAB of tissue spheroids in a yield-stress support bath (Ayan, Celik, et al., 2020). Just as in suspended EBB, in AAfB, the building blocks of the bioprinted construct are placed in a slurry of jammed microgel particles, which undergo solid-liquid transition when the shear stress exceeds the yield stress and becomes solid right after the passage of the print nozzle. To minimize the disturbance in the support bath, Ayan et al. used a 27 G blunt needle (413 μm outer diameter and 210 μm inner diameter) as an aspiration nozzle and demonstrated the feasibility of AAfB in 1.2% Carbopol and 0.5% alginate microgel slurries. Bioprinted constructs were built from spheroids made of 2×10^4 or 5×10^4 mesenchymal stem cells (MSCs). The spheroids were placed in a medium-filled reservoir and transferred into a square Petri dish. About half of the Petri dish was occupied by the support bath, whereas the other half was filled with cell culture medium, which also covered the bioink reservoir. The medium was separated from the support bath by a vertical polydimethylsiloxane slice. Prior to bioprinting, the separator was removed, leaving behind a vertical interface (Fig. 5.20A) (Ayan, Celik, et al., 2020).

As shown in Fig. 5.20B and C, the freeform version of AAB does not require extracting the captured spheroid from its culture medium. Hence, it circumvents the most challenging event in AAB, but it still requires an aspiration pressure of tens of mmHg to maintain the spheroid attached to the nozzle despite the drag force exerted by the gel. Actually, AAfB involves several forces, including the elastic indentation of the gel as the spheroid bumps into the interface (Fig. 5.20B), and the interfacial tension force as it traverses the interface, but, according to the careful analysis of (Ayan, Celik, et al., 2020), the drag force exerted by the gel onto the spheroid is dominant. Combined with rheological measurements, their theoretical analysis provided an estimate of the minimum aspiration pressure:

$$P_a \approx \frac{\sqrt{R^2 - r^2}}{r^3} 6Rv\left(\frac{\tau_0}{\dot{\gamma}} + K\dot{\gamma}^{n-1}\right) \tag{5.1}$$

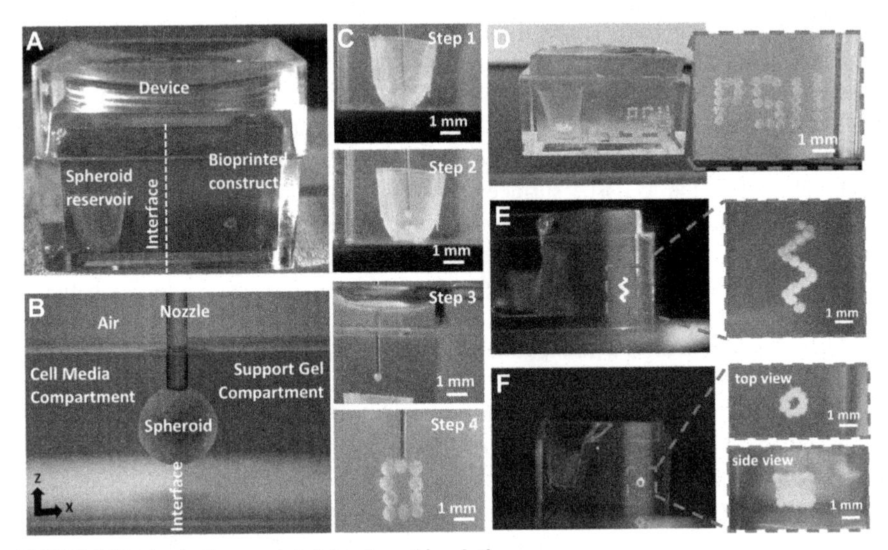

FIGURE 5.20 Aspiration-assisted freeform bioprinting.

(A) Picture of the bioprinting chamber—a square Petri dish partially filled with yield-stress gel (*right*) and cell culture medium containing the bioink reservoir (*left*); (B) schematic representation of the passage of an aspirated tissue spheroid through the medium–gel interface; (C) snapshots of the AAfB process captured by the inspection camera: Step 1—the aspiration nozzle approaches the spheroids from the medium-filled reservoir, Step 2—a spheroid is captured, Step 3—the nozzle has left the reservoir and is heading toward the support bath (opaque region), Step 4—the spheroid is deposited to complete a rectangular tissue construct firmly embedded in the support bath; (D–F) photographs of tissue constructs built from MSC spheroids using AAfB: (D) initials of Penn State University, (E) helicoidal construct, and (F) tubular structure.

Adapted from Ayan, B., Celik, N., Zhang, Z., Zhou, K., Kim, M.H., Banerjee, D., Wu, Y., Costanzo, F. & Ozbolat, I.T. (2020). Aspiration-assisted freeform bioprinting of pre-fabricated tissue spheroids in a yield-stress gel. Communications Physics, 3, *183. https://doi.org/10.1038/s42005-020-00449-4 Reprinted under the terms of the Creative Commons Attribution 4.0 International License (https://creativecommons.org/licenses/by/4.0/legalcode).*

where R is the spheroid radius, r is the internal radius of the aspiration nozzle, v is the nozzle velocity (2.5 mm/s), $\dot{\gamma}$ is the shear rate ($\dot{\gamma} \approx v/R$), τ_0 is the yield stress (25.7 Pa for 1.2% carbopol and 21.9 Pa for 0.5% alginate), n is the flow behavior index (smaller than one for a shear-thinning fluid), and K is the consistency index ($n = 0.3$ and $K = 44$ Pa s$^{0.3}$ for 1.2% Carbopol, whereas $n = 0.18$ and $K = 13.8$ Pa s$^{0.18}$ for 0.5% alginate) (Ayan, Celik, et al., 2020). Eq. (5.1) results from the balance of torques with respect to the trailing edge of the aspiration nozzle: the aspiration force, $P_a \pi r^2$, has the arm equal to the nozzle radius, r, whereas the drag force, $6\pi R v \eta_v$, has the arm equal to the distance between the nozzle tip and the spheroid center, $\sqrt{R^2 - r^2}$. Hence, the balance of torques reads $P_a \pi r^3 = 6\pi R v \eta_v \sqrt{R^2 - r^2}$, yielding Eq. (5.1) if one assumes that the support bath is a Herschel-Bulkley fluid, for which the shear stress is related to the shear rate according to the following relationship $\tau = \tau_0 + K\dot{\gamma}^n$. The Herschel-Bulkley model implies that the shear rate is zero unless the shear stress, τ, exceeds the yield stress, τ_0. Then, the viscosity of the support bath can be expressed as $\eta_v = \tau/\dot{\gamma} = \tau_0/\dot{\gamma} + K\dot{\gamma}^{n-1}$, which is the last factor in Eq. (5.1).

The 0.5% alginate microgel performed better in the context of AAfB than the 1.2% carbopol yield-stress gel. The medium−gel interface remained stable for 24 h in the case of alginate, whereas the Carbopol microgel particles migrated into the adjacent medium within a few hours. The bioprinted structure could be rapidly removed from alginate by sodium citrate treatment (within ½ hour). By contrast, the printout could be released from Carbopol by dissolving the microgel in cell culture medium. What is most important, during a 3-day culture of MSC spheroids in a 1.2% carbopol microgel cell viability dropped to 74%, whereas a similar experiment performed in 0.5% alginate microgel resulted in 93% cell viability (Ayan, Celik, et al., 2020).

In conclusion, AAB is a relatively affordable, cell-friendly bioprinting technique capable of building live structures of high cell density both layer by layer or in a freeform manner.

References

Akkouch, A., Yu, Y., & Ozbolat, I. T. (2015). Microfabrication of scaffold-free tissue strands for three-dimensional tissue engineering. *Biofabrication, 7*, 031002. https://doi.org/10.1088/1758-5090/7/3/031002

Arcaute, K., Mann, B. K., & Wicker, R. B. (2006). Stereolithography of three-dimensional bioactive poly(ethylene glycol) constructs with encapsulated cells. *Annals of Biomedical Engineering, 34*, 1429−1441. https://doi.org/10.1007/s10439-006-9156-y

Ashkin, A. (1970). Acceleration and trapping of particles by radiation pressure. *Physical Review Letters, 24*, 156−159. https://doi.org/10.1103/PhysRevLett.24.156

Ayan, B., Celik, N., Zhang, Z., Zhou, K., Kim, M. H., Banerjee, D., Wu, Y., Costanzo, F., & Ozbolat, I. T. (2020). Aspiration-assisted freeform bioprinting of pre-fabricated tissue spheroids in a yield-stress gel. *Communications on Physics, 3*, 183. https://doi.org/10.1038/s42005-020-00449-4

Ayan, B., Heo, D. N., Zhang, Z., Dey, M., Povilianskas, A., Drapaca, C., & Ozbolat, I. T. (2020). Aspiration-assisted bioprinting for precise positioning of biologics. *Science Advances, 6*, 5111. https://doi.org/10.1126/sciadv.aaw5111

Bhattacharjee, T., Gil, C. J., Marshall, S. L., Urueña, J. M., O'Bryan, C. S., Carstens, M., Keselowsky, B., Palmer, G. D., Ghivizzani, S., Gibbs, C. P., Sawyer, W. G., & Angelini, T. E. (2016). Liquid-like solids support cells in 3D. *ACS Biomaterials Science & Engineering, 2*, 1787−1795. https://doi.org/10.1021/acsbiomaterials.6b00218

Bhattacharjee, T., Zehnder Steven, M., Rowe Kyle, G., Jain, S., Nixon Ryan, M., Sawyer, W. G., & Angelini Thomas, E. (2015). Writing in the granular gel medium. *Science Advances, 1*, e1500655. https://doi.org/10.1126/sciadv.1500655

Bhusal, A., Dogan, E., Nguyen, H. A., Labutina, O., Nieto, D., Khademhosseini, A., & Miri, A. K. (2022). Multi-material digital light processing bioprinting of hydrogel-based microfluidic chips. *Biofabrication, 14*, 014103. https://doi.org/10.1088/1758-5090/ac2d78

Bohandy, J., Kim, B. F., & Adrian, F. J. (1986). Metal deposition from a supported metal film using an excimer laser. *Journal of Applied Physics, 60*, 1538–1539. https://doi.org/10.1063/1.337287

Brassard, J. A., Nikolaev, M., Hübscher, T., Hofer, M., & Lutolf, M. P. (2021). Recapitulating macro-scale tissue self-organization through organoid bioprinting. *Nature Materials, 20*, 22–29. https://doi.org/10.1038/s41563-020-00803-5

Brass, E. P., Forman, W. B., Edwards, R. V., & Lindan, O. (1978). Fibrin formation: Effect of calcium ions. *Blood, 52*, 654–658. https://doi.org/10.1182/blood.V52.4.654.654

Breslin, S., & O'Driscoll, L. (2013). Three-dimensional cell culture: The missing link in drug discovery. *Drug Discovery Today, 18*, 240–249. https://doi.org/10.1016/j.drudis.2012.10.003

Chan, V., Zorlutuna, P., Jeong, J. H., Kong, H., & Bashir, R. (2010). Three-dimensional photo-patterning of hydrogels using stereolithography for long-term cell encapsulation. *Lab on a Chip, 10*, 2062–2070. https://doi.org/10.1039/C004285D

De Moor, L., Merovci, I., Baetens, S., Verstraeten, J., Kowalska, P., Krysko, D. V., De Vos, W. H., & Declercq, H. (2018). High-throughput fabrication of vascularized spheroids for bioprinting. *Biofabrication, 10*, 035009. https://doi.org/10.1088/1758-5090/aac7e6

De Moor, L., Smet, J., Plovyt, M., Bekaert, B., Vercruysse, C., Asadian, M., De Geyter, N., Van Vlierberghe, S., Dubruel, P., & Declercq, H. (2021). Engineering microvasculature by 3D bioprinting of prevascularized spheroids in photo-crosslinkable gelatin. *Biofabrication, 13*, 045021. https://doi.org/10.1088/1758-5090/ac24de

Demirci, U. (2006). Acoustic picoliter droplets for emerging applications in semiconductor industry and biotechnology. *Journal of Microelectromechanical Systems, 15*, 957–966. https://doi.org/10.1109/JMEMS.2006.878879

Demirci, U., & Montesano, G. (2007). Single cell epitaxy by acoustic picolitre droplets. *Lab on a Chip, 7*, 1139–1145. https://doi.org/10.1039/B704965J

Derby, B. (2010). Inkjet printing of functional and structural materials: Fluid property requirements, feature stability, and resolution. *Annual Review of Materials Research, 40*, 395–414. https://doi.org/10.1146/annurev-matsci-070909-104502

Derby, B. (2012). Printing and prototyping of tissues and scaffolds. *Science, 338*, 921–926. https://doi.org/10.1126/science.1226340

Dhariwala, B., Hunt, E., & Boland, T. (2004). Rapid prototyping of tissue-engineering constructs, using photopolymerizable hydrogels and stereolithography. *Tissue Engineering, 10*, 1316–1322. https://doi.org/10.1089/ten.2004.10.1316

Duocastella, M., Patrascioiu, A., Dinca, V., Fernández-Pradas, J. M., Morenza, J. L., & Serra, P. (2011). Study of liquid deposition during laser printing of liquids. *Applied Surface Science, 257*, 5255–5258. https://doi.org/10.1016/j.apsusc.2010.10.148

Ebrahimi Orimi, H., Hosseini Kolkooh, S. S., Hooker, E., Narayanswamy, S., Larrivée, B., & Boutopoulos, C. (2020). Drop-on-demand cell bioprinting via laser induced side transfer (LIST). *Scientific Reports, 10*, 9730. https://doi.org/10.1038/s41598-020-66565-x

Eggers, J., & Villermaux, E. (2008). Physics of liquid jets. *Reports on Progress in Physics, 71*, 036601. https://doi.org/10.1088/0034-4885/71/3/036601

Fogarassy, E., Fuchs, C., Kerherve, F., Hauchecorne, G., & Perrière, J. (1989). Laser-induced forward transfer: A new approach for the deposition of high Tc superconducting thin films. *Journal of Materials Research, 4*, 1082–1086. https://doi.org/10.1557/JMR.1989.1082

Gauvin, R., Chen, Y.-C., Lee, J. W., Soman, P., Zorlutuna, P., Nichol, J. W., Bae, H., Chen, S., & Khademhosseini, A. (2012). Microfabrication of complex porous tissue engineering scaffolds using 3D projection stereolithography. *Biomaterials, 33*, 3824–3834. https://doi.org/10.1016/j.biomaterials.2012.01.048

Gudapati, H., Dey, M., & Ozbolat, I. (2016). A comprehensive review on droplet-based bioprinting: Past, present and future. *Biomaterials, 102*. https://doi.org/10.1016/j.biomaterials.2016.06.012

Guillemot, F., Souquet, A., Catros, S., & Guillotin, B. (2010). Laser-assisted cell printing: Principle, physical parameters versus cell fate and perspectives in tissue engineering. *Nanomedicine, 5*, 507–515. https://doi.org/10.2217/nnm.10.14

Gutzweiler, L., Kartmann, S., Troendle, K., Benning, L., Finkenzeller, G., Zengerle, R., Koltay, P., Stark, G. B., & Zimmermann, S. (2017). Large scale production and controlled deposition of single HUVEC spheroids for bioprinting applications. *Biofabrication, 9*, 025027. https://doi.org/10.1088/1758-5090/aa7218

Hay, I. D., Ur Rehman, Z., Moradali, M. F., Wang, Y., & Rehm, B. H. A. (2013). Microbial alginate production, modification and its applications. *Microbial Biotechnology, 6*(6), 637–650. https://doi.org/10.1111/1751-7915.12076

Highley, C. B., Rodell, C. B., & Burdick, J. A. (2015). Direct 3D printing of shear-thinning hydrogels into self-healing hydrogels. *Advanced Materials, 27*, 5075–5079. https://doi.org/10.1002/adma.201501234

Hinton, T. J., Jallerat, Q., Palchesko, R. N., Park, J. H., Grodzicki, M. S., Shue, H.-. J., Ramadan, M. H., Hudson, A. R., & Feinberg, A. W. (2015). Three-dimensional printing of complex biological structures by freeform reversible embedding of suspended hydrogels. *Science Advances, 1*. https://doi.org/10.1126/sciadv.1500758

Hopp, B., Smausz, T., Antal, Z., Kresz, N., Bor, Z., & Chrisey, D. (2004). Absorbing film assisted laser induced forward transfer of fungi (Trichoderma conidia). *Journal of Applied Physics, 96*, 3478–3481. https://doi.org/10.1063/1.1782275

Hospodiuk, M., Dey, M., Sosnoski, D., & Ozbolat, I. T. (2017). The bioink: A comprehensive review on bioprintable materials. *Biotechnology Advances, 35*, 217–239. https://doi.org/10.1016/j.biotechadv.2016.12.006

Hu, W., Wang, Z., Xiao, Y., Zhang, S., & Wang, J. (2019). Advances in crosslinking strategies of biomedical hydrogels. *Biomaterials Science, 7*, 843–855. https://doi.org/10.1039/C8BM01246F

Itoh, M., Nakayama, K., Noguchi, R., Kamohara, K., Furukawa, K., Uchihashi, K., Toda, S., Oyama, J., Node, K., & Morita, S. (2015). Scaffold-free tubular tissues created by a bio-3D printer undergo remodeling and endothelialization when implanted in rat Aortae. *PLoS One, 10*, e0136681. https://doi.org/10.1371/journal.pone.0136681

Jakab, K., Damon, B., Neagu, A., Kachurin, A., & Forgacs, G. (2006). Three-dimensional tissue constructs built by bioprinting. *Biorheology, 43*, 509–513.

Jeon, O., Song, S. J., Lee, K.-J., Park, M. H., Lee, S.-H., Hahn, S. K., Kim, S., & Kim, B.-S. (2007). Mechanical properties and degradation behaviors of hyaluronic acid hydrogels cross-linked at various cross-linking densities. *Carbohydrate Polymers, 70*, 251–257. https://doi.org/10.1016/j.carbpol.2007.04.002

Khalil, S., Nam, J., & Sun, W. (2005). Multi-nozzle deposition for construction of 3D biopolymer tissue scaffolds. *Rapid Prototyping Journal, 11*, 9–17. https://doi.org/10.1108/13552540510573347

Klebe, R. J. (1988). Cytoscribing: A method for micropositioning cells and the construction of two- and three-dimensional synthetic tissues. *Experimental Cell Research, 179*, 362–373. https://doi.org/10.1016/0014-4827(88)90275-3

Kleinman, H. K., & Martin, G. R. (2005). Matrigel: Basement membrane matrix with biological activity. *Seminars in Cancer Biology, 15*, 378–386. https://doi.org/10.1016/j.semcancer.2005.05.004

Koch, L., Gruene, M., Unger, C., & Chichkov, B. (2013). Laser assisted cell printing. *Current Pharmaceutical Biotechnology, 14*, 91–97. https://doi.org/10.2174/1389201011314010012

Kolesky, D. B., Homan, K. A., Skylar-Scott, M. A., & Lewis, J. A. (2016). Three-dimensional bioprinting of thick vascularized tissues. *Proceedings of the National Academy of Sciences, 113*, 3179–3184. https://doi.org/10.1073/pnas.1521342113

Kolesky, D. B., Truby, R. L., Gladman, A. S., Busbee, T. A., Homan, K. A., & Lewis, J. A. (2014). 3D bioprinting of vascularized, heterogeneous cell-laden tissue constructs. *Advanced Materials, 26*, 3124–3130. https://doi.org/10.1002/adma.201305506

Kryou, C., Theodorakos, I., Karakaidos, P., Klinakis, A., Hatziapostolou, A., & Zergioti, I. (2021). Parametric study of jet/droplet formation process during LIFT printing of living cell-laden bioink. *Micromachines, 12*. https://doi.org/10.3390/mi12111408

Landers, R., & Mülhaupt, R. (2000). Desktop manufacturing of complex objects, prototypes and biomedical scaffolds by means of computer-assisted design combined with computer-guided 3D plotting of polymers and reactive oligomers. *Macromolecular Materials and Engineering, 282*, 17–21. https://doi.org/10.1002/1439-2054(20001001)282:1<17::AID-MAME17>3.0.CO;2-8

Lee, A., Hudson, A. R., Shiwarski, D. J., Tashman, J. W., Hinton, T. J., Yerneni, S., Bliley, J. M., Campbell, P. G., & Feinberg, A. W. (2019). 3D bioprinting of collagen to rebuild components of the human heart. *Science, 365*, 482. https://doi.org/10.1126/science.aav9051

Lee, K.-S., Kim, R. H., Yang, D.-Y., & Park, S. H. (2008). Advances in 3D nano/microfabrication using two-photon initiated polymerization. *Progress in Polymer Science, 33*, 631–681. https://doi.org/10.1016/j.progpolymsci.2008.01.001

Lee, S.-H., Moon, J. J., & West, J. L. (2008). Three-dimensional micropatterning of bioactive hydrogels via two-photon laser scanning photolithography for guided 3D cell migration. *Biomaterials, 29*, 2962–2968. https://doi.org/10.1016/j.biomaterials.2008.04.004

Lippens, E., Swennen, I., Gironès, J., Declercq, H., Vertenten, G., Vlaminck, L., Gasthuys, F., Schacht, E., & Cornelissen, R. (2011). Cell survival and proliferation after encapsulation in a chemically modified Pluronic® F127 hydrogel. *Journal of Biomaterials Applications, 27*, 828–839. https://doi.org/10.1177/0885328211427774

Malda, J., Visser, J., Melchels, F. P., Jüngst, T., Hennink, W. E., Dhert, W. J. A., Groll, J., & Hutmacher, D. W. (2013). 25th anniversary article: Engineering hydrogels for biofabrication. *Advanced Materials, 25*, 5011–5028. https://doi.org/10.1002/adma.201302042

McCormack, A., Highley, C. B., Leslie, N. R., & Melchels, F. P. W. (2020). 3D printing in suspension baths: Keeping the promises of bioprinting afloat. *Trends in Biotechnology, 38*, 584–593. https://doi.org/10.1016/j.tibtech.2019.12.020

Mekhileri, N. V., Lim, K. S., Brown, G. C. J., Mutreja, I., Schon, B. S., Hooper, G. J., & Woodfield, T. B. F. (2018). Automated 3D bioassembly of micro-tissues for biofabrication of hybrid tissue engineered constructs. *Biofabrication, 10*, 024103. http://stacks.iop.org/1758-5090/10/i=2/a=024103.

Miao, S., Cui, H., Nowicki, M., Xia, L., Zhou, X., Lee, S.-J., Zhu, W., Sarkar, K., Zhang, Z., & Zhang, L. G. (2018). Stereolithographic 4D bioprinting of multiresponsive architectures for neural engineering. *Advanced Biosystems, 2*, 1800101. https://doi.org/10.1002/adbi.201800101

Mironov, V., Boland, T., Trusk, T., Forgacs, G., & Markwald, R. R. (2003). Organ printing: Computer-aided jet-based 3D tissue engineering. *Trends in Biotechnology, 21*, 157–161. http://www.sciencedirect.com/science/article/pii/S0167779903000337.

Mironov, V., Visconti, R. P., Kasyanov, V., Forgacs, G., Drake, C. J., & Markwald, R. R. (2009). Organ printing: Tissue spheroids as building blocks. *Biomaterials, 30*, 2164–2174. http://www.sciencedirect.com/science/article/pii/S0142961209000052.

Moldovan, N. I., Hibino, N., & Nakayama, K. (2017). Principles of the kenzan method for robotic cell spheroid-based three-dimensional bioprinting. *Tissue Engineering Part B Reviews, 23*, 237–244. https://doi.org/10.1089/ten.teb.2016.0322

Morley, C. D., Ellison, S. T., Bhattacharjee, T., O'Bryan, C. S., Zhang, Y., Smith, K. F., Kabb, C. P., Sebastian, M., Moore, G. L., Schulze, K. D., Niemi, S., Sawyer, W. G., Tran, D. D., Mitchell, D. A., Sumerlin, B. S., Flores, C. T., & Angelini, T. E. (2019). Quantitative characterization of 3D bioprinted structural elements under cell generated forces. *Nature Communications, 10*, 3029. https://doi.org/10.1038/s41467-019-10919-1

Murata, D., Arai, K., & Nakayama, K. (2020). Scaffold-free bio-3D printing using spheroids as "bio-inks" for tissue (Re-)Construction and drug response tests. *Advanced Healthcare Materials, 9*, 1901831. https://doi.org/10.1002/adhm.201901831

Nahmias, Y., & Odde, D. J. (2006). Micropatterning of living cells by laser-guided direct writing: Application to fabrication of hepatic–endothelial sinusoid-like structures. *Nature Protocols, 1*, 2288–2296. https://doi.org/10.1038/nprot.2006.386

Nahmias, Y., Schwartz, R. E., Verfaillie, C. M., & Odde, D. J. (2005). Laser-guided direct writing for three-dimensional tissue engineering. *Biotechnology and Bioengineering, 92*, 129–136. https://doi.org/10.1002/bit.20585

Nakatsu, M. N., Sainson, R. C. A., Aoto, J. N., Taylor, K. L., Aitkenhead, M., Pérez-del-Pulgar, S., Carpenter, P. M., & Hughes, C. C. W. (2003). Angiogenic sprouting and capillary lumen formation modeled by human umbilical vein endothelial cells (HUVEC) in fibrin gels: The role of fibroblasts and angiopoietin-1. *Microvascular Research, 66*, 102–112. https://doi.org/10.1016/S0026-2862(03)00045-1

Nishiyama, Y., Nakamura, M., Henmi, C., Yamaguchi, K., Mochizuki, S., Nakagawa, H., & Takiura, K. (2008). Development of a three-dimensional bioprinter: Construction of cell supporting structures using hydrogel and state-of-the-art inkjet technology. *Journal of Biomechanical Engineering, 131*. https://doi.org/10.1115/1.3002759

Noor, N., Shapira, A., Edri, R., Gal, I., Wertheim, L., & Dvir, T. (2019). 3D printing of personalized thick and perfusable cardiac patches and hearts. *Advanced Science, 6*, 1900344. https://doi.org/10.1002/advs.201900344

Norotte, C., Marga, F. S., Niklason, L. E., & Forgacs, G. (2009). Scaffold-free vascular tissue engineering using bioprinting. *Biomaterials, 30*, 5910–5917. http://www.sciencedirect.com/science/article/pii/S0142961209006401.

O'Bryan, C. S., Bhattacharjee, T., Marshall, S. L., Gregory Sawyer, W., & Angelini, T. E. (2018). Commercially available microgels for 3D bioprinting. *Bioprinting, 11*, e00037. https://doi.org/10.1016/j.bprint.2018.e00037

Odde, D. J., & Renn, M. J. (1999). Laser-guided direct writing for applications in biotechnology. *Trends in Biotechnology, 17*, 385–389. https://doi.org/10.1016/S0167-7799(99)01355-4

Odde, D. J., & Renn, M. J. (2000). Laser-guided direct writing of living cells. *Biotechnology and Bioengineering, 67*, 312–318. https://doi.org/10.1002/(SICI)1097-0290(20000205)67:3<312::AID-BIT7>3.0.CO;2-F

Ovsianikov, A., Gruene, M., Pflaum, M., Koch, L., Maiorana, F., Wilhelmi, M., Haverich, A., & Chichkov, B. (2010). Laser printing of cells into 3D scaffolds. *Biofabrication, 2*, 014104. https://doi.org/10.1088/1758-5082/2/1/014104

Ozbolat, I. T., & Hospodiuk, M. (2016). Current advances and future perspectives in extrusion-based bioprinting. *Biomaterials, 76*, 321–343. https://doi.org/10.1016/j.biomaterials.2015.10.076

Park, J.-U., Hardy, M., Kang, S. J., Barton, K., Adair, K., Mukhopadhyay, D. kishore, Lee, C. Y., Strano, M. S., Alleyne, A. G., Georgiadis, J. G., Ferreira, P. M., & Rogers, J. A. (2007). High-resolution electrohydrodynamic jet printing. *Nature Materials, 6*, 782–789. https://doi.org/10.1038/nmat1974

Pawelec, K. M., Best, S. M., & Cameron, R. E. (2016). Collagen: A network for regenerative medicine. *Journal of Materials Chemistry B, 4*, 6484–6496. https://doi.org/10.1039/C6TB00807K

Pfister, A., Landers, R., Laib, A., Hübner, U., Schmelzeisen, R., & Mülhaupt, R. (2004). Biofunctional rapid prototyping for tissue-engineering applications: 3D bioplotting versus 3D printing. *Journal of Polymer Science Part A: Polymer Chemistry, 42*, 624–638. https://doi.org/10.1002/pola.10807

Raman, R., & Bashir, R. (2015). Chapter 6 - stereolithographic 3D bioprinting for biomedical applications A2 - atala, anthony. In A. Atala, & J. J. Yoo (Eds.), *Essentials of 3D biofabrication and translation* (pp. 89–121). Academic Press. https://doi.org/10.1016/B978-0-12-800972-7.00006-2

Ringeisen, B. R., Kim, H., Barron, J. A., Krizman, D. B., Chrisey, D. B., Jackman, S., Auyeung, R. Y. C., & Spargo, B. J. (2004). Laser printing of pluripotent embryonal carcinoma cells. *Tissue Engineering, 10*, 483–491. https://doi.org/10.1089/107632704323061843

Ringeisen, B. R., Othon, C. M., Barron, J. A., Young, D., & Spargo, B. J. (2006). Jet-based methods to print living cells. *Biotechnology Journal, 1*, 930–948. https://doi.org/10.1002/biot.200600058

Roversi, K., Ebrahimi Orimi, H., Falchetti, M., Lummertz da Rocha, E., Talbot, S., & Boutopoulos, C. (2021). Bioprinting of adult dorsal root ganglion (DRG) neurons using laser-induced side transfer (LIST). *Micromachines, 12*. https://doi.org/10.3390/mi12080865

Schiele, N. R., Corr, D. T., Huang, Y., Raof, N. A., Xie, Y., & Chrisey, D. B. (2010). Laser-based direct-write techniques for cell printing. *Biofabrication, 2*, 032001. https://doi.org/10.1088/1758-5082/2/3/032001

Seo, B. R., Chen, X., Ling, L., Song, Y. H., Shimpi, A. A., Choi, S., Gonzalez, J., Sapudom, J., Wang, K., Andresen Eguiluz, R. C., Gourdon, D., Shenoy, V. B., & Fischbach, C. (2020). Collagen microarchitecture mechanically controls myofibroblast differentiation. *Proceedings of the National Academy of Sciences, 117*, 11387. https://doi.org/10.1073/pnas.1919394117

Shiwarski, D. J., Hudson, A. R., Tashman, J. W., & Feinberg, A. W. (2021). Emergence of FRESH 3D printing as a platform for advanced tissue biofabrication. *APL Bioengineering, 5*, 010904. https://doi.org/10.1063/5.0032777

Skylar-Scott, M. A., Uzel, S. G. M., Nam, L. L., Ahrens, J. H., Truby, R. L., Damaraju, S., & Lewis, J. A. (2019). Biomanufacturing of organ-specific tissues with high cellular density and embedded vascular channels. *Science Advances, 5*, eaaw2459. https://doi.org/10.1126/sciadv.aaw2459

Sun, W., Starly, B., Daly, A. C., Burdick, J. A., Groll, J., Skeldon, G., Shu, W., Sakai, Y., Shinohara, M., Nishikawa, M., Jang, J., Cho, D.-W., Nie, M., Takeuchi, S., Ostrovidov, S., Khademhosseini, A., Kamm, R. D., Mironov, V., Moroni, L., & Ozbolat, I. T. (2020). The bioprinting roadmap. *Biofabrication, 12*, 022002. https://doi.org/10.1088/1758-5090/ab5158

Tarassoli, S. P., Jessop, Z. M., Jovic, T., Hawkins, K., & Whitaker, I. S. (2021). Candidate bio-inks for extrusion 3D bioprinting—a systematic review of the literature. *Frontiers in Bioengineering and Biotechnology, 9*. https://doi.org/10.3389/fbioe.2021.616753

Wilson, W. C., Jr., & Boland, T. (2003). Cell and organ printing 1: Protein and cell printers. *The Anatomical Record Part A: Discoveries in Molecular, Cellular, and Evolutionary Biology, 272A*, 491–496. https://doi.org/10.1002/ar.a.10057

Wu, W., DeConinck, A., & Lewis, J. A. (2011). Omnidirectional printing of 3D microvascular networks. *Advanced Materials, 23*, H178–H183. https://doi.org/10.1002/adma.201004625

Wu, P. K., Ringeisen, B. R., Callahan, J., Brooks, M., Bubb, D. M., Wu, H. D., Piqué, A., Spargo, B., McGill, R. A., & Chrisey, D. B. (2001). The deposition, structure, pattern deposition, and activity of biomaterial thin-films by matrix-assisted pulsed-laser evaporation (MAPLE) and MAPLE direct write. *Thin Solid Films, 398–399*, 607–614. https://doi.org/10.1016/S0040-6090(01)01347-5

Xiong, R., Zhang, Z., Chai, W., Huang, Y., & Chrisey, D. B. (2015). Freeform drop-on-demand laser printing of 3D alginate and cellular constructs. *Biofabrication, 7*, 045011. https://doi.org/10.1088/1758-5090/7/4/045011

Xu, T., Jin, J., Gregory, C., Hickman, J. J., & Boland, T. (2005). Inkjet printing of viable mammalian cells. *Biomaterials, 26*, 93–99. https://doi.org/10.1016/j.biomaterials.2004.04.011

Xu, Tao, Kincaid, H., Atala, A., & Yoo, J. J. (2008). High-throughput production of single-cell microparticles using an inkjet printing technology. *Journal of Manufacturing Science and Engineering, 130*. https://doi.org/10.1115/1.2903064

Xu, Tao, Gregory, C. A., Molnar, P., Cui, X., Jalota, S., Bhaduri, S. B., & Boland, T. (2006). Viability and electrophysiology of neural cell structures generated by the inkjet printing method. *Biomaterials, 27*, 3580–3588. https://doi.org/10.1016/j.biomaterials.2006.01.048

Yao, H., Wang, J., & Mi, S. (2018). Photo processing for biomedical hydrogels design and functionality: A review. *Polymers, 10*, 11. https://www.mdpi.com/2073-4360/10/1/11.

Yue, K., Trujillo-de Santiago, G., Alvarez, M. M., Tamayol, A., Annabi, N., & Khademhosseini, A. (2015). Synthesis, properties, and biomedical applications of gelatin methacryloyl (GelMA) hydrogels. *Biomaterials, 73*, 254–271. https://doi.org/10.1016/j.biomaterials.2015.08.045

Yu, Y., Moncal, K. K., Li, J., Peng, W., Rivero, I., Martin, J. A., & Ozbolat, I. T. (2016). Three-dimensional bioprinting using self-assembling scalable scaffold-free "tissue strands" as a new bioink. *Scientific Reports, 6*, 28714. https://doi.org/10.1038/srep28714

Zhu, W., Qu, X., Zhu, J., Ma, X., Patel, S., Liu, J., Wang, P., Lai, C. S. E., Gou, M., Xu, Y., Zhang, K., & Chen, S. (2017). Direct 3D bioprinting of prevascularized tissue constructs with complex microarchitecture. *Biomaterials, 124*, 106–115. https://doi.org/10.1016/j.biomaterials.2017.01.042

CHAPTER

Theoretical methods for the optimization of 3D bioprinting: printability, formability, and cell survival

Bioprinting, whether in 3D or 4D, seeks to build live structures. Therefore, material properties and process parameters should be carefully balanced according to the physiological needs of the embedded cells. Biological requirements pose different challenges depending on the employed bioprinting technique and stage of processing. Some of them pertain to bioink handling before printing, others to the printing process itself, and yet others to the manipulation of the printout. In hybrid bioprinting approaches, which also involve the use of cell-free biomaterials, there are less stringent requirements for material processing, but the integration of cellular and inanimate components of the output is of special concern. Given the vast variety of bioprinting methods and materials, as well as many options for construct geometry and microarchitecture, mathematical and computational models are increasingly used to aid experimental efforts aimed at optimizing the outcome of bioprinting.

What aspects need to be considered in the course of optimization? There is compelling evidence that biological function goes hand in hand with the proper structuring of a multicellular construct, which mimics the organizational complexity of native tissues (Moroni et al., 2018). Hence, the printing *resolution* is of crucial importance. Single-cell resolution, however, usually comes with low *throughput* (number of cells delivered in unit time), which leads to prolonged printing, with adverse effects on cells. Another important feature of a bioprinted construct is its *formability*, also known as *shape stability* or *print accuracy*, defined as the ability to maintain its as-printed conformation in good agreement with the original digital model. Also, the rather intuitive but complex concept of *printability* refers to the suitability of a bioink to be handled by a given bioprinting technique for achieving desirable outcomes. Over the years, the term printability has been used to designate "the properties that facilitate handling and deposition by the bioprinter," which include rheological and gelation properties (Murphy & Atala, 2014), as well as "relationships between bioinks and substrates that result in printing an accurate, high-quality pattern" (Mandrycky et al., 2016). According to the most general definition proposed to date, *printability is* "the ability of a material, when subjected to a certain set of printing conditions, to be printed in a way which results in printing outcomes

Towards 4D Bioprinting. https://doi.org/10.1016/B978-0-12-818653-4.00007-3

which are desirable for a given application" (Gillispie et al., 2020). In this definition, printing conditions mean well-defined parameters specifying printer settings and environmental conditions, whereas printing outcomes are (preferably quantitative) measures of successful printing.

The optimization problem at hand depends on the employed bioprinting method. Therefore, this chapter is organized accordingly, dealing with extrusion-based bioprinting, droplet-based, and photopolymerization-based bioprinting in separate sections.

1. Theoretical tools in the optimization of extrusion-based bioprinting

1.1 Rheological properties of bioinks

In extrusion-based bioprinting (EBB), the bioink is typically a cell-laden hydrogel. It is loaded into a syringe, dispensed through a narrow nozzle under high shear conditions, and comes to rest in the form of a specific arrangement of struts. The rheological parameters involved in these processes are the bioink's viscosity and viscoelastic shear moduli (Cooke & Rosenzweig, 2021).

Viscosity describes a fluid's resistance to flow under the action of an applied force. Defined as the ratio of shear stress to shear rate, viscosity is a constant for many fluids, including water; these are known as Newtonian fluids (see Fig. 6.1A, *black line*). By contrast, polymer solutions, such as hydrogel precursor solutions with or without suspended cells, are non-Newtonian because their viscosity depends on flow conditions. Most of them are shear-thinning materials, displaying a decrease in viscosity above a threshold shear rate because dissolved polymer chains become disentangled and elongated. Once they are fully stretched, the solution's viscosity does not change despite an increase in shear rate (Fig. 6.1B, *solid line*).

Yield stress materials, such as Bingham plastics and Herschel—Bulkley fluids, exhibit a smooth transition between the solid and fluid states as soon as the applied shear stress reaches a critical value, σ_y, called yield stress (Fig. 6.1A). Such materials are especially useful as support baths in freeform bioprinting, but they are desirable also in bioink formulations because they prevent the cell suspension from settling in the print head and enable extrusion when the applied pressure results in a large-enough shear stress. Ideally, the extruded bioink filament should quickly undergo elastic recovery to ensure the shape fidelity of the printout. High yield stress favors shape fidelity (the extruded bioink filament being able to sustain multiple layers without collapsing), but it demands a high extrusion pressure associated with harsh biomechanical conditions for the included cells (see the next subsection).

Hence, EBB relies on changes in the viscoelastic behavior of the bioink, quantified in terms of the viscoelastic shear moduli G' (the shear storage modulus) and G'' (the shear loss modulus); G' is also known as shear elastic modulus because it is a measure of the energy of elastic deformation stored in the material; the

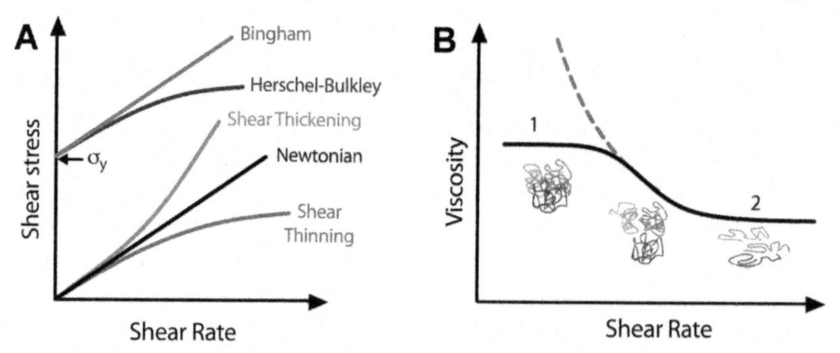

FIGURE 6.1 Rheological properties of fluids used in extrusion-based bioprinting.

(A) Plots of shear stress versus shear rate for various types of fluids and yield-stress materials and (B) plot of viscosity versus shear rate for a yield-stress fluid (*dashed line*) as well as for a polymer solution exhibiting a shear thinning behavior at intermediate shear rates (*solid line*); 1 (2) denote the Newtonian plateaus corresponding to low (high) shear rates.

From Cooke, M. E. & Rosenzweig, D. H. (2021). The rheology of direct and suspended extrusion bioprinting. APL Bioengineering, 5, *011502. https://doi.org/10.1063/5.0031475. Reprinted under the terms of the Creative Commons Attribution 4.0 International License (https://creativecommons.org/licenses/by/4.0/legalcode).*

alternative name of G'' is shear viscous modulus because it accounts for the energy dissipated by the material due to internal friction. Their ratio, G''/G', goes by the names of loss tangent or damping factor and is denoted by $\tan(\delta)$. These quantities are measured by dynamic mechanical analysis (oscillatory rheology), whereby a sinusoidally varying shear strain is applied to the material and the resulting stress is recorded. In an elastic material, stress is proportional to the strain, so they oscillate in phase. In a viscoelastic material, there is a phase lag δ, between stress and strain because the material's response is partly viscous and partly elastic. Amplitude sweep experiments show a linear viscoelastic region for low shear stress levels (in which $G' > G''$ and both are independent of the stress), the yield point (yield stress) at which G' starts to drop with increasing stress (i.e., the material suffers permanent deformation because its structure is affected by micro-ruptures), and the flow point (flow stress), above which the viscous modulus exceeds the elastic modulus, $G'' > G'$—i.e., the viscous behavior dominates and the entire material flows. For certain materials, which undergo brittle fracturing, the yield stress and flow stress are roughly equal. It is important to note, however, that the rheological parameters derived from dynamic mechanical analysis depend on the measuring conditions, such as the preset frequency of the oscillations (Schwab et al., 2020).

1.2 Fluid dynamics description of bioink extrusion

An extrusion-based bioprinter applies pressure to transfer the bioink from the printer's cartridge onto the print bed. Along the way, the bioink passes through the nozzle's orifice, of a few hundred micrometers in diameter, experiencing high shear stress, which might affect the embedded cells. To characterize the biomechanical and geometrical aspects of EBB, it is useful to know the velocity of the bioink in various regions of the nozzle. The bioink velocity profile has been derived for a rigid, cylindrical extrusion nozzle (a needle), and employed for the study of bioink printability (Lee & Yeong, 2015; Paxton et al., 2017).

Shear thinning hydrogels are well described by the Power Law model, which expresses the viscosity, η, of the material as a function of the shear rate, $\dot{\gamma}$:

$$\eta = K\,\dot{\gamma}^{n-1} \tag{6.1}$$

Here, K is the consistency index and n is the power law index, which takes values between zero and one; the more shear-thinning is the material, the closer is n to zero. The consistency index is expressed in $Pa \cdot s^n$ and is numerically equal to the viscosity of the fluid when the shear rate is 1 s^{-1}.

Consider a shear-thinning bioink dispensed through a nozzle of radius R and length L under the action of the applied pressure Δp (Fig. 6.2A). To a good approximation, the pressure gradient along the extrusion nozzle is equal to $\Delta p/L$ because the pressure drop along the body of the printer cartridge is much smaller than the one encountered along the nozzle. As usual in continuum hydrodynamics, we assume that the nonslip boundary condition is satisfied: the bioink layer located next to the nozzle wall is at rest. More distant layers move due to the applied pressure, but their movement is resisted by viscosity. Let us focus on the bioink located next to the nozzle's axis, in a cylinder of radius $r < R$, represented schematically as a light green cylinder in Fig. 6.2B. It is propelled by the force resulting from the applied pressure, $F_p = \Delta p \pi r^2$, while it is pulled back by the force of viscous drag, $F_v = \tau\, 2\pi r\, L$ (i.e., the shear stress, τ, multiplied by the lateral area of the cylinder).

The definition of viscosity relates shear stress to the shear rate, $\tau = \eta\dot{\gamma}$. Taking into account Eq. (6.1), the condition of stationary flow, assured by the equilibrium of forces, can be written as follows:

$$\pi r^2 \cdot \Delta p = K\,\dot{\gamma}^{n-1} \cdot \dot{\gamma} \cdot 2\pi r\, L \tag{6.2}$$

which can be solved for the shear rate to obtain

$$\dot{\gamma} = \left(\frac{\Delta p}{2KL}\right)^{\frac{1}{n}} r^{\frac{1}{n}} \tag{6.3}$$

The velocity of the hydrogel is highest along the nozzle axis, where the influence of the stationary wall is the smallest. Hence, the derivative of the flow velocity, v, with respect to the radial distance, r, is negative, and the absolute value of the shear rate is given by $\dot{\gamma} = -\frac{dv}{dr}$. Substituting this expression into Eq. (6.3) results in a differential equation for satisfied by the unknown function $v(r)$:

$$dv = -\left(\frac{\Delta p}{2KL}\right)^{\frac{1}{n}} r^{\frac{1}{n}}\, dr \tag{6.4}$$

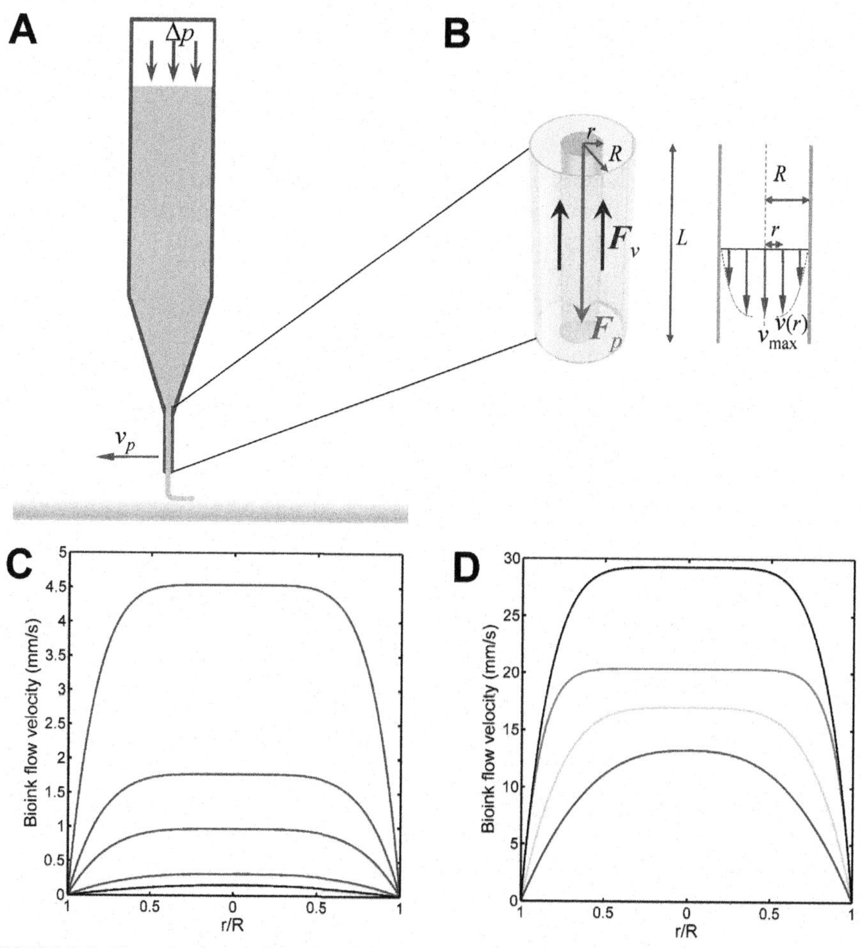

FIGURE 6.2 The velocity profile of a bioink thread during extrusion.

(A) Schematics of an extrusion print head moving with respect to the collector plate; (B) a perspective representation of the nozzle with a cylindrical volume element of bioink highlighted along with the forces acting on it (*center*), as well as a cross-sectional drawing of the nozzle showing the velocity field of a typical shear-thinning hydrogel (*right*); (C) theoretical velocity profiles computed from Eq. (6.6) for a Newtonian fluid ($n = 1$, black line) and for a set of hypothetical shear-thinning materials with power law indices, $n = 0.2$, 0.25, 0.3, and 0.5 (*blue lines*, from *top* to *bottom*, respectively)—for all five curves, $K = 400$ Pa s n, $\Delta p = 3$ bar, $R = 100$ μm, and $L = 1.22$ cm; and (D) theoretical velocity profiles calculated for $R = 125$ μm, $L = 1.22$ cm, and rheological parameters obtained experimentally for four different hydrogels: (i) 25 wt% poloxamer 407 (*red line*, gray in print); (ii) 8%/1% alginate (*green line*, light gray in print); (iii) 4% alginate + 20% gelatin (*blue line*, dark gray in print), and (iv) 10% polyethylene glycol (PEG) + 4% Laponite XLG (*black line*); the compositions and rheological parameters of these hydrogels are given in Table 6.1, lines (i)–(iv), respectively.

The integration of this equation gives

$$v(r) = -\left(\frac{\Delta p}{2KL}\right)^{\frac{1}{n}} \frac{r^{1+\frac{1}{n}}}{1+\frac{1}{n}} + C \tag{6.5}$$

The integration constant, C can be obtained from the nonslip boundary condition, $v(R) = 0$. Thus, the velocity profile of the shear-thinning bioink flowing through a cylindrical needle will be given by (Paxton et al., 2017):

$$v(r) = v_{\max}\left[1 - \left(\frac{r}{R}\right)^{1+\frac{1}{n}}\right], \text{ where } v_{\max} = \left(\frac{\Delta p}{2KL}\right)^{\frac{1}{n}} \frac{R^{1+\frac{1}{n}}}{1+\frac{1}{n}} \tag{6.6}$$

In Eq. (6.6), $v_{\max}$ is the maximum velocity—the velocity of bioink flow along the nozzle axis.

The average extrusion velocity can be computed by integrating over the cross-sectional area of the nozzle and dividing the result by that area:

$$\overline{v} = \frac{1}{S}\int_S v\,dS = \frac{1}{\pi R^2}\int_0^R v(r)2\pi r dr = \frac{2v_{\max}}{R^2}\int_0^R r\left(1 - \frac{r^{1+\frac{1}{n}}}{R^{1+\frac{1}{n}}}\right)dr = v_{\max}\frac{1+\frac{1}{n}}{3+\frac{1}{n}} \tag{6.7}$$

Finally, the volumetric flow rate, $Q = S\cdot\overline{v} = \pi R^2\cdot\overline{v}$ defined as the volume of fluid dispensed during one second, is given by

$$Q = \pi\left(\frac{\Delta p}{2KL}\right)^{\frac{1}{n}} \frac{R^{3+\frac{1}{n}}}{3+\frac{1}{n}} \tag{6.8}$$

In the limit of a Newtonian fluid ($n = 1$ and $K = \eta$), Eq. (6.8) becomes the well-known Hagen–Poiseuille formula of the rate of volumetric flow of an incompressible viscous liquid through a rigid tube—see (Splinter, 2010), Chapter 17.

The velocity profile given by Eq. (6.6) is visualized in Fig. 6.2C for a set of hypothetical hydrogels that share the same viscosity, 400 Pa s, at the reference shear rate of 1 s^{-1}, but differ in their shear-thinning properties. The bottom curve, represented as a black solid line, corresponds to a Newtonian fluid, whereas the blue curves arching above it correspond to more and more shear-thinning materials, with power-law indices $n = 0.5, 0.3, 0.25$, and 0.2, from bottom to top, respectively. All these plots were obtained for the same printing parameters (given in the caption of Fig. 6.2), demonstrating that, indeed, shear thinning facilitates extrusion.

In contrast to Newtonian fluids, which have a parabolic velocity profile, shear-thinning fluids display a velocity plateau near the nozzle axis, resulting in an average extrusion velocity, $\overline{v}$, larger than $v_{\max}/2$. This feature is apparent in Fig. 6.2 not only in *panel* C but also in *panel* D, obtained for shear thinning parameters of actual hydrogels commonly used in bioprinting (Table 6.1).

Table 6.1 Rheological parameters of representative hydrogels used in bioprinting.

Composition, consistency index, power law index, and typical extrusion pressure are listed for representative 3D printable hydrogels.

Number	Hydrogel composition	K (Pa s n)	n	Δp (bar)	References
(i)	25 wt% poloxamer 407 (also known as pluronic F127)	406	0.127	2	Paxton et al. (2017)
(ii)	8%/1% alginate (8% alginate precrosslinked with 1% CaCl2 mixed in 7:3 volume mixing ratio)	254	0.307	3.5	Paxton et al. (2017)
(iii)	4% alginate + 20% gelatin	13.3	0.608	0.8	Paxton et al. (2017)
(iv)	10% PEG + 4% laponite XLG	60	0.2	0.5	Peak et al. (2018)

The thickness of the hydrogel filament that descends from the nozzle tip depends on the relative velocity of the print head with respect to the collector plate (the so-called print velocity, v_p). If the hydrogel has the right rheology to leave the nozzle as a uniform cylinder (see next subsection), volume conservation can be applied to infer the geometry of the extruded filament. If the print speed matches the average extrusion velocity $\overline{v}$, given by Eq. (6.7), the filament thickness is equal to the inner diameter of the nozzle. If $v_p > \overline{v}$ the descending filament is stretched on the fly, and its diameter d, becomes smaller. From volume conservation, $\pi(d/2)^2 \cdot v_p = \pi R^2 \cdot \overline{v}$, we obtain:

$$d = 2R\left(\frac{\overline{v}}{v_p}\right)^{\frac{1}{2}} \tag{6.9}$$

In this case, the feature sizes of the printed construct can be smaller than the nozzle diameter. Conversely, when the collector plate receives more material per unit length than present in the lame length of extruded filament (i.e., the deposited bioink strand is overfed), the result is a larger and often nonuniform strand diameter.

Besides the theoretical diameter given by Eq. (6.9), the print resolution depends on the shape stability of the extruded bioink filament. Ideally, it should remain cylindrical upon deposition, but in practice, it flattens due to the combined effects of gravity and interfacial tension forces. Lee and Yeong computed the width of the printed struts assuming that their cross-section was either a circle or half an ellipse, but in both cases, the theoretical strut width turned out to be smaller than its experimental counterpart (Lee & Yeong, 2015). To improve print accuracy, they suggest to take advantage of buoyancy by printing in a liquid medium, which would also diminish interfacial forces.

The shear rate can be computed from Eq. (6.6) by taking the derivative of the velocity with respect to the distance from the nozzle axis (with opposite sign):

$$\dot{\gamma} = \frac{v_{\max}}{R}\left(1 + \frac{1}{n}\right)\left(\frac{r}{R}\right)^{\frac{1}{n}} \tag{6.10}$$

To calculate the shear stress, one could rely on the definition of viscosity $\tau = \eta\dot{\gamma} = K\dot{\gamma}^n$, and substitute $\dot{\gamma}$ from Eq. (6.10), but it is simpler to return to the expression of mechanical equilibrium (Equation 6.2):

$$\tau = \frac{r\Delta p}{2L} \tag{6.11}$$

The shear rate and shear stress in usual hydrogels under typical bioprinting conditions are plotted in Fig. 6.3.

It is clear that the vicinity of the extrusion nozzle wall is a harsh medium for the encapsulated cells because the shear stress is highest there (Fig. 6.3B). Additionally, the flow velocity is small next to the wall (Fig. 6.2D), resulting in a long residence time under high shear conditions. The shear stress experienced by cells is relatively high for the 8%/1% alginate bioink (Table 6.1 ii), but this translates into a better shape fidelity, as discussed in the following subsection. Fortunately, biological cells seem to be sturdy enough to withstand shear stress levels of up to 5 kPa (Blaeser et al., 2016).

Mathematical modeling combined with shear viscosity measurements aid the optimization of bioink formulation, making sure it can be extruded at the desired flow rate using a given set of nozzles and applying pressures within reach of the employed bioprinter.

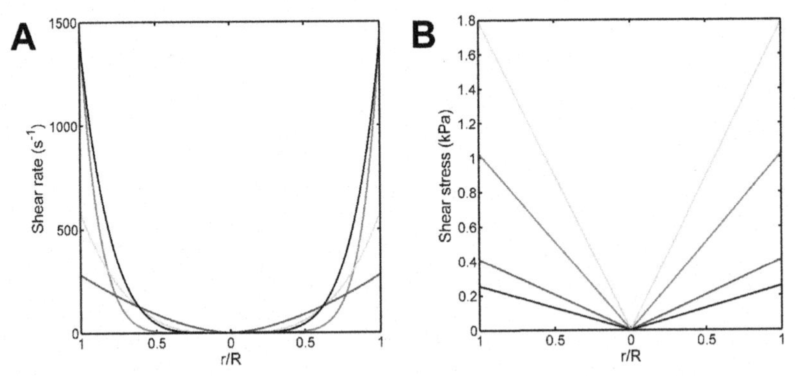

FIGURE 6.3 The distribution of the shear rate and shear stress in a bioink as it passes through the extrusion nozzle.

(A) The shear rate plotted versus the distance, *r*, from the nozzle axis expressed as a fraction of the nozzle radius, *R*; (B) the shear stress plotted versus the normalized radial distance; calculations were performed for a nozzle of radius $R = 125\ \mu m$, and length $L = 1.22$ cm; red, green, blue, and black lines refer to the bioinks listed in Table 6.1, lines (i)–(iv), respectively.

Nevertheless, the mathematical model presented here is not free from limitations. First, it is anchored in the cylindrical geometry and assumes that the bioink flow is laminar and steady. The latter is clearly not true near the needle's entrance. The model also assumes that the bioink does not slip along the nozzle wall, which has been challenged in certain studies (Lee & Yeong, 2015; Sarker & Chen, 2017). It also considers that bioink rheology does not change with time. Therefore, the above analysis does not apply to thixotropic hydrogels, whose viscosity decreases in time at constant shear rate and recovers as soon as the flow ceases. Although it is appealing for extrudability and subsequent structural integrity, thixotropy is not of special interest in bioink design because, to control the extrusion rate, it asks for real-time adjustment of the applied pressure (Schwab et al., 2020).

The above limitations were partially addressed in the literature. Sarker and Chen, for example, developed a model to describe the flow of hydrogels extruded through conical nozzles (Sarker & Chen, 2017). Their formalism is based on the Herschel–Bulkley model, $\tau = \tau_0 + K\dot{\gamma}^n$, taking into account a potential yield stress behavior, but in the context of EBB the applied pressure leads to shear stress levels way beyond the bioink's yield stress, making the Herschel–Bulkley model equivalent to the power law model, $\tau = K\dot{\gamma}^n$. Comparing their model predictions with experimental data on medium viscosity alginate hydrogels, they came to the conclusion that slip flow needs to be taken into account in order to assure a good fit between theory and experiment (Sarker & Chen, 2017).

1.3 Quantitative assessment of printability

The optimization of bioprinting is facilitated by reproducible methods designed for evaluating printability. Excellent review papers are available on this topic (Cooke & Rosenzweig, 2021; Gillispie et al., 2020; Schwab et al., 2020); therefore, this subsection aims at presenting the core concepts and illustrate them instead of providing a comprehensive overview of this highly dynamic field.

Paxton et al. proposed a systematic, repeatable evaluation of bioink printability in two steps (Paxton et al., 2017) (Fig. 6.4): (1) a low-cost manual extrusion experiment looking at filament formation and layer stacking and (2) shear-viscosity rheometry combined with mathematical modeling of bioink flow within the nozzle, discussed in the previous subsection.

Upon extrusion, the bioink is expected to exit the nozzle as a continuous, straight filament. Some bioinks emerge instead as droplets because their shear recovery is too slow and/or their yield stress is too small to resist surface tension. This is the case, for instance, with dilute polymer solutions (Fig. 6.5B and C). At the other extreme, when precrosslinking is applied for improving shape fidelity, the emergent filament may be irregular (fractured) because of excessive gelation. Inconsistent filament thickness is another common problem, mainly associated with nozzle clogging. It may arise even in the absence of clogging if composite bioinks, with solid microparticles or cell aggregates in their composition, are dispensed at constant pressure. A wavy filament may result also from an inadequate distance between

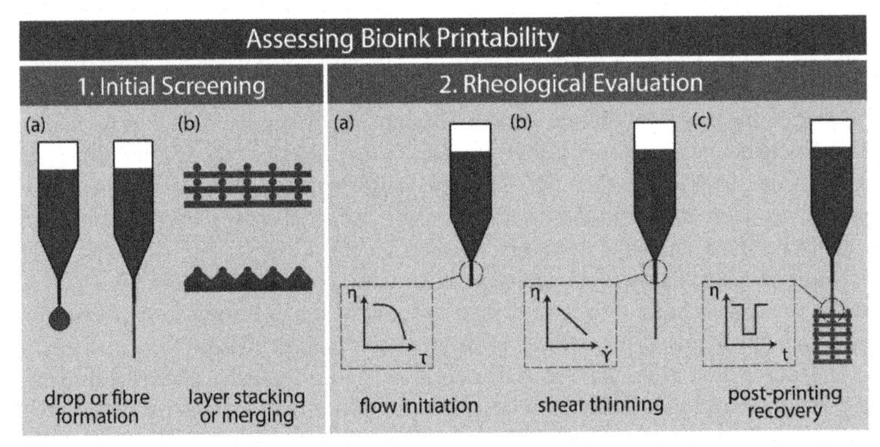

FIGURE 6.4 Two-step procedure for assessing bioink printability.

Step 1 involves a manual extrusion test of (a) filament formation and (b) layer stacking; step 2 consists of rheological measurements that characterize (a) bioink flow initiation and yield stress, (b) shear-thinning behavior, and (c) elastic recovery after extrusion; step 2 also includes a theoretical analysis of bioink flow based on the measured rheological parameters.

From Paxton, N., Smolan, W., Böck, T., Melchels, F., Groll, J. & Jungst, T. (2017). Proposal to assess printability of bioinks for extrusion-based bioprinting and evaluation of rheological properties governing bioprintability. Biofabrication, 9, *044107. https://doi.org/10.1088/1758-5090/aa8dd8. Reprinted under the terms of the Creative Commons Attribution 3.0 International License (https://creativecommons.org/licenses/by/3.0/legalcode).*

the tip of the nozzle and the collector plate (nozzle offset) or from a mismatch between the printing speed and the average extrusion velocity (Zhang et al., 2021).

In their manual dispensing experiments, Paxton et al. used Nivea Creme as a control material because of its notorious extrudability and formability; fibers printed from this material remain cylindrical, and do not fuse when they are dispensed on top of each other (Paxton et al., 2017). Poloxamer 407 samples of up to 20 wt% polymer concentration failed the fiber formation test and the overlaid hydrogel fibers merged, suggesting that a woodpile structure made of them would collapse (Fig. 6.5B and C). The most printable poloxamer 407 hydrogel was that of 25 wt % (Fig. 6.5D). Higher concentrations still lead to uniform, cohesive, and stackable fibers (Fig. 6.5E), but they require high extrusion pressures associated with high shear stress levels.

Achieving a homogeneous microstructure is difficult in the case of multimaterial bioinks, but it is worth the effort because they feature all the benefits of their individual components. For example, a 10% PEG hydrogel displays high compliance but no elastic recovery. The addition of 4% Laponite results in a shear-thinning hydrogel of good extrudability at 0.5 bar (see Fig. 6.3 and Table 6.1) and optimal elastic recovery upon relaxation, presumably because Laponite particles rapidly form bridges

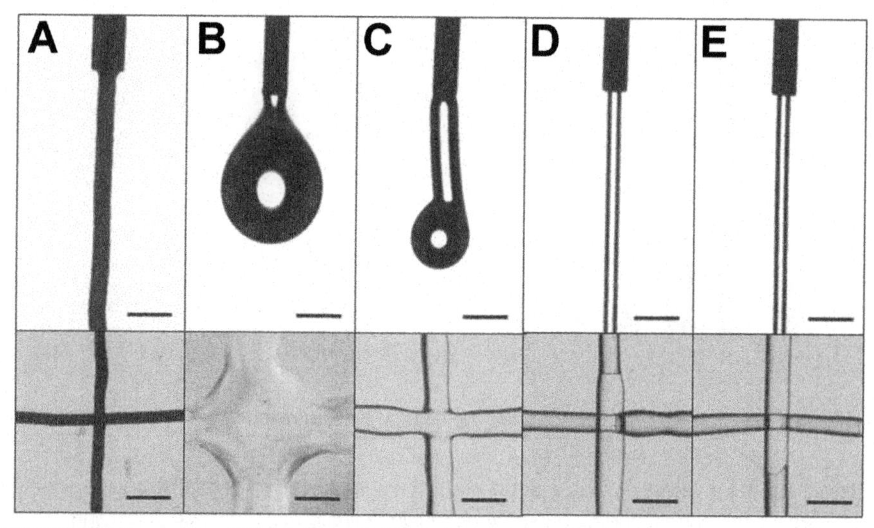

FIGURE 6.5 Examples of bioink filament formation and layer stacking.

The photographs from the top row show lateral views of fiber formation, whereas those from the bottom row represent top views of the region where two mutually perpendicular fibers overlap; tests were performed using (A) Nivea Creme, (B) 15 wt% poloxamer 407, (C) 20 wt% poloxamer 407, (D) 25 wt% poloxamer 407, and (E) 30 wt% poloxamer 407; scale bar = 1 mm.

From Paxton, N., Smolan, W., Böck, T., Melchels, F., Groll, J. & Jungst, T. (2017). Proposal to assess printability of bioinks for extrusion-based bioprinting and evaluation of rheological properties governing bioprintability. Biofabrication, 9, *044107. https://doi.org/10.1088/1758-5090/aa8dd8. Reprinted under the terms of the Creative Commons Attribution 3.0 International License (https://creativecommons.org/licenses/by/3.0/legalcode).*

between polymer chains. Indeed, this material enabled the 3D printing of constructs over 20 layers in height, which did not collapse (Peak et al., 2018).

1.4 Extrusion uniformity

Extrusion uniformity was characterized in qualitative terms, as well as by the uniformity ratio U, defined as the ratio of the lateral contour length of the filament divided by what it would have been for a perfectly straight and uniform filament (Gao et al., 2018). To calculate U, the filament contour needs to be outlined manually or using image analysis software, the length of each longitudinal contour line should be computed in pixels, and their sum should be divided by twice the length of the straight segment that runs along the nozzle axis from the beginning of the filament to its end. Ideally, the uniformity ratio should be as close to 1 as possible. Gao et al. conducted oscillatory rheometry experiments on composite bioinks prepared from alginate and gelatin and correlated rheological quantities with printability metrics. Nearly ideal uniformity was observed for tan $(\delta) \geq 0.43$ (Fig. 6.6A). Nevertheless,

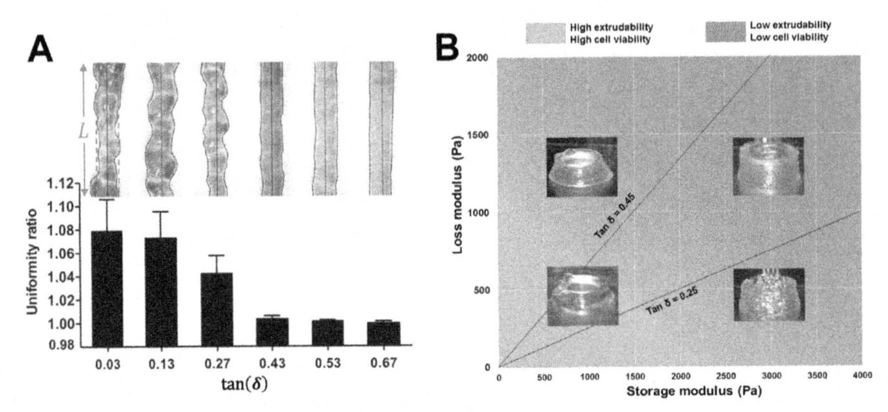

FIGURE 6.6 Prediction of printability from rheological measurements.

(A) Bar plot of filament uniformity ratios corresponding to gelatin–alginate mixtures with different loss tangent values; representative pictures of extruded filament fragments are shown above the corresponding bars; the rectangle delimited by a dashed green line represents the ideal shape of a filament of the same length, L (*top-left* image) and (B) printability diagram of gelatin-alginate bioinks.

Republished from Gao, T., Gillispie, G. J., Copus, J. S., Pr, A. K., Seol, Y.-J., Atala, A., Yoo, J. J. & Lee, S. J. (2018). Optimization of gelatin–alginate composite bioink printability using rheological parameters: a systematic approach. Biofabrication, 10, *034106. https://doi.org/10.1088/1758-5090/aacdc7 with permission from IOP Publishing, Inc.*

too high a loss tangent affected the structural integrity of the bioprinted structure (Fig. 6.6B) because a large loss modulus is associated with a fluid-like character. Hence, gelatin–alginate composite bioinks feature a printability window expressed in terms of the loss tangent, $0.25 \leq \tan(\delta) \leq 0.45$, which ensures extrusion uniformity without compromising shape fidelity.

Besides being used for mapping printability, the experimentally measured values of the storage modulus, G', and loss modulus, G'', served as independent variables to predict the required extrusion pressure via multiple linear regression with interaction: $p = 26.6 + 5 \times 10^{-2} \times G' + 7 \times 10^{-2} \times G'' - 3.5 \times 10^{-6} \times G' \times G''$ (kPa), valid for a 25 G nozzle (Gao et al., 2018).

1.5 Shape fidelity metrics

An extruded bioink fiber is expected to undergo rapid elastic recovery, retaining its shape despite numerous forces acting on it, such as gravity, surface tension, or traction forces exerted by subsequently deposited filaments. Experiments specifically designed to quantify the shape stability of extruded fibers are useful in bioink development and in bioprinting study design. People seeking to devise new bioink formulations can monitor formability versus composition. Researchers interested in applicative aspects of bioprinting can evaluate a set of commercially available bioinks beforehand, saving the time and money needed for trial and error experiments.

Quantitative tests of shape fidelity can look at the ability of an extruded fiber to bridge gaps between underlying supporting structures. They require to print straight filaments over a set of pillars with different separations and measure the sagging of overhang fiber portions.

One option for quantifying **filament collapse** is to measure the area below the curve that runs along the filament's axis. The collapse area factor, $C_f = \left(A_a^c / A_t^c\right) \cdot 100\%$, is defined as the percentage of the actual area below the curve A_a^c, compared to the theoretical value A_t^c that would correspond to an ideal filament capable of forming a straight bridge between two consecutive pillars (Habib et al., 2018). Hence, C_f tells us what percentage of the theoretically calculated lateral pore area is still present under the sagging filament.

Another option is to measure the angle of deflection θ_s, of the filament at the point of contact with the support—more precisely, the margin of the pillar facing the gap spanned by the filament (Ribeiro et al., 2018). If the filament is uniform, it collapses symmetrically so it is sufficient to focus on the forces that act on half of it (Fig. 6.7A, *top panel*). Let us denote the gap distance by $2L$ and the length of the overhang portion of the filament by $2L_f$. The mass of half the filament can be expressed as $m = \rho A_f L_f$, where ρ is the density of the bioink (e.g., $\rho = 1020\ kg/m^3$ for poloxamer 407) and A_f is the cross-sectional area of the filament. The mechanical equilibrium of this portion is ensured by the bioink's resistance to yield at the edge of the pillar, F_σ, the horizontal elastic force, F_{σ_0}, exerted by the other half of the filament and the weight force $F_g = mg$ (Fig. 6.7A).

Along the horizontal direction, $F_\sigma \cos\theta_s - F_{\sigma_0} = 0$, which shows that the stress is higher at the pillar's edge than at the lowest point of the filament. Along the vertical direction, $F_\sigma \sin\theta_s - mg = 0$, so the angle of deflection can be expressed as follows (Ribeiro et al., 2018):

$$\theta_s = \sin^{-1}\left(\frac{\rho A_f L_f g}{F_\sigma}\right) \cong \sin^{-1}\left(\frac{\rho g L}{F_\sigma / A_f}\right) = \sin^{-1}\left(\frac{\rho g L}{\sigma_y}\right) \tag{6.12}$$

This equation contains one approximation, $L_f \cong L$, justified by experiments showing that a satisfactory bioink fiber has a small relative elongation ($L_f - L$ is much smaller than L). Eq. (6.12) indicates that a small deflection angle is possible only if the bioink's yield stress is high. An initially horizontal fiber elongates slightly under its own weight until its resistance to yield cancels the vector sum of the other two forces that act on half of the bioink arch.

Note that the derivation of Eq. (6.12) does not depend on the precise shape of the sagging filament. The only underlying assumptions are that (i) the overhang portion of the filament is symmetrical with respect to the middle of the gap and (ii) internal forces are capable of eventually assuring mechanical equilibrium (i.e., the filament assumes a stable geometry). The lateral view images shown in Fig. 6.7A, *panels* (i)–(vi), were acquired 20 s after printing because filament collapse practically ceased after this time interval. More precisely, the sagging filaments kept on elongating, but at a much slower rate (Ribeiro et al., 2018). From the practical point

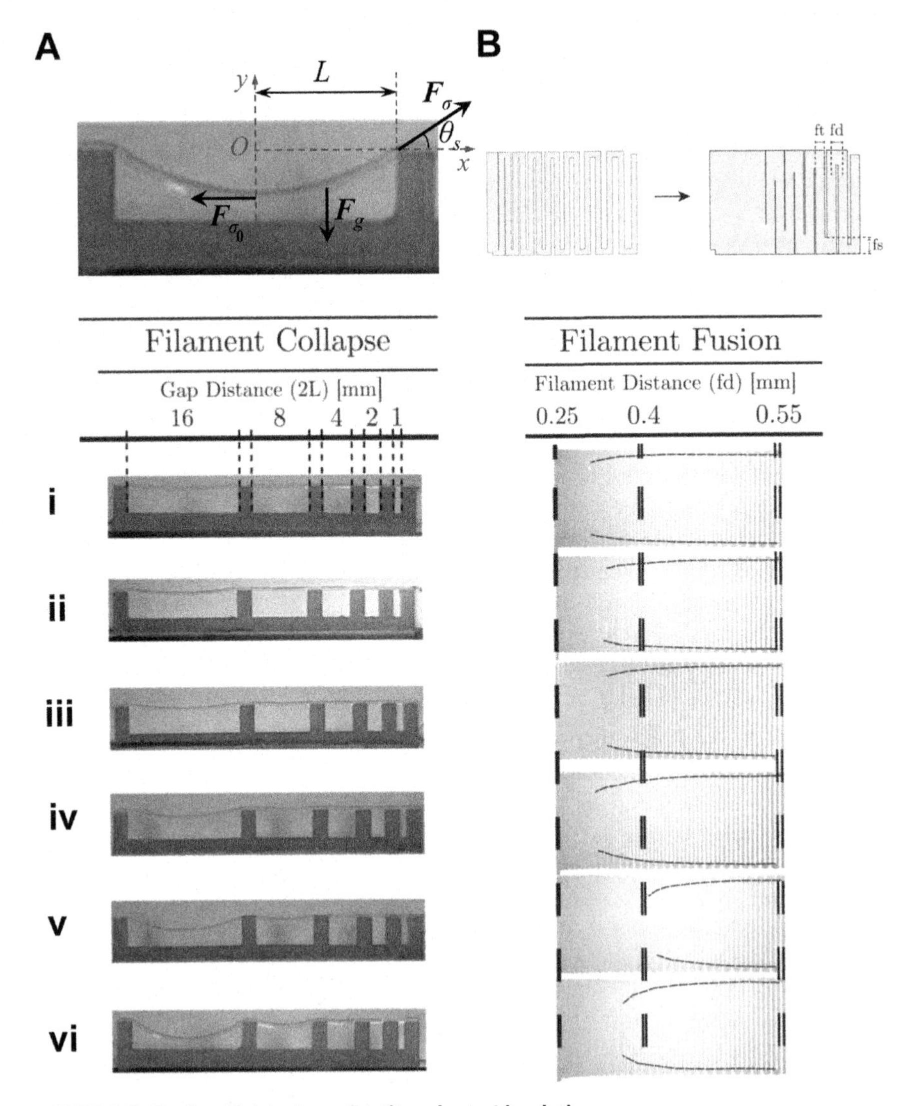

FIGURE 6.7 Testing bioink shape fidelity prior to bioprinting.

(A) Filament collapse test: forces that ensure the equilibrium of half of the sagging portion of the filament are shown in the *top panel* and experimental results obtained for different bioink compositions are depicted in rows (i)–(vi); (B) filament fusion test based on printing the gradient pattern shown in the *top panel* (*left*); the evaluation is based on measuring the ratio of fused segment length (*fs*) to filament thickness (*ft*) for various values of the filament distance (*fd*)—the fused pattern and the parameters *fs, ft,* and *fd* are shown schematically in the *top right* drawing; in the stereomicroscopy images of a single layer printed on a glass substrate, shown in rows (i)–(vi), *blue lines* join the points where the adjacent filaments separate; in both (A) and (B), experimental images refer to hydrogels prepared from (i) 30% poloxamer 407 (px), (ii) 29% px + 1% PEG, (iii) 28% px + 2% PEG, (iv) 27% px + 3% PEG, (v) 26% px + 4% PEG, and (vi) 20% px.

Reprinted from Ribeiro, A., Blokzijl, M. M., Levato, R., Visser, C. W., Castilho, M., Hennink, W. E., Vermonden, T. & Malda, J. (2018). Assessing bioink shape fidelity to aid material development in 3D bioprinting. Biofabrication, 10, *014102. https://doi.org/10.1088/1758-5090/aa90e2 Reproduced with permission from IOP Publishing.*

of view, that evolution is unimportant, since bioprinted constructs are stabilized post-printing by crosslinking the extruded strands after every layer deposition, which takes minutes. Gelation mechanisms employed for this task rely on various physicochemical factors, including temperature change, addition of ionic solutes, light exposure, enzymatic reactions, host–guest interaction, and pH change (Schwab et al., 2020).

A more accurate analysis of the filament collapse problem is required to derive, under the same assumptions, the equation of the filament's shape. It turns out that this problem is similar to finding the shape of a chain hanging between two posts (Lockwood, 1961), Chapter 13. Its solution is the equation of the catenary arch

$$y(x) = a\left[\cosh\left(\frac{x}{a}\right) - \cosh\left(\frac{L}{a}\right)\right], \tag{6.13}$$

where a is a length scale, defined as $a \equiv F_{\sigma_0} L_f/(mg) = F_{\sigma_0}/(\rho g A_f)$, which can be obtained numerically from the y coordinate of the deepest point, $y(0) = -h$.

The slope of the curve at any point is the derivative of the function: $\tan\theta(x) = y'(x) = \sinh(x/a)$, so the angle of deflection at the support ($x = L$) can be expressed as $\theta_s = \tan^{-1}[\sinh(L/a)]$. The length of the half filament results from the implicit equation of the catenary, $L_f = a \cdot \tan\theta_s = a \cdot \sinh(L/a)$. The elastic force exerted by the other half is given by $F_{\sigma_0} = \rho g a A_f$, which leads to the expression of the elastic force acting at the pillar's edge, $F_\sigma = F_{\sigma_0}/\cos\theta_s = F_{\sigma_0} \cdot cosh(L/a)$. Finally, the bioink's yield stress is related to its density and the catenary curve's parameters by the following formula: $\sigma_y = F_\sigma/A_f = \rho g a \cdot \cosh(L/a)$. Note that Eq. (6.12) leads to the same formula: $\sigma_y = \rho g L_f/\sin\theta = \rho g a \tan\theta/\sin\theta = \rho g a/\cos\theta = \rho g a \cdot \cosh(L/a)$.

The catenary equation, named after the Latin word catena (= chain), has numerous applications. It was used, for instance, by Eero Saarinen in his design of the suspended rooftop of Dulles International Airport (https://rachitect.wordpress.com/tag/dulles-airport). Its most intriguing feature is that a catenary curve turned upside down makes an especially stable arch. Many famous buildings comprise catenary curves, including some by Antoni Gaudi, and the Gateway Arch in St. Louis, designed by Eero Saarinen (Osserman, 2010).

The theoretical analysis based on Eqs. (6.12) and (6.13) is less accurate at small angles of deflection (Ribeiro et al., 2018). A hydrogel is a viscoelastic medium; therefore, elastic forces contribute to a hydrogel fiber's resistance to bending, making it capable of spanning gap distances comparable with the fiber thickness without noticeable deflection. A chain, on the other hand, does not possess bending stiffness and, thus, requires an infinite force of tension to stay strictly straight between two posts. Indeed, Eq. (6.12) was found to overestimate the angle of deflection, and the relative error was especially large for high yield stress hydrogels and small separations between pillars. The theory of elastic beams might provide further insights into the shape fidelity of structures created by extrusion-based bioprinting.

The **filament fusion** test evaluates the ability to print structures that involve sharp corners and nearby filaments (small pores). Such attempts are especially challenging because surface tension forces work against the yield stress of the material, causing deformation and pore closure. The filament fusion test is based on printing the meander pattern with gradually growing filament distance (*fd*) shown in the top-left scheme of Fig. 6.7B. The fused filament length (*fs*) observed in the printout is larger than the filament thickness (*ft*). Experiments indicate that their ratio, fs/ft, is well fitted by an exponential function, $fs/ft = 1 + \exp(-fd/b)$, where the constant b is the characteristic filament distance, which depends on the bioink formulation (Ribeiro et al., 2018).

The experimental results depicted in Fig. 6.7B, *panels* (i)–(vi), were obtained by printing a single winding pattern on a glass substrate. Since the bioink filaments are pinned down by adhesive forces onto the receiving substrate, the resulting pattern depends on the interplay of bioink-substrate adhesion and forces of cohesion acting between bioink components. When several layers are printed on top of each other, the winding pattern reflects the intrinsic properties of the bioink. Further research will lead to a deeper understanding of the concerted action of interfacial forces, yield stress, gravity, and potential changes in material properties during and after dispensing, providing guidelines for choosing the right materials and printing parameters for the desired outcome.

In a multilayered construct, one can measure the filament width (in-plane diameter), as well as filament height (measured along the axial direction) to infer whether the bioink properties and printer settings enable woodpile stacking or they lead to a structure unable to sustain itself. When the bioink is in an ideal state of gelation, the extruded filament is uniform, smooth, and circular in cross-section even after deposition. Taken together, these features encompass all aspects of printability (extrudability, filament formation, and shape stability). To quantify deviations from this ideal scenario, Ouyang et al. proposed to print a rectangular grid and analyze the shapes of the in-plane pores. They also proposed a printability index, *Pr*, defined by the formula $Pr = (\text{Perimeter})^2/(16 \times \text{Area})$ (Ouyang et al., 2016). These quantities are expressed in pixels and can be measured using image analysis software, such as ImageJ (freely available at https://imagej.nih.gov/ij). The pore perimeter is the number of pixels located at the pore boundary, whereas its area is the number of pixels that compose the pore's image. For perfectly square pores, $Pr = 1$, for pores with rounded corners, $Pr < 1$, whereas for irregular pores delimited by bumpy filaments, $Pr > 1$. Note that the definition of *Pr* is related to a well-known metrics of compactness, called circularity, $C = 4\pi \times \text{Area}/(\text{Perimeter})^2$, which is 1 (highest) for a circle. For a square, $C = \pi/4$, so $Pr = (\pi/4)(1/C)$ is equal to one for a perfect square-shaped pore. As a rule of thumb, shape fidelity was deemed poor if it resulted in *Pr* values below 0.9 (Ouyang et al., 2016).

Habib et al. modified the printability test by printing a rectangular pattern of several pore sizes and observed that *Pr* tends to increase with the pore size (Habib et al., 2018), perhaps because the relative contribution of the rounded corners is smaller in the case of larger pores. This observation suggests that the pore sizes of cross-hatch patterns designed for printability assessment should be scaled according to the filament widths measured in preliminary tests. Also, it is recommended to dismiss the first two layers to minimize the impact of the print bed on filament spreading (Ribeiro et al., 2018).

An equally important measure of successful bioprinting is printing accuracy, which refers to the extent to which the output matches its intended shape, size, and microarchitecture (Gillispie et al., 2020). The in-plane geometric accuracy can be characterized by the relative deviations of pore sizes from their intended values, called the rate of material spreading, $R_s = 100\% \times (A_t - A_a)/A_t$, where, A_t denotes the theoretical pore area and A_a is the actual pore area (Habib et al., 2018). The mean value of R_s for all the pores from a given layer is a good estimate of printing accuracy. Another option is to measure the total area occupied by the construct viewed from above—i.e., to look at the filaments instead of looking at the holes (Gillispie et al., 2020).

Once the chosen materials and printer settings enable a reliable printing of a few test layers, the question arises what is the maximal height of a multilayered structure that can be achieved without compromising the construct's stability. Taking side view pictures of the printout is one way of assessing the ratio of actual height to intended height, which is a widely used index of shape fidelity. When it is nearly 1, layer stacking is optimal. Smaller values, which are quite common, indicate filament flattening and/or merging.

A detailed evaluation of printing accuracy can be performed by microcomputed tomography (micro-CT), or by optical coherence tomography (OCT). Both of these techniques provide 3D volumetric views of micron-scale resolution. Moreover, they are bundled with software applications that allow for the morphological analysis of filaments and pores. Micro-CT even reveals defects within filaments (e.g., air bubbles), but it is harmful to the embedded cells because it relies on the attenuation of X-rays. OCT, on the other hand, is non-invasive and, therefore, it might develop into a real-time monitoring tool, providing feedback to the 3D printing software to adapt the printing parameters on-the-fly (Schwab et al., 2020).

2. Mathematical and computational methods for improving droplet-based bioprinting

Trading bioink strands for droplets boosts printing resolution. Therefore, droplet-based bioprinting (DBB) has attracted much attention as an alternative or complementary technique to EBB. Droplet formation has been investigated to ensure control over bioink dispensing rate and droplet volume and to limit the impact of DBB on the cells present in the bioprinted construct.

2.1 The biological impact of shear stress

A fluid-dynamics model allowed for tuning microvalve-based bioprinting to achieve high resolution without exposing the cells from the bioink to harmful levels of shear

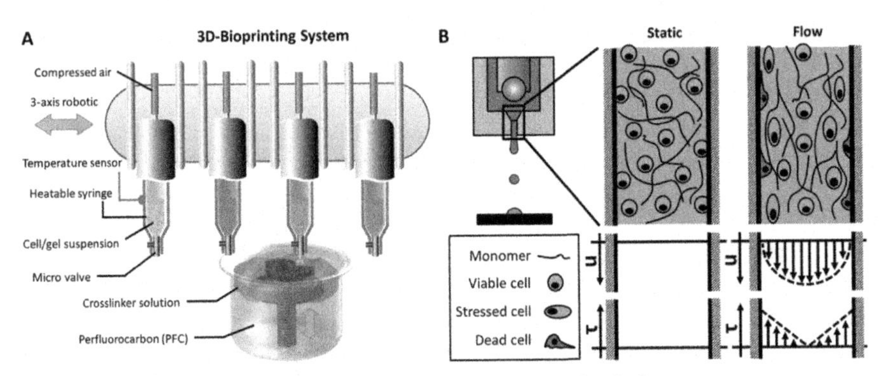

FIGURE 6.8 Schematics of microvalve-based multimaterial bioprinting.

(A) The working principle of the bioprinter capable of dispensing four types of bioink onto a print bed immersed into a crosslinking solution that floats on top of a perfluorocarbon support fluid (Fluorinert Electronic Liquids FC-43) and (B) schematic representation of the velocity and shear stress profile of a bioink column advancing along the print nozzle and their impact on the embedded cells.

Republished from Blaeser, A., Duarte Campos, D. F., Puster, U., Richtering, W., Stevens, M. M. & Fischer, H. (2016). Controlling shear stress in 3D bioprinting is a key factor to balance printing resolution and stem cell integrity. Advanced Healthcare Materials, 5, *326–333. https://doi.org/10.1002/adhm.201500677, with permission from John Wiley and Sons.*

stress (Blaeser et al., 2016). The bioprinter shown in Fig. 6.8A generates individual droplets of cell-laden hydrogel via jetting controlled by electromagnetic microvalves. While a valve is open, bioink is pneumatically expelled from the narrow nozzle (150–300 μm in diameter), and the transient viscous flow brings about considerable shear stress, especially near the nozzle wall (Fig. 6.8B).

The theoretical analysis conducted by Blaeser et al. sought to identify cell-friendly printing parameters and verify them in ample tests of cell viability, proliferative potential, and cell type-specific protein synthesis. Their continuum fluid dynamics model relied on the Bernoulli equation for unsteady flow, the Hagen–Poiseuille equation of the volumetric flow rate of a viscous fluid, and the Power Law model of shear thinning behavior (Equation 6.1); these equations were combined to derive the differential equation satisfied by the average drop speed v (Blaeser et al., 2016),

$$\frac{\partial v}{\partial t} = \frac{1}{\rho L}\left[p + \rho g h - \frac{1}{2}\rho v^2 - 2K\frac{(3+1/n)^n}{R^{n+1}}v^n\right], \tag{6.14}$$

where p is the applied pressure, L and R are the extrusion nozzle's length and inner radius, respectively, whereas ρ, K, and n are the material constants of the shear-thinning bioink (ρ– density, K– consistency index, and n– power-law exponent). Blaeser et al. validated their model by predicting the volumes of droplets of 0.5%, 1%, and 1.5% (w/v) alginate expelled during a gating time of 0.5 ms through two different nozzle types (150 and 300 μm inner diameter) at different printing pressures ranging from 0.5 to 1.5 bar. To determine the droplet volume for a given set of printing parameters, 1000 droplets were printed into a glass container, and their weight was measured and divided by 1000 times the ink's density (approximated to 10^3 kg/m^3).

Then, the theoretical model was used to calculate shear stress levels encountered while printing alginate droplets without cells and with them (1 million and 10 million cells/mL). Experiments conducted on L292 mouse fibroblasts suspended in alginate demonstrated that low shear stress levels, of up to 5 kPa, had no significant impact on the embedded cells, leading to 96% cell viability. For medium shear stress, 5–10 kPa, cell viability was 91%, whereas for high shear stress, over 10 kPa, it dropped to 76%. Similar tests, performed at three particular shear stress levels (4, 9, and 18 kPa) on human mesenchymal stem cells (hMSC) harvested from five donors revealed significant differences in cell viability (94%, 92%, and 86%, respectively) right after bioprinting, but they waned after 1 week in culture (Fig. 6.9A).

The proliferative capacity of hMSC at various stress levels is illustrated in Fig. 6.9B. Cell numbers, expressed in relative fluorescence units, indicate that low shear stress caused a small, statistically insignificant loss of cells compared to the unprinted controls. This conclusion is valid at all time points. High shear stress diminished significantly the proliferation potential of the bioprinted cells compared to control cells, but it did not inhibit proliferation altogether: the number of cells printed at high shear stress increased 1.9-fold from day 1 to day 7 (Fig. 6.9B). In contrast, cells dispensed at low-to-medium shear stress levels displayed a slightly higher proliferation rate than control cells (indeed, their number increased 2.8 times from day 1 to day 7, while the control cell population increased 2.4-fold).

During 1 week in culture, all printed cells preserved their phenotypes. Regardless of the shear stress (of up to 18 kPa) experienced in the print nozzle, vimentin expression was not altered during microvalve-based bioprinting, as demonstrated by Fig. 6.9E. Furthermore, printed cells remained CD34-negative, just as control cells (Fig. 6.9F).

2.2 Theoretical model of the landing of cell-laden droplets on solid collector plates

The impact and spreading of viscous droplets on a flat target is a problem of interest in most DBB techniques. As cell-encapsulating droplets land on the receiving substrate, the enclosed cells undergo fast and ample deformation. Also, the transient flow of the embedding prepolymer solution past the cell membrane during droplet landing can lead to harmful levels of shear stress. An accurate prediction of the extent of cell deformation and detailed characterization of the flow velocity field next to the cell are desirable for minimizing cell damage.

Muradoglu and Tasoglu tackled the problem of viscous droplet landing on dry solid walls using computational fluid dynamics (Muradoglu & Tasoglu, 2010). They devised a front-tracking method to solve the axisymmetric Navier–Stokes equations discretized on a stationary Eulerian grid. The interface between the

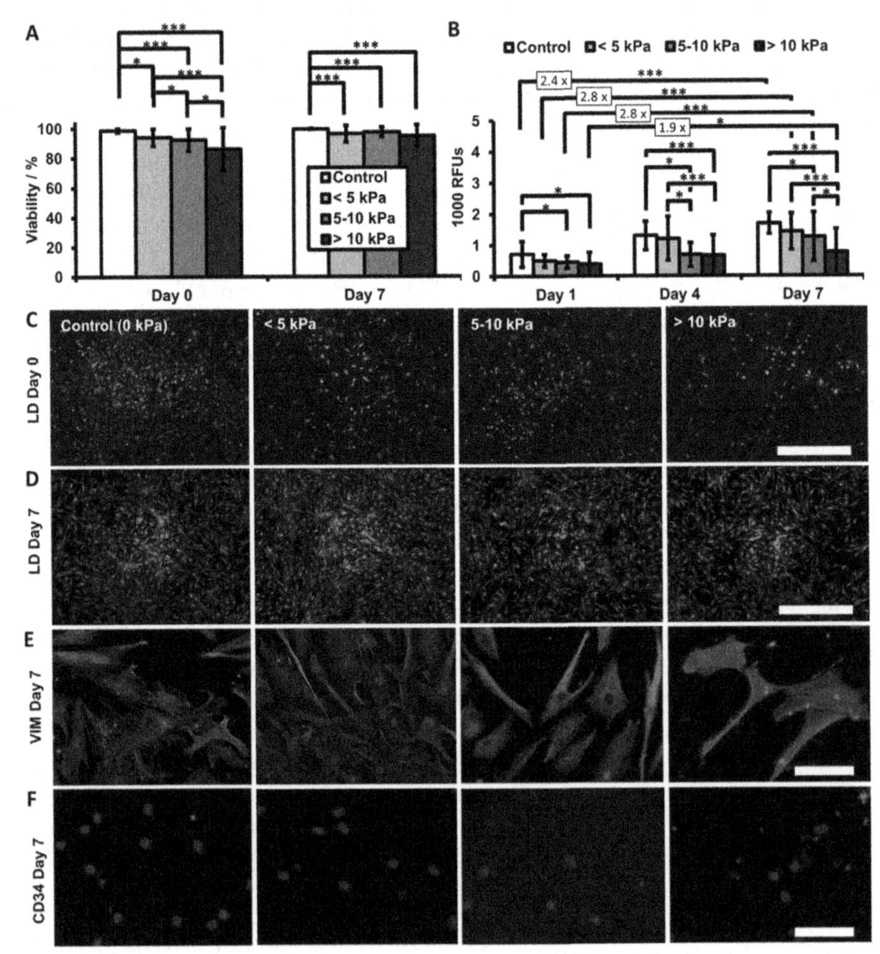

FIGURE 6.9 Human mesenchymal stem cell viability, proliferation, and protein expression after microvalve-bioprinting at different shear stress levels.

(A) Cell viability plotted versus the shear stress calculated from the theoretical model; cell viability was evaluated in triplicate, on days 0 and 7, by fluorescence staining with propidium iodide and fluorescein diacetate; (B) total cell numbers, expressed in relative fluorescence units (RFUs), plotted against shear stress; (C and D) representative images showing cell viability and growth on days 0 and 7, respectively; scale bars = 1 mm; (E) immunofluorescence staining with vimentin a mesenchymal stem cell marker; scale bar = 100 μm; and (F) immunofluorescence staining with the hematopoietic cell marker CD34, demonstrating that the printed cells did not express this protein; scale bar = 100 μm; in *panels* E and F, cell nuclei are stained blue with DAPI.

Reproduced from Blaeser, A., Duarte Campos, D. F., Puster, U., Richtering, W., Stevens, M. M. & Fischer, H. (2016). Controlling shear stress in 3D bioprinting is a key factor to balance printing resolution and stem cell integrity. Advanced Healthcare Materials, 5, *326–333. https://doi.org/10.1002/adhm.201500677, with permission from John Wiley and Sons.*

droplet and the surrounding fluid is represented by a Lagrangian grid—a "necklace" of marker points joined by front elements that move with the velocity of the local flow inferred by interpolation from the Eulerian grid. The Lagrangian grid is used to compute the surface tension, which is transferred to the adjacent Eulerian grid points via the cosine distribution function proposed by Peskin in his computational analysis of blood flow (Peskin, 1977). Hence, surface tension is represented as a body force included in the fluid momentum equations. The Lagrangian grid needs to be restructured at every time step by removing front elements that are too short and by splitting those that are too long, thereby keeping the front elements comparable in size with the Eulerian grid spacing (Tryggvason et al., 2001).

As the droplet approaches the target wall, a challenging problem arises when one tries to describe fluid motion next to the contact disk: the no-slip boundary condition gives a stress singularity because the fluid velocity is zero at the target and finite at the free surface of the droplet. In their axisymmetric analysis, Muradoglu and Tasoglu described the evolution of the contact line (the radius of the contact disk) by receding from the wall by a threshold distance, h_{th}, equal to four Euclidean grid spacings, and imposing a dynamic contact angle determined iteratively. They assumed that the droplet connects the wall as soon as it gets closer to it than h_{th}. The marker points that crossed the threshold surface were eliminated, and the droplet's profile was extrapolated by a cubic spline fit to find the true point of contact Fig. 6.10. For technical details, the reader is referred to the original paper (Muradoglu & Tasoglu, 2010).

The front-tracking method was first tested against formulas that describe the equilibrium shape of a viscous droplet placed on a dry surface. This problem has simple analytic solutions at extreme values of the Eötvös number, *Eo* (named for the Hungarian physicist Loránd Eötvös), defined as the gravitational force to surface tension force ratio. Consider a droplet of initial radius R_o and density ρ_d, immersed in a medium of density ρ_o, and placed in touch with a horizontal plate. In this context, $Eo = (\rho_d - \rho_o) g R_0^2 / \sigma$, where σ denotes the surface tension (droplet–medium interfacial tension) and g is the gravitational acceleration. When gravity is negligible compared to surface tension ($Eo << 1$), the equilibrium shape of the droplet is a spherical cap whose height, H_0, can be computed from volume conservation by taking into account the static contact angle, θ_e:

$$H_0 = 2^{2/3} R_0 (1 - \cos\theta_e)^{1/3} (2 + \cos\theta_e)^{-1/3}. \tag{6.15}$$

In the asymptotic limit, when gravity is overwhelming ($Eo >> 1$), the equilibrium shape of the droplet resembles a pancake of height

$$H_\infty = H_0 \cdot 2Eo^{-1/2} \cos(\theta_e / 2), \tag{6.16}$$

where H_0 is given by Eq. (6.15) (Muradoglu & Tasoglu, 2010).

The predictions of the front-tracking model were found in agreement with experimental data concerning the landing and subsequent spreading of glycerin droplets on wax, as well as glass plates (Muradoglu & Tasoglu, 2010).

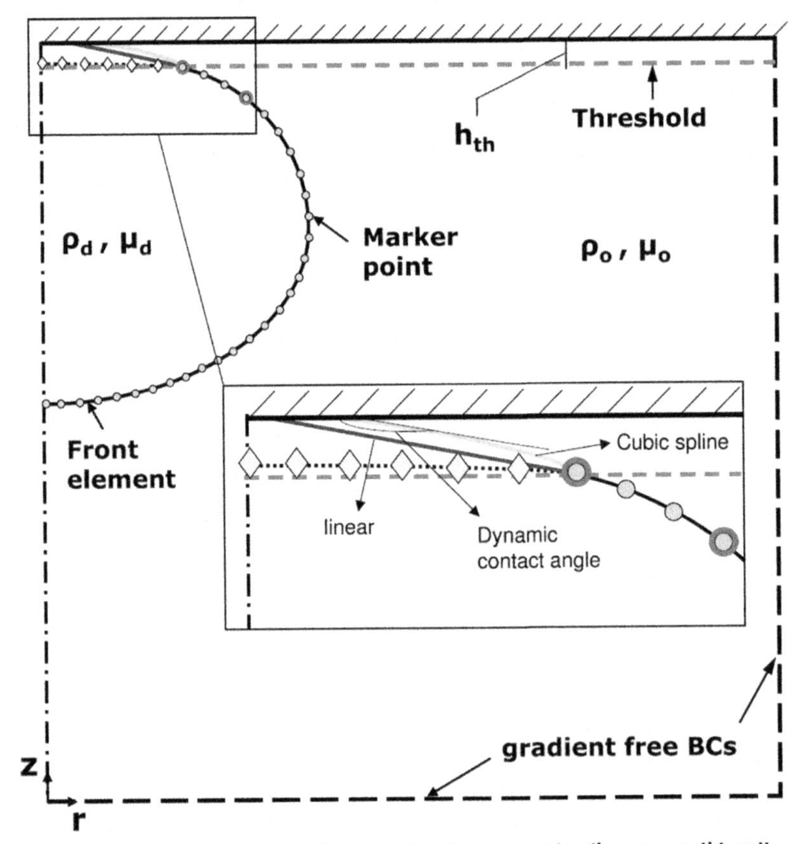

FIGURE 6.10 Front-tracking computational model of a droplet landing on a solid wall.

Circles stand for marker points, whereas the segments delimited by them represent front elements that correspond to the interface between the droplet and the surrounding fluid. The equations of fluid dynamics are solved numerically to describe the radial cross-section of the axisymmetric droplet, except for the region located between the wall and a nearby threshold surface. *Marker* points that move beyond this surface are eliminated (these are represented by *yellow diamonds*). The dynamic contact angle is computed iteratively and imposed while the marker points highlighted by large, *red circles* are fitted by a cubic spline to extrapolate the droplet profile (*cyan curve*, light gray in print). The dark blue (dark gray in print) line depicts the less precise, linear extrapolation of the droplet profile.

Reprinted from Muradoglu, M. & Tasoglu, S. (2010). A front-tracking method for computational modeling of impact and spreading of viscous droplets on solid walls. Computers & Fluids, 39, *615–625. https://doi.org/10.1016/j.compfluid.2009.10.009, with permission from Elsevier.*

Tasoglu et al. extended the above front-tracking approach to treat a problem of practical interest in DBB: the landing of a cell-laden viscous droplet on a solid collector plate. More precisely, they considered a compound droplet composed of a highly viscous droplet, representing the cell, embedded in a larger droplet of about

10 times lower viscosity. For tractability, the inner droplet, the wrapping droplet, and the surrounding air were all treated as Newtonian fluids of different material properties (Tasoglu et al., 2010). The numerical methods developed for dealing with a homogeneous droplet were effective also in this case, which involves the tracking of two interfaces (Lagrangian grids).

The validation of the extended model was performed by solving the problem of droplet shape relaxation and considering the limiting cases described by Eqs. (6.15) and (6.16). Also, the predicted contact line dynamics was compared with experimental findings on glycerin droplets spreading on flat wax and glass surfaces, as before.

To bring the theoretical predictions closer to the biological realm, model parameters were inferred from experimental work on acoustic bioprinting of single cells enclosed in droplets of about 40 μm in diameter (Demirci & Montesano, 2007). In these experiments, various cell types were suspended in an aqueous solution of 8.5% sucrose and 0.3% dextrose, and tiny droplets were expelled from an open pool by an acoustic field (see also Chapter 5, Section 2). Since the ejection process is gentle and nozzle-free, droplet landing is the sole event that could cause cell damage.

Fig. 6.11 plots velocity vectors (*left*) and shear stress levels (*right*) in successive snapshots of the droplet's shape evolution after landing on the receiving substrate (Tasoglu et al., 2010).

The shear stress is maximal near the contact line right after the moment of landing and decreases steadily as the droplet spreads on the target surface. Interestingly, the maximum shear stress is encountered at the droplet—air interface because the velocity gradient is highest there. Negative shear stress emerges next to the contact line, where both the velocity and pressure gradients are large (although their assessment might be affected by numerical errors, which are also highest in the vicinity of the contact line) (Tasoglu et al., 2010).

The front-tracking model suggests that the encapsulating droplet provides mechanical protection to the enclosed cell, hampering its deformation and shielding it from shear stress.

3. Optimization of photopolymerization-based bioprinting

Photopolymerization-based bioprinters do not expose cells to harsh physical factors, such as shear stress or high temperatures. They rely on spatially modulated light exposure of a photocrosslinkable prepolymer solution mixed with a photoinitiator and live cells, often called a bioresin by analogy with the photocurable materials used in traditional stereolithography (SL). Nonetheless, light irradiation and the chemical environment required for photopolymerization can affect the cells present in the bioink.

Bioprinting technologies based on vat photopolymerization are nozzle-free, so they do not require filament or droplet formation. Hence, they pose different

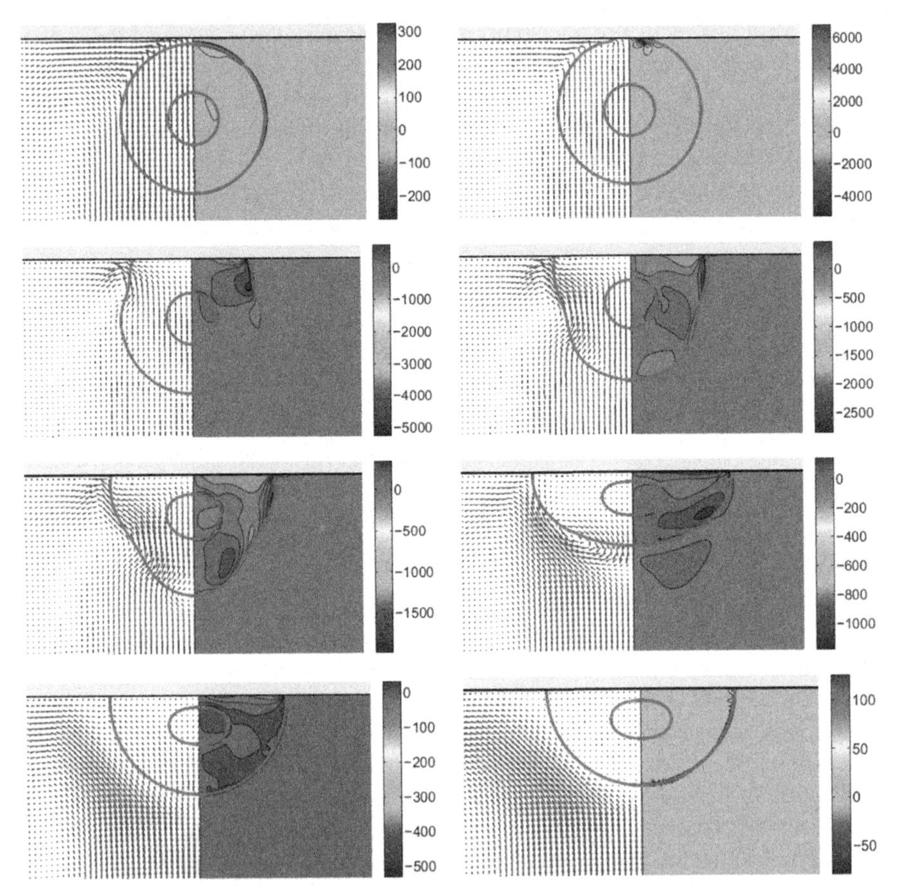

FIGURE 6.11 Velocity field and shear stress distribution during compound droplet spreading on a solid wall.

Snapshots of the droplet's cross-section at various stages of spreading, corresponding to dimensionless times $t^* = 2.7 \times 10^{-4}$, 5.4×10^{-2}, 0.135, 0.216, 0.270, 0.514, 1.03, and 3.84, from *left* to *right* and from *top* to *bottom*, respectively; in each *panel*, the left half represents fluid velocity vectors, whereas the right half is a heatmap plot of shear stress levels; model parameters were as follows: Weber number = 0.5, Reynolds number = 30, the ratio of the surface tension coefficients of the droplet-air interface and the droplet-droplet interface = 2541, and the ratio of inner droplet viscosity and embedding droplet viscosity = 10.

Reprinted from Tasoglu, S., Kaynak, G., Szeri, A. J., Demirci, U. & Muradoglu, M. (2010). Impact of a compound droplet on a flat surface: A model for single cell epitaxy. Physics of Fluids, 22, *082103. https://doi.org/10.1063/1.3475527, with the permission of AIP Publishing.*

rheological requirements than other bioprinting techniques. Digital light processing (DLP) and SL require low viscosity bioresins, which rapidly flow under a lifted part in a bottom-up setup or flood the top surface of the hardened part in a top-down setup. The bottom-up approach has been adapted also for relatively viscous, composite resins (ceramic suspensions) by covering the vat bottom with an elastic membrane to reduce the lifting force. As the hardened part is lifted, the membrane detaches from it progressively, starting from the periphery, and the neighboring uncured resin infiltrates into the cleft formed between them (Hu et al., 2018). Low-viscosity hydrogel precursors are desirable also for printing tissue constructs with narrow channels and/or small pores because they help to drain uncured bioink. Low viscosity, however, promotes the sedimentation of the suspended cells. The periodic movement of the build platform was found sufficient to prevent cell settling in a bioresin composed of 10 wt% methacrylated poly(vinyl alcohol) (PVA-MA) and 1 wt% gelatin-methacryloyl (GelMA) (Lim et al., 2018), whereas a 10% PEGDA solution had to be supplemented with 37.5% (v/v) Percoll to keep cells suspended by means of buoyancy (Lin et al., 2013).

3.1 The Jacobs equation

The theoretical resolution of SL and DLP (i.e., the smallest feature size outlined by the light pattern) is of the order of 5 μm (Schwab et al., 2020). The corresponding voxel size, however, is larger because the photopolymerization is not strictly confined within the irradiated spot. Light scattering and the diffusion of the reactive species released as a result of the irradiation of the photoinitiator cause fortuitous polymerization of nearby bioresin. These phenomena spoil the printing resolution both in the x, y plane, orthogonal to the direction of the incident light beam, as well as along the z axis, parallel to it. The resolution in the z direction is mainly determined by the curing depth, C_d, defined as the thickness of the hardened bioink layer. To make sure that successive layers stick firmly to each other, C_d should be larger than the distance covered by the build platform after each photocuring step (the designed layer height). It should not be much larger, though, because in constructs designed to have pores it would cause inadvertent polymerization of the bioink planned to drain away. It is important to prevent the overcuring of the resin located beyond the focal layer. Thus, printing accuracy depends on a careful control of the curing depth warranting a closer look at the physicochemical factors that determine C_d.

In additive manufacturing based on photopolymerization, it is assumed that the prepolymer solution obeys the Beer–Lambert law (Jacobs, 1992),

$$I(z) = I_0 \exp\left(-\frac{z}{D_p}\right) \tag{6.17}$$

where I_0 is the irradiance at the surface of the prepolymer solution, and $I(z)$ is the irradiance at a depth z. By definition, *irradiance* is the radiant energy that crosses within one second the unit area in a plane normal to the beam. Irradiance is also called *light intensity.* In Eq. (6.17), D_p denotes the penetration depth, defined as the depth at which the irradiance is $1/e \cong 0.368$ times the surface irradiance (e is Euler's number, 2.71,828 ...).

The time integral of the irradiance, $E = \int Idt$, is called *exposure* or *energy dose*, and represents the energy delivered to the unit surface oriented normally to the incident light beam. It is an experimental fact that a photocurable prepolymer solution remains liquid when the exposure remains below a critical value, E_c. Gelation begins when the critical exposure is achieved.

Consider, for simplicity, a uniformly illuminated patch on the resin's surface, which receives a surface exposure $E_0 = \int I_0 dt > E_c$—e.g., from the projector of a DLP system. From the Beer–Lambert law, the exposure at depth z is given by $E(z) = E_0 exp(-z/D_p)$. That is, remote layers get less energy, and the resin undergoes polymerization up to the depth at which the exposure barely reaches the critical level—i.e., $E(z) = E_c$ at $z = C_d$. Hence, the cure depth is related to the surface exposure, the penetration depth, and the critical exposure according to the relationship $E_c = E_0 exp(-C_d/D_p)$, which yields the Jacobs Equation:

$$C_d = D_p \ln\left(\frac{E_0}{E_c}\right) \tag{6.18}$$

This equation conveys the idea that the higher the energy dose delivered to the bioresin surface, the larger the cure depth—but the relationship is logarithmic, not linear. It is known as the equation of the working curve of SL or DLP, obtained by plotting the observed cure depth versus the surface exposure expressed in logarithmic scale. According to Eq. (6.18), the corresponding experimental points can be fitted by a straight line, whose x intercept is the critical exposure, whereas its slope is the penetration depth (Fig. 6.12). More precisely, D_p can be calculated by taking two points on the line, (C_{d_1}, E_{0_1}) and (C_{d_2}, E_{0_2}), and computing $D_p = \Delta C_d / \ln(E_{0_2}/E_{0_1})$.

The cure depth can be measured by stereomicroscopy, whereas the bioprinter allows for setting the surface exposure by specifying the surface irradiance (W/m^2, or rather mW/cm^2) and the curing time (s). Their product is the surface exposure, $E_0 = I_0 \cdot t$, in the most common case when the power of the light source does not change with time—see, for example (Huh et al., 2021); otherwise, E_0 is the integral of the surface irradiance over the curing time interval.

Fig. 6.13A exemplifies two working curves obtained by Lim et al. for a bioresin composed of 10 wt% methacrylated poly(vinyl alcohol) (PVA-MA) mixed with a photoinitiator (PI), 0.2 mM tris-bipyridylruthenium (II) hexahydrate (Ru) combined with 2 mM sodium persulfate (SPS), abbreviated as 0.2 mM/2 mM Ru/SPS (Lim et al., 2018). The working curve of this bioresin is represented by solid squares, whereas open triangles plot the working curve obtained when the same bioresin was supplemented with 1 wt% Ponceau 4R, a photoabsorber (PA)—nonreactive dye molecules that have high absorbance at the wavelength of the light source. A PA dye competes with the PI in absorbing the incident light, so it lowers the penetration depth and, thereby, the extent of overcuring. Moreover, they also limit light

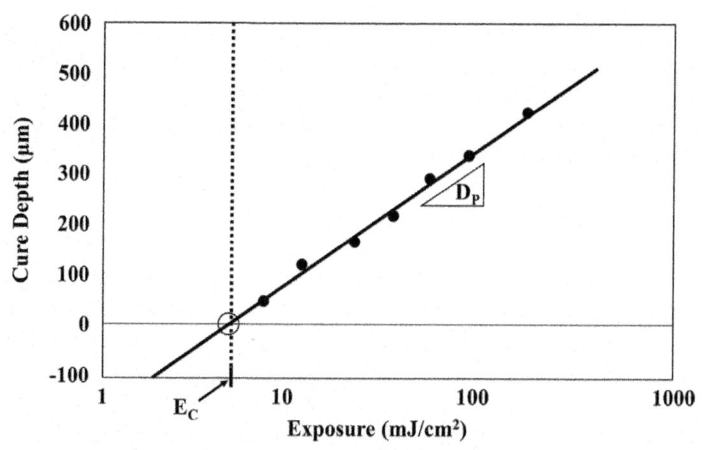

FIGURE 6.12 The working curve of photopolymerization-based 3D (bio)printing.

The extrapolated line, obtained from the least squares fit of the data, meets the horizontal axis at the smallest exposure capable of triggering the resin's gelation, the critical exposure, E_C; the slope of the line is a measure of the light penetration depth.

Reprinted from Chartrain, N. A., Williams, C. B. & Whittington, A. R. (2018). A review on fabricating tissue scaffolds using vat photopolymerization. Acta Biomaterialia, *74, 90–111. https://doi.org/10.1016/j.actbio.2018.05.010, with permission from Elsevier.*

scattering, which improves the printing resolution in the x, y plane. Indeed, the DLP printing accuracy was drastically improved in the presence of the PA (compare Fig. 6.13Biii with the digital model Fig. 6.13Bi); without it, the stray light caused undesired curing far beyond the designed lateral surfaces of the cube (Fig. 6.13Bii).

Panels C and D demonstrate the capabilities of DLP when light absorption is carefully balanced. Constructs with intertwined filaments are beyond the reach of EBB and even for structures whose topology is compatible with continuous filament extrusion the feature sizes are an order of magnitude lower than those of created using EBB.

The spectacular gain in accuracy, of course, comes at the cost of a lower throughput (since the layer thickness should remain below the curing depth) and a higher surface exposure (since it needs to surpass the critical exposure, which increased by an order of magnitude because of the PA). In the experiments that furnished the results from Fig. 6.13C and D, the vertical step size was in the interval 25–50 µm, the surface irradiance was 7.25 mW/cm^2, and the curing time was 10 s (Lim et al., 2018).

A slight variation of the Jacobs equation was used by Li et al. to express the UV exposure time needed to achieve a desired cure depth (Li et al., 2019). Working at a preset light intensity ($I_0 = 2.25$ mW/cm^2), they chose to express the critical exposure as $E_c = I_0 t_T$, where t_T is the exposure time needed to bring the closest resin layer in a pregelled state—i.e., t_T is the threshold curing time. Also writing the

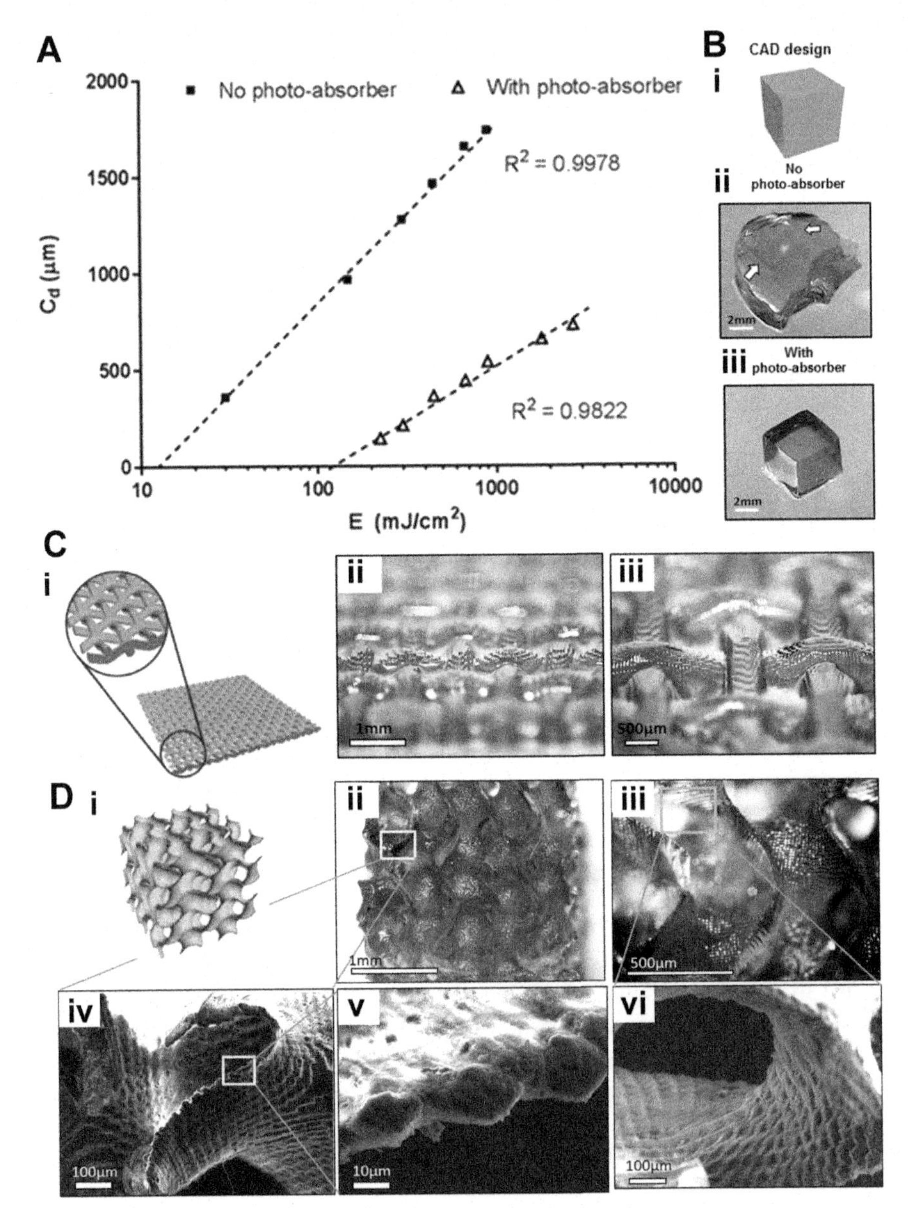

FIGURE 6.13 High-resolution digital light processing (DLP) printing.

(A) Working curves derived from the Jacobs Equation for a PVA-MA + Ru/SPS bioresin with and without Ponceau 4R photoabsorber (PA) (empty triangles and solid squares, respectively); (B) (i) computer-aided design (CAD) digital model of a cube, (ii) the printout obtained by DLP printing of the cube in the absence of PA in the bioresin, (iii) the output of the DLP printer in the presence of the PA; (C) (i) CAD model of a woven mat, (ii) optical microscopy image of the DLP printed mat, (iii) close-up view of representative struts of the mat; (D) (i) CAD model of a gyroid pattern commonly used in 3D printing infills and tissue engineering scaffolds, (ii) side view of the DLP printed pattern, (iii) optical microscopy close-up view of the gyroid construct's surface features, and (iv)–(vi) scanning electron microscopy images of the gyroid construct microscopic features captured at different magnifications.

Republished from Lim, K. S., Levato, R., Costa, P. F., Castilho, M. D., Alcala-Orozco, C. R., van Dorenmalen, K. M. A., Melchels, F. P. W., Gawlitta, D., Hooper, G. J., Malda, J. & Woodfield, T. B. F. (2018). Bio-resin for high resolution lithography-based biofabrication of complex cell-laden constructs. Biofabrication, 10, *034101. https://doi.org/10.1088/1758-5090/aac00c, with permission granted by IOP Publishing, Inc.*

applied energy dose as $E_0 = I_0 t$, the Jacobs equation takes the form $C_d = D_p \ln (t/t_T)$. When $t = t_T$, the depth of cure is zero because the surface layer (that next to the light source) has barely reached the gel point. To assure a target cure depth, an extra time Δt is necessary beyond t_T, i.e., $t = t_T + \Delta t$, and the Jacobs equation can be solved for the extra time: $\Delta t = t_T \left[exp\left(C_d / D_p \right) - 1 \right]$. Their theoretical analysis served as a background for a new DLP approach, called precuring DLP. The prepolymer solution is first precured by uniform illumination of duration t_T. Then, the image pattern corresponding to the first layer is projected onto the pregelled solution for time Δt to cure the first layer. As the print bed is lifted, pregelled solution streams in from the construct's periphery, and the next layer is cured, with patterned light in time Δt, and so on. Precuring DLP provides increased throughput, especially when the slice thickness is very small; e.g., for 10 μm-thick slices, it gave an 18-fold boost in printing speed compared to conventional DLP. Furthermore, the smaller time of UV exposure was shown to increase the viability of PC12 cells embedded in GelMA bioink from 54% (for conventional DLP) to 90.2% (for precuring DLP) (Li et al., 2019).

DLP has limitations, too. Finding the optimal combination of PI and PA is a subject of intense research because they provide control over print resolution and are mainly responsible for the cytotoxicity of photopolymerization-based bioprinting (Grigoryan et al., 2019; Huh et al., 2021). Ru/SPS is a particularly advantageous PI because it has a large molar extinction coefficient $\varepsilon = 14.6 \times 10^3\ M^{-1}\ cm^{-1}$, i.e., it displays high light absorption, at the wavelength of 450 nm. By comparison, other widely used PIs that are well tolerated by live cells have about 100 times smaller molar absorptivities—e.g., $\varepsilon = 4\ M^{-1}\ cm^{-1}$ for Irgacure 2959 and $\varepsilon = 218\ M^{-1}\ cm^{-1}$ for LAP at 365 nm (Lim et al., 2020). The result is poor photoreactivity, which needs to be compensated by higher light exposure to ensure proper crosslinking. The low light absorption gives chance for light scattering, pushing the print resolution into the 100 μm range (Lim et al., 2018).

3.2 Phenomenological models of light curing

To control bioprinting techniques that take advantage of light curing, it is desirable to express the depth of penetration and the critical energy dose as a function of the system's composition and the physicochemical parameters that characterize their involvement in free radical photopolymerization.

One phenomenon shared by all these techniques is the attenuation of the light beam that penetrates the liquid resin. It is described by the Beer–Lambert law, which can be expressed in several (mathematically equivalent) ways. Besides Eq. (6.14), it is also written as $I(z) = I_0 \exp(-\alpha z)$, where $\alpha = 1/D_p$ goes by the name of attenuation coefficient, or $I(z) = I_0 exp(-\varepsilon c z)$, where ε is called molar extinction coefficient and c is the molar concentration of a solute capable of absorbing light of the given wavelength. When several absorbing species are present in a solution, the irradiance at depth z is given by, $I(z) = I_0 \exp\left[-(\varepsilon_1 c_1 + \varepsilon_2 c_2 + \ldots) z\right]$.

The attenuation coefficient can be expressed as the sum of attenuation coefficients associated with each mechanism that lowers the number of photons encountered at a certain depth below the resin surface: $\alpha = \alpha_{PI} + \alpha_{PA} + \alpha_{cells}$. If suspended cells occupy a volume fraction f, and neither the PI nor the PA are able to cross the cell membrane, they reside in a fraction $1-f$ of the total resin volume. Therefore, $\alpha_{PI} = (1-f)\varepsilon c$ where ε is the molar extinction coefficient of the photoinitiator and c is its molar concentration, and $\alpha_{PA} = (1-f)\varepsilon' c'$, where ε' and c' refer to the photoabsorber. The factor$(1-f)$ should be omitted in the case of chemical species that penetrate the suspended cells. The contribution of cells to the attenuation coefficient, $\alpha_{cells} = f\mu_a\overline{\langle L\rangle(\mu_a)} + f\beta$, takes into account the attenuation of light due to absorption and scattering. In the first term, $\overline{\langle L\rangle(\mu_a)}$ is the mean average path length of detected photons over the absorption coefficient range $[0, \mu_a]$—here μ_a is the absorption coefficient of a tissue-like, compact cell cluster (Sassaroli & Fantini, 2004). The second term is a first-order approximation of the contribution of light scattering caused by cells. The scattering term has been proposed and validated experimentally for photocurable suspensions of nonabsorbing ceramic particles at low volume fractions (Tomeckova & Halloran, 2010c). Typical cell densities used in photopolymerization-based bioprinting do not exceed 10^7 cells/mL, which corresponds to a cell volume fraction of the order of 10^{-2}. (Indeed, the cell size is of the order of 10^{-5} m, so the cell volume $V_{cell} \approx 10^{-15}$ m^3; hence, an aggregate of cohesive cells ($f = 1$) contains about 10^9 cells/mL). Experiments conducted on NIH 3T3 mouse fibroblasts show that a cell density of 10^7 cells/mL in 5% (w/v) GelMA corresponds to a volume fraction $f = 0.0176$,and lowers the cure depth by 10% in comparison to the cure depth measured in the absence of cells (Wadnap et al., 2019). Thus, at commonly used, low cell fractions, the cure depth can be approximated by the following formula:

$$D_p = [(1-f)\varepsilon c + (1-f)\varepsilon' c' + f\lambda]^{-1} \tag{6.19}$$

where λ is a constant that incorporates both absorption and scattering induced by the suspended cells (a rough, first-order approximation). At cell densities of 10^6 cells/mL or lower, the above equation can be further simplified by taking $f \cong 0$; i.e., $D_p \cong (\varepsilon c + \varepsilon' c')^{-1}$.

Eq. (6.19) is based on the works of Tomeckova and Halloran concerning photocurable ceramic suspensions (Tomeckova & Halloran, 2010b; 2010c). A more accurate treatment of the impact of suspended cells on the curing depth might take advantage of the theoretical concepts developed in the field of continuous-wave near-infrared tissue spectroscopy (Kocsis et al., 2006; Mallet et al., 2021). Near-infrared light is of special interest because it is weakly absorbed by mammalian cells, and, therefore, it penetrates deeply into biological tissues without affecting them. It was recently shown that near-infrared light can be used to induce the photopolymerization of common UV-curable bioresins. The idea behind this new technology is to coat up-conversion nanoparticles with a UV-sensitive photoinitiator (i.e., LAP) (Chen et al., 2020). Upon near-infrared irradiation, the nanoparticles emit

UV photons; these are absorbed by the photoinitiator, which triggers the polymerization of the monomer solution. In their seminal paper, Chen et al. injected subcutaneously, in nude mice, a bioresin composed of 15 wt% GelMA and 1 wt% nanoinitiator and demonstrated DLP printing beneath the skin.

In the context of ceramic suspensions, Tomeckova and Halloran derived a theoretical formula also for the critical energy dose, E_c. They started from the hypothesis that photopolymerization cannot begin unless the number of free radicals generated by the photoinitiator exceeds the number of free radicals destroyed by the inhibitors present in the liquid resin (Tomeckova & Halloran, 2010a; 2010c). Most photocurable resins contain added inhibitors to prevent accidental polymerization. Besides these, dissolved oxygen also acts as an inhibitor, maintaining a dead zone (i.e., unpolymerized resin layer) in the vicinity of the atmosphere or next to an oxygen-permeable membrane (Suh et al., 2011). Furthermore, the PA, an inert dye added to the prepolymer mixture, also contributes to free radical depletion as it absorbs part of the light and, therefore, a number of free radicals never get created. Hence, the number of free radicals annihilated in the unit volume of resin can be written as $n_{\text{depletion}} = \gamma' c' + \sum_i \gamma_i c_i$, where c_i are the concentrations of various inhibitor species (including O_2) and γ_i are model parameters that characterize their ability to destroy free radicals; c' and γ' refer to the contribution of the PA.

The number of free radicals created in the unit volume can be expressed in terms of the number of photons absorbed by PI molecules present in that volume and the quantum yield, ϕ, defined as the number of free radicals created per absorbed photon: $n_{\text{creation}} = \phi\, n_{\text{phAbsPI}}/(1-f)$. The last factor takes into account that the generated free radicals reside in the extracellular space. When an energy dose E is delivered to a resin patch of area S, the number of photons that cross the surface is $S \cdot E/(h\nu)$, where h is Planck's constant and ν is the photons' frequency. Most of these photons are absorbed in a layer of thickness $1/(\varepsilon c)$, which is a characteristic length scale of UV absorption by the PI (Tomeckova & Halloran, 2010c). Consequently, the number of photons absorbed in the unit volume is given by $n_{phAbs} = [S\,E/(h\nu)]/[S/(\varepsilon c)] = E\varepsilon c/(h\nu)$, but only a fraction α_{PI}/α of these photons is absorbed by the PI. Therefore, $n_{\text{phAbsPI}} = E\varepsilon c\alpha_{PI}/(h\nu\alpha)$ and $n_{\text{creation}} = E\phi\varepsilon^2 c^2/(h\nu\alpha)$, where $\alpha = 1/D_p$ can be substituted from Eq. (6.19).

At the critical energy dose, $E = E_c$, the creation of free radicals is balanced by their depletion ($n_{\text{creation}} = n_{\text{depletion}}$), which can be solved for E_c to obtain:

$$E_c = \frac{h\nu}{\phi}\left(\gamma' c' + \sum_i \gamma_i c_i\right)\frac{(1-f)\varepsilon c + (1-f)\varepsilon' c' + f\lambda}{\varepsilon^2 c^2}. \tag{6.20}$$

Retaining only the first-order term in $1/(\varepsilon c)$, one obtains the linearized form of the critical energy derived from the inhibitor exhaustion model (Tomeckova & Halloran, 2010c):

$$E_c = (1-f)\frac{h\nu}{\phi}\frac{\gamma' c' + \sum_i \gamma_i c_i}{\varepsilon c}. \tag{6.21}$$

Eq. (6.21) has been validated against a vast set of experimental data obtained for photocurable suspensions of UV-transparent ceramic microparticles (Tomeckova & Halloran, 2010a). A similar equation, $E_c = K + K'c'$, was found in very good agreement with data derived from working curves obtained, in the absence of cells, for 20% (w/v) PEGDA hydrogel with 0.5% (w/v) LAP, augmented with various concentrations of brilliant blue PA in the interval 0.1%–0.25% (w/v) (Li et al., 2019).

Assuming that cells are not actively involved in the photopolymerization process, the analogy between bioprinting and the photopolymerizaiton-based 3D printing of ceramic suspensions seems reasonable, providing quantitative tools for the analysis of experimental data and process optimization. For example, Eq. (6.16) predicts that the penetration depth (visualized in Fig. 6.12 as the slope of the working curve) decreases when a photoabsorber is added to the bioink. At the same time, the critical energy increases, as shown by Eq. (6.21). These predictions agree with the experimental results plotted in Fig. 6.13A (Lim et al., 2018).

Another interesting consequence of these phenomenological equations can be derived by inserting Eqs. (6.19) and (6.21) into Eq. (6.18) (the Jacobs equation) and taking the first and second derivatives of the curing depth with respect to the photoinitiator concentration, c. It turns out that the first derivative vanishes at a particular concentration, obtained by solving the equation $\ln(u) = 1 + k/u$, where $u = E_0/E_c$ and $k = (1-f)^{-1}E_0\phi[\varepsilon'c' + \lambda f/(1-f)] \Big/ \left[h\nu\left(\gamma'c' + \sum_i \gamma_i c_i\right)\right]$ depends on model parameters, dye concentration, inhibitor concentrations, and surface exposure. Note that the $f/(1-f)$ factor is divergent as the cell fraction approaches 1, which is another indication that the scattering term, $f\lambda$, in Eq. (6.19) is only valid at very low cell densities. Even if scattering would be dwarfed by absorption (which is not the case in a multicellular system—a crowded medium with numerous interfaces), the limit $f \rightarrow 1$ does not make sense in the framework of this model—indeed, k would still diverge due to the factor $(1-f)^{-1}$. Since the logarithm increases monotonously toward infinity, and $1 + k/u$ decreases monotonously from infinity toward the asymptotic value 1, this equation is satisfied by just one value, u_m, which can be obtained numerically if k is known. At that particular value, $u = u_m$, the second derivative is negative, indicating that the depth of cure is maximal at that point. When no photoabsorber is present ($c' = 0$) and the cell volume fraction of cells is negligibly low ($f \cong 0$), also $k = 0$ and the equation is satisfied for $u_m = e$, Euler's number. Then, the photoinitiator's molar concentration, c_m, for which the depth of cure is maximal can be computed from $E_0/E_c(c_m) = e$, with E_c given by Eq. (6.18): $c_m = e\,h\nu\left(\sum_i \gamma_i c_i\right) \Big/ (\phi\varepsilon E_0)$; the corresponding maximum curing depth is given by $C_{d_{max}} = D_p = 1 \Big/ (\varepsilon c_m) = \phi E_0 \Big/ \left[eh\nu\left(\sum_i \gamma_i c_i\right)\right]$. These results are in qualitative agreement with kinetic model calculations and experimental results (Lee et al., 2001).

Provided that the rate constants are known, kinetic models are preferable over phenomenological ones because they provide more insight into the photochemical reactions at play during light curing. For instance, a kinetic model was found in excellent agreement with experimental data on the inhibition of free radical photopolymerization by dissolved oxygen (Dendukuri et al., 2008), and the model was extended to take into account the influence of photoabsorbers present in the prepolymer mixture (Suh et al., 2011).

The main strengths of phenomenological models are simplicity and flexibility. They can be formulated analytically and depend on a relatively small number of model parameters, which can be inferred by least squares fitting of data acquired from calibration experiments—see, e.g, (Tomeckova & Halloran, 2010a; 2010b). Once the parameters are determined for a given bioink composition, they can be used to optimize the bioprinting process, providing control over layer height and printing speed by tuning the curing light intensity and exposure time. For instance, overcuring of undercut and overhang portions of the printout can be avoided by reducing the laser power in tandem with the z-step size (to keep it smaller than the layer thickness calculated from the predictive model) until frontal walls are hardened. Then, the exposure can be increased to make the printing faster.

When it comes to adjusting in-plane (x, y) resolution, however, a limitation of the phenomenological approach becomes apparent: it does not account for the diffusion of chemical species (inhibitors or free radicals) and does not assess the amount of scattered light. In this respect, kinetic models and computer simulations are more powerful.

Finally, phenomena that are only just beginning to be deciphered concern the interplay between cell biochemistry and the photochemistry of light curing. It turns out that cells are not just passive players that somewhat alter light propagation through the bioink. Recently, chondrocytes were found to interact with several types of PEG hydrogel precursors, via thiol side chains of their cell surface proteins (Poole, 2015), depleting the dithiol crosslinker pool in their vicinity (Chu et al., 2020). Also, it is known that the cell membrane is capable of quenching free radicals by lipid peroxidation, so it inhibits polymerization. Consequently, the hydrogel was "rarefied" near the encapsulated chondrocytes, as revealed by fluorescence microscopy, and this feature could be prevented by treating the cells with an antioxidant (estradiol) before including them in the bioink. This effect was unsurprising since estradiol was shown previously to increase superoxide dismutase, and to accumulate in the cell membrane, protecting against lipid peroxidation. The accumulated evidence suggests that cells interfere with photopolymerization through multiple mechanisms, which also depend on cell maturity; adult chondrocytes were more effective than juvenile ones in altering the surrounding hydrogel, presumably because of their increased levels of antioxidant proteins and free thiols in their membrane (Chu et al., 2020).

3.3 Printability assays for light curing-based bioprinting

As SL and DLP approach maturity as bioprinting techniques, with an ever-growing choice of bioresin formulations, there is an increased need for quantitative measures of printability. They are needed to quantify the output of experiments meant to optimize printer settings and bioinks. Objective measures of printing outcomes also help to expedite the experimental design through machine learning. Mathematical models predict certain outcomes, such as the cure depth, but they do not describe the ability to print constructs with specific features, such as cavities of predefined geometry or sub-millimetric channels of circular cross-sections. A machine learning algorithm can be trained to predict a variety of outcomes by taking the printing parameters as input and minimizing the differences between the planned and the real outcome (Ng et al., 2020)—provided that the latter is described quantitatively.

To measure cure depth in a top-down DLP setup, a specific 3D structure was designed, which resembles the Stonehenge monument from Wiltshire, England (Wadnap et al., 2019). It consists of four vertical, 13-layers high pillars, connected at the top by four horizontal beams of a single layer in thickness, resulting in a square-shaped frame sustained by four columns. The designed layer height was 0.2 mm: the z-stage of the print bed was lowered into the bioink by this distance after each photocuring step. During each step, the prepolymer solution was exposed for 15–45 s to UV light ranging from 7 to 16 mW/cm^2 in intensity. The cure depth was assessed as the maximum thickness of the horizontal beams. For a surface exposure of 0.3 J/cm^2 of the 5% (w/v) GelMA hydrogel, C_d was about 1 mm and dropped to 0.9 mm when the hydrogel was loaded with 10^7 cells/mL (Wadnap et al., 2019). A similar structure, shown in Fig. 6.14, was used for the measurement of the cured

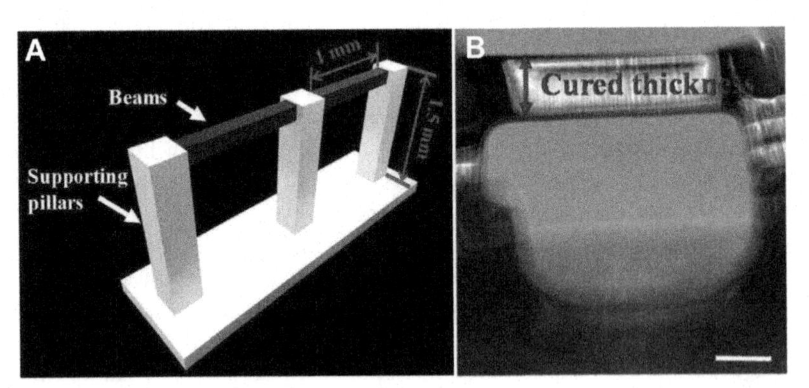

FIGURE 6.14 Cure depth assessment.

(A) Digital model of a structure specifically designed for the measurement of the depth of cure in photopolymerization-based 3D (bio)printing and (B) a representative printout obtained using a digital light processing (DLP) instrument to print the model of *panel* A from 20% (w/v) PEGDA hydrogel.

Reprinted from Li, Y., Mao, Q., Li, X., Yin, J., Wang, Y., Fu, J. & Huang, Y. (2019). High-fidelity and high-efficiency additive manufacturing using tunable pre-curing digital light processing. Additive Manufacturing, 30, *100889. https://doi.org/10.1016/j.addma.2019.100889, with permission from Elsevier.*

layer thickness in PEGDA hydrogels with various PA (brilliant blue food dye) concentrations, and the results were analyzed to find material constants needed for the calculation of printing parameters in precuring DLP (Li et al., 2019, 2021).

Yu et al. conducted an ample study of GelMA hydrogel printability using a bottom-up DLP bioprinter (Yu et al., 2022). In their prepolymer mixture, they varied the GelMA concentration from 5% to 20% (w/v), added 0.5% (w/v) LAP photoinitiator as well as 0%−1% tartrazine (a yellow food dye) as PA. Besides testing the printing accuracy for a variety of geometries, they also performed photorheological studies in oscillatory mode while the mixture was exposed to 405 nm light at 4 mW/cm^2 irradiance. As a reference hydrogel, 20% (w/v) PEGDA was used. Recording the storage modulus, G', and loss modulus, G'', over time, they identified the gel point for each composition as the point at which G' overcomes G'', so that the solid-like behavior becomes prevalent. PEGDA displayed the fastest gelling (within 13 s), followed by 20% GelMA (within 20 s). The more dilute formulations required progressively more time to reach their gel point: a twofold decrease in GelMA concentration resulted in about twice as long gelation time (Yu et al., 2022).

In a previous paper (Sun et al., 2021), the same research group proposed a standardized model for the assessment of DLP printing accuracy and resolution. The motivation for their endeavor reportedly originated from difficulties encountered in trying to apply contact-based measurement methods on hydrogel structures, which are notoriously soft and transparent. Hydrogel constructs are delicate and slippery; hence, they are hard to manipulate. Being wet and transparent, they beget light reflection and refraction, so their boundaries are hard to distinguish. Attempts have been made to circumvent these hurdles with the help of advanced imaging. Scanning electron microscopy (SEM) gives high-resolution images of construct surface morphology, but sample preparation is prone to affect its geometry (Sun et al., 2021). Viable methods for assessing shape fidelity include micro-CT and OCT, with certain limitations, as discussed in the context of EBB (first section of this chapter) (Schwab et al., 2020). Therefore, a standardized model suitable for quantitative evaluation by optical microscopy is highly appealing to whoever intends to optimize photocuring parameters and/or candidate bioresins.

The model proposed by Sun et al. consists of a disk, akin to a coin, decorated with straight lines (spokes) of a single voxel in thickness that diverge from the center of the disk. Since the projected line width, h, is finite, adjacent lines overlap next to the center and form a small disk that is concentric with the cylindrical base (Fig. 6.15). The actual printed line is wider than the projected one because of the inadvertent hardening of the bioresin caused by stray light and free radicals that diffused away from the irradiated area. The resolution is defined as the smallest distance between two points of the digital model that can be distinguished from each other on the printout. In the context of their model of Fig. 6.15$A_{i\text{-}iii}$, the printing resolution p, has been defined as the distance between the edges of the projected lines at the point of separation of the actual printed lines (Fig. 6.15A_{iii}) (Sun et al., 2021).

Consider a model with n uniformly distributed spokes. The angle between adjacent spokes is $2\pi/n$ radians. The bisector of this angle contains the point where the

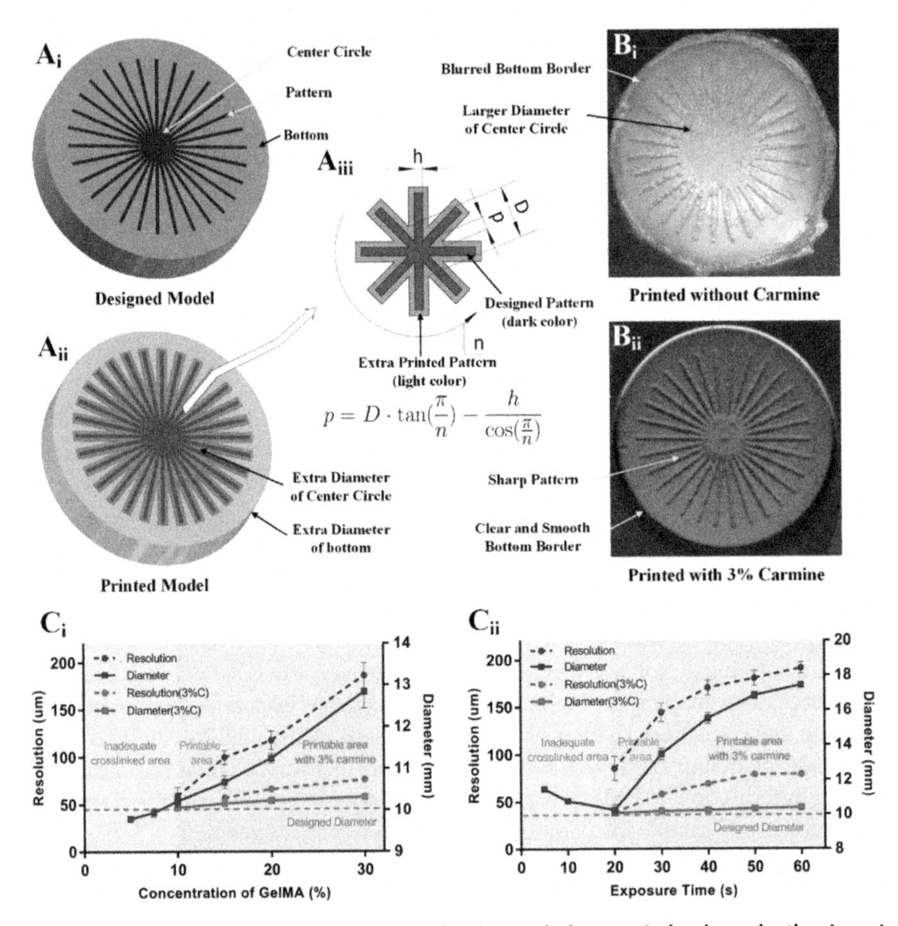

FIGURE 6.15 Definition and measurement of in-plane printing resolution in projection-based 3D (bio)printing.

(A_i) Digital representation of the spokes model and (A_{ii}) a schematic drawing of the corresponding printed structure; (A_{iii}) scheme of the projected pattern, with spokes of width *h* (*dark-blue*) and the resulting printout with wider spokes (*dark-* and *light-blue*); a central disk of diameter *D* is formed as adjacent lines overlap; the printing resolution, *p*, can be expressed in terms of the number of spokes, *n*, and the accurately measurable diameter, *D*; (B_i) digital photography of a representative printout made of 30% (w/v) GelMA hydrogel; (B_{ii}) picture of the printout obtained when 3% carmine PA was added to the 30% (w/v) GelMA precursor solution; (C_i) plot of the printing resolution and central disk diameter versus polymer concentration; (C_{ii}) plot of the impact of light exposure time on printing resolution. For interpretation of the references to color in this figure legend, please refer online version of this title.

Reprinted from Sun, Y., Yu, K., Nie, J., Sun, M., Fu, J., Wang, H. & He, Y. (2021). Modeling the printability of photocuring and strength adjustable hydrogel bioink during projection-based 3D bioprinting. Biofabrication, 13, *035032. https://doi.org/10.1088/1758-5090/aba413, with permission from Elsevier.*

two spokes separate from each other, at a distance R from the center; $R = D/2$ is the radius of the central circle on the printed sample — its diameter, D, can be measured accurately by light microscopy. The tangent to the circle at the point of separation intersects the symmetry axes of the spokes in two points that delimit a segment of length $2R\tan(\pi/n)$. The distance between the edges of the projected lines is smaller by $2(h/2)/\cos(\pi/n)$. Thus, the in-plane (x, y) printing resolution of projection-based 3D (bio)printing is given by (Sun et al., 2021; Yu et al., 2022):

$$p = D\tan(\pi / n) - \frac{h}{\cos(\pi/n)}. \tag{6.22}$$

The printing resolution p, is an estimate of the smallest achievable feature size that can be created using DLP.

Although the benefits of augmenting the GelMA solution with 3% carmine PA are clear from the visual inspection of *panels* B_i and B_{ii} of Fig. 6.15, the quantitative test based on the spoke model provides deeper insight into the set of factors that influence the accuracy of DLP. The printability window of pure GelMA is limited to the polymer concentration range of 10%–15% (w/v), and even in this range, p increases twofold as the GelMA concentration grows. In the presence of 3% carmine, the printability window expands to 10%–30% (w/v) GelMA with a moderate (less than 50%) increase in p over this range of polymer concentrations (Fig. 6.15C_i). For a given prepolymer solution (e.g., 15% GelMA), larger exposure times result in higher crosslinking density (better mechanical properties), but also larger values of the minimum printable feature size; fortunately, exposure has less impact on print resolution when PA is present in the ink (Fig. 6.15C_{ii}). Interestingly, p also depends on the projected line width, h, presumably because wider lines correspond to a larger number of incident photons and a proportionally larger number of scattered photons. For example, in the case of a 10% GelMA ink with 0.5% tartrazine PA, the resolution was about 500 μm for $h = 0.25$ mm and 800 μm for $h = 0.5$ mm (Yu et al., 2022).

To quantify the printing resolution along the z axis, the direction of the incident light beam, Yu et al. proposed to print a spiral structure capped by a horizontal platform and analyze the length of the contact line between the spiral and the platform (Yu et al., 2022).

The objective measures of printing accuracy pointed out the optimal printing parameters for the fabrication of solid constructs, perfusable tissue structures, nerve conduits, porous scaffolds as well as microneedle patches, covering a wide variety of potential applications. For instance, solid constructs, such as cartilaginous structures of the ear and nose, are best printed from a bioink of low GelMA concentration (5%–10%) and high PA concentration, at low exposure. An organoid with perfusable branching channels, on the other hand, requires a bioink of medium GelMA concentration (10%–15%), a carefully chosen PA concentration for preventing the overcuring of narrow tubes, and high exposure, based on high irradiance (5 mW/cm^2), for assuring fast crosslinking of the bioink (Yu et al., 2022). Further research will explore the influence of the embedded cells on DLP bioprinting parameters.

References

Blaeser, A., Duarte Campos, D. F., Puster, U., Richtering, W., Stevens, M. M., & Fischer, H. (2016). Controlling shear stress in 3D bioprinting is a key factor to balance printing resolution and stem cell integrity. *Advanced Healthcare Materials, 5*, 326–333. https://doi.org/10.1002/adhm.201500677

Chen, Y., Zhang, J., Liu, X., Wang, S., Tao, J., Huang, Y., Wu, W., Li, Y., Zhou, K., Wei, X., Chen, S., Li, X., Xu, X., Cardon, L., Qian, Z., & Gou, M. (2020). Noninvasive in vivo 3D bioprinting. *Science Advances, 6*, eaba7406. https://doi.org/10.1126/sciadv.aba7406

Chu, S., Maples, M. M., & Bryant, S. J. (2020). Cell encapsulation spatially alters crosslink density of poly(ethylene glycol) hydrogels formed from free-radical polymerizations. *Acta Biomaterialia, 109*, 37–50. https://doi.org/10.1016/j.actbio.2020.03.033

Cooke, M. E., & Rosenzweig, D. H. (2021). The rheology of direct and suspended extrusion bioprinting. *APL Bioengineering, 5*, 011502. https://doi.org/10.1063/5.0031475

Demirci, U., & Montesano, G. (2007). Single cell epitaxy by acoustic picolitre droplets. *Lab on a Chip, 7*, 1139–1145. https://doi.org/10.1039/B704965J

Dendukuri, D., Panda, P., Haghgooie, R., Kim, J. M., Hatton, T. A., & Doyle, P. S. (2008). Modeling of oxygen-inhibited free radical photopolymerization in a PDMS microfluidic device. *Macromolecules, 41*, 8547–8556. https://doi.org/10.1021/ma801219w

Gao, T., Gillispie, G. J., Copus, J. S., Pr, A. K., Seol, Y.-J., Atala, A., Yoo, J. J., & Lee, S. J. (2018). Optimization of gelatin–alginate composite bioink printability using rheological parameters: A systematic approach. *Biofabrication, 10*, 034106. https://doi.org/10.1088/1758-5090/aacdc7

Gillispie, G., Prim, P., Copus, J., Fisher, J., Mikos, A. G., Yoo, J. J., Atala, A., & Lee, S. J. (2020). Assessment methodologies for extrusion-based bioink printability. *Biofabrication, 12*, 022003. https://doi.org/10.1088/1758-5090/ab6f0d

Grigoryan, B., Paulsen Samantha, J., Corbett Daniel, C., Sazer Daniel, W., Fortin Chelsea, L., Zaita Alexander, J., Greenfield Paul, T., Calafat Nicholas, J., Gounley John, P., Ta Anderson, H., Johansson, F., Randles, A., Rosenkrantz Jessica, E., Louis-Rosenberg Jesse, D., Galie Peter, A., Stevens Kelly, R., & Miller Jordan, S. (2019). Multivascular networks and functional intravascular topologies within biocompatible hydrogels. *Science, 364*, 458–464. https://doi.org/10.1126/science.aav9750

Habib, A., Sathish, V., Mallik, S., & Khoda, B. (2018). 3D printability of alginate-carboxymethyl cellulose hydrogel. *Materials, 11*. https://doi.org/10.3390/ma11030454

Huh, J., Moon, Y.-W., Park, J., Atala, A., Yoo, J. J., & Lee, S. J. (2021). Combinations of photoinitiator and UV absorber for cell-based digital light processing (DLP) bioprinting. *Biofabrication, 13*, 034103. https://doi.org/10.1088/1758-5090/abfd7a

Hu, K., Wei, Y., Lu, Z., Wan, L., & Li, P. (2018). Design of a shaping system for stereolithography with high solid loading ceramic suspensions. *3D Printing and Additive Manufacturing, 5*, 311–318. https://doi.org/10.1089/3dp.2017.0065

Jacobs, P. F. (1992). Fundamentals of stereolithography. In *Solid freeform fabrication conference proceedings, 1992, Austin, Texas* (pp. 196–211).

Kocsis, L., Herman, P., & Eke, A. (2006). The modified Beer–Lambert law revisited. *Physics in Medicine and Biology, 51*, N91–N98. https://doi.org/10.1088/0031-9155/51/5/n02

Lee, J. H., Prud'homme, R. K., & Aksay, I. A. (2001). Cure depth in photopolymerization: Experiments and theory. *Journal of Materials Research, 16*, 3536–3544. https://doi.org/10.1557/JMR.2001.0485

Lee, J. M., & Yeong, W. Y. (2015). A preliminary model of time-pressure dispensing system for bioprinting based on printing and material parameters. *Virtual and Physical Prototyping, 10*, 3–8. https://doi.org/10.1080/17452759.2014.979557

Li, Y., Mao, Q., Li, X., Yin, J., Wang, Y., Fu, J., & Huang, Y. (2019). High-fidelity and high-efficiency additive manufacturing using tunable pre-curing digital light processing. *Additive Manufacturing, 30*, 100889. https://doi.org/10.1016/j.addma.2019.100889

Li, Y., Mao, Q., Yin, J., Wang, Y., Fu, J., & Huang, Y. (2021). Theoretical prediction and experimental validation of the digital light processing (DLP) working curve for photocurable materials. *Additive Manufacturing, 37*, 101716. https://doi.org/10.1016/j.addma.2020.101716

Lim, K. S., Galarraga, J. H., Cui, X., Lindberg, G. C. J., Burdick, J. A., & Woodfield, T. B. F. (2020). Fundamentals and applications of photo-cross-linking in bioprinting. *Chemical Reviews, 120*, 10662–10694. https://doi.org/10.1021/acs.chemrev.9b00812

Lim, K. S., Levato, R., Costa, P. F., Castilho, M. D., Alcala-Orozco, C. R., van Dorenmalen, K. M. A., Melchels, F. P. W., Gawlitta, D., Hooper, G. J., Malda, J., & Woodfield, T. B. F. (2018). Bio-resin for high resolution lithography-based biofabrication of complex cell-laden constructs. *Biofabrication, 10*, 034101. https://doi.org/10.1088/1758-5090/aac00c

Lin, H., Zhang, D., Alexander, P. G., Yang, G., Tan, J., Cheng, A. W.-M., & Tuan, R. S. (2013). Application of visible light-based projection stereolithography for live cell-scaffold fabrication with designed architecture. *Biomaterials, 34*, 331–339. https://doi.org/10.1016/j.biomaterials.2012.09.048

Lockwood, E. H. (1961). *A book of curves*. Cambridge University Press.

Mallet, A., Tsenkova, R., Muncan, J., Charnier, C., Latrille, É., Bendoula, R., Steyer, J.-P., & Roger, J.-M. (2021). Relating near-infrared light path-length modifications to the water content of scattering media in near-infrared spectroscopy: Toward a new Bouguer–Beer–lambert law. *Analytical Chemistry, 93*, 6817–6823. https://doi.org/10.1021/acs.analchem.1c00811

Mandrycky, C., Wang, Z., Kim, K., & Kim, D.-H. (2016). 3D bioprinting for engineering complex tissues. *Biotechnology Advances, 34*(4), 422–434. https://doi.org/10.1016/j.biotechadv.2015.12.011

Moroni, L., Boland, T., Burdick, J. A., De Maria, C., Derby, B., Forgacs, G., Groll, J., Li, Q., Malda, J., Mironov, V. A., Mota, C., Nakamura, M., Shu, W., Takeuchi, S., Woodfield, T. B. F., Xu, T., Yoo, J. J., & Vozzi, G. (2018). Biofabrication: A guide to technology and terminology. *Trends in Biotechnology, 36*, 384–402. https://doi.org/10.1016/j.tibtech.2017.10.015

Muradoglu, M., & Tasoglu, S. (2010). A front-tracking method for computational modeling of impact and spreading of viscous droplets on solid walls. *Computers & Fluids, 39*, 615–625. https://doi.org/10.1016/j.compfluid.2009.10.009

Murphy, S. V., & Atala, A. (2014). 3D bioprinting of tissues and organs. *Nature Biotechnology, 32*, 773–785. https://doi.org/10.1038/nbt.2958

Ng, W. L., Chan, A., Ong, Y. S., & Chua, C. K. (2020). Deep learning for fabrication and maturation of 3D bioprinted tissues and organs. *Virtual and Physical Prototyping, 15*, 340–358. https://doi.org/10.1080/17452759.2020.1771741

Osserman, R. (2010). How the gateway arch got its shape. *Nexus Network Journal, 12*, 167–189. https://doi.org/10.1007/s00004-010-0030-8

Ouyang, L., Yao, R., Zhao, Y., & Sun, W. (2016). Effect of bioink properties on printability and cell viability for 3D bioplotting of embryonic stem cells. *Biofabrication, 8*, 035020. https://doi.org/10.1088/1758-5090/8/3/035020

Paxton, N., Smolan, W., Böck, T., Melchels, F., Groll, J., & Jungst, T. (2017). Proposal to assess printability of bioinks for extrusion-based bioprinting and evaluation of rheological properties governing bioprintability. *Biofabrication, 9*, 044107. https://doi.org/10.1088/1758-5090/aa8dd8

Peak, C. W., Stein, J., Gold, K. A., & Gaharwar, A. K. (2018). Nanoengineered colloidal inks for 3D bioprinting. *Langmuir, 34*, 917–925. https://doi.org/10.1021/acs.langmuir.7b02540

Peskin, C. S. (1977). Numerical analysis of blood flow in the heart. *Journal of Computational Physics, 25*, 220–252. https://doi.org/10.1016/0021-9991(77)90100-0

Poole, L. B. (2015). The basics of thiols and cysteines in redox biology and chemistry. *Free Radical Biology & Medicine, 80*, 148–157. https://doi.org/10.1016/j.freeradbiomed.2014.11.013

Ribeiro, A., Blokzijl, M. M., Levato, R., Visser, C. W., Castilho, M., Hennink, W. E., Vermonden, T., & Malda, J. (2018). Assessing bioink shape fidelity to aid material development in 3D bioprinting. *Biofabrication, 10*, 014102. https://doi.org/10.1088/1758-5090/aa90e2

Sarker, M., & Chen, X. B. (2017). Modeling the flow behavior and flow rate of medium viscosity alginate for scaffold fabrication with a three-dimensional bioplotter. *Journal of Manufacturing Science and Engineering, 139*. https://doi.org/10.1115/1.4036226

Sassaroli, A., & Fantini, S. (2004). Comment on the modified Beer–Lambert law for scattering media. *Physics in Medicine and Biology, 49*, N255–N257. https://doi.org/10.1088/0031-9155/49/14/n07

Schwab, A., Levato, R., D'Este, M., Piluso, S., Eglin, D., & Malda, J. (2020). Printability and shape fidelity of bioinks in 3D bioprinting. *Chemical Reviews, 120*, 11028–11055. https://doi.org/10.1021/acs.chemrev.0c00084

Splinter, R. (2010). *Handbook of physics in medicine and biology* (1st ed.). CRC Press, Taylor & Francis Group. https://doi.org/10.1201/9781420075250

Suh, S. K., Bong, K. W., Hatton, T. A., & Doyle, P. S. (2011). Using stop-flow lithography to produce opaque microparticles: Synthesis and modeling. *Langmuir, 27*, 13813–13819. https://doi.org/10.1021/la202796b

Sun, Y., Yu, K., Nie, J., Sun, M., Fu, J., Wang, H., & He, Y. (2021). Modeling the printability of photocuring and strength adjustable hydrogel bioink during projection-based 3D bioprinting. *Biofabrication, 13*, 035032. https://doi.org/10.1088/1758-5090/aba413

Tasoglu, S., Kaynak, G., Szeri, A. J., Demirci, U., & Muradoglu, M. (2010). Impact of a compound droplet on a flat surface: A model for single cell epitaxy. *Physics of Fluids, 22*, 082103. https://doi.org/10.1063/1.3475527

Tomeckova, V., & Halloran, J. W. (2010a). Critical energy for photopolymerization of ceramic suspensions in acrylate monomers. *Journal of the European Ceramic Society, 30*, 3273–3282. https://doi.org/10.1016/j.jeurceramsoc.2010.08.003

Tomeckova, V., & Halloran, J. W. (2010b). Cure depth for photopolymerization of ceramic suspensions. *Journal of the European Ceramic Society, 30*, 3023–3033. https://doi.org/10.1016/j.jeurceramsoc.2010.06.004

Tomeckova, V., & Halloran, J. W. (2010c). Predictive models for the photopolymerization of ceramic suspensions. *Journal of the European Ceramic Society, 30*, 2833–2840. https://doi.org/10.1016/j.jeurceramsoc.2010.01.027

Tryggvason, G., Bunner, B., Esmaeeli, A., Juric, D., Al-Rawahi, N., Tauber, W., Han, J., Nas, S., & Jan, Y. J. (2001). A front-tracking method for the computations of multiphase flow. *Journal of Computational Physics, 169*, 708–759. https://doi.org/10.1006/jcph.2001.6726

Wadnap, S., Krishnamoorthy, S., Zhang, Z., & Xu, C. (2019). Biofabrication of 3D cell-encapsulated tubular constructs using dynamic optical projection stereolithography. *Journal of Materials Science: Materials in Medicine, 30*, 36. https://doi.org/10.1007/s10856-019-6239-5

Yu, K., Zhang, X., Sun, Y., Gao, Q., Fu, J., Cai, X., & He, Y. (2022). Printability during projection-based 3D bioprinting. *Bioactive Materials, 11*, 254–267. https://doi.org/10.1016/j.bioactmat.2021.09.021

Zhang, Y. S., Haghiashtiani, G., Hübscher, T., Kelly, D. J., Lee, J. M., Lutolf, M., McAlpine, M. C., Yeong, W. Y., Zenobi-Wong, M., & Malda, J. (2021). 3D extrusion bioprinting. *Nature Reviews Methods Primers, 1*, 75. https://doi.org/10.1038/s43586-021-00073-8

CHAPTER

Multicellular self-assembly

7

How do cells give rise to functional tissues and organs? This question preoccupied developmental biologists for more than a century. Their progress in understanding embryonic morphogenesis is increasingly important for regenerative medicine and tissue engineering (Ingber & Levin, 2007). This chapter will present principles of developmental biology that serve as a foundation for computational tools devised to predict the evolution of tissue constructs built in the laboratory.

1. The differential adhesion hypothesis

Morphogenesis is a complex process that leads to the formation of functional tissues and organs. Although it is genetically controlled, embryonic morphogenesis is ultimately carried out by physical forces that guide the spatial arrangement of multiple cell types. Developmental biologists investigated morphogenetic mechanisms for decades (Steinberg, 1996), and their findings led to the formulation of the differential adhesion hypothesis (DAH) (Steinberg, 1963; 1970). DAH states that a population of cohesive and motile cells evolves toward the state of minimum free energy of adhesion—that is, cells rely on their mobility to maximize the number of strong bonds with their neighbors.

DAH originated from an analogy between clusters of cohesive cells and ordinary liquids (Steinberg, 1963). Both of them consist of discrete building blocks able to move while adhering to each other. Liquids are made of interacting molecules subject to thermal motion, whereas multicellular clusters consist of cells that stick to each other and move on the account of energy-releasing biochemical reactions. The cohesive forces between cells are intermediated by cell adhesion molecules (Gumbiner, 1996)—membrane proteins that interdigitate, in an antiparallel alignment, as the membranes of adjacent cells approach each other (Zhu et al., 2003).

Several experiments indicate that, on a time scale of hours, embryonic tissues behave as highly viscous liquids (Steinberg, 1996). Irregular tissue fragments become spherical within 1–2 days of incubation. Two contiguous tissue fragments of different type fuse such that the less cohesive one spreads over the surface of the other, eventually enveloping it. Moreover, the same equilibrium configuration emerges when dissociated cells of those two tissue types are randomly intermixed and allowed to sort out (Steinberg, 1963, 1970). Such findings motivated numerous studies that aimed to predict the behavior of cell populations in light of the physics of fluids.

Towards 4D Bioprinting. https://doi.org/10.1016/B978-0-12-818653-4.00001-2

The close resemblance between liquids and embryonic tissues has also shaped the terminology that concerns cells capable of interacting with each other or with a substrate. Interactions between cells of the same type are called cohesive, whereas those between different cell types or between cells and a substrate are called adhesive. Moreover, physical quantities commonly used to describe the behavior of liquids, such as surface tension and viscosity, have been adopted also in the context of tissues. Experimental techniques devised to measure them are presented in the next two subsections.

2. Tissue surface tension

Surface tension characterizes a liquid in equilibrium. By definition, it is the free energy change when the surface area of a liquid is increased by the unit area (Israelachvili, 2011). Surface tension is a positive quantity, reflecting that molecules on the surface are less connected to their neighbors than those in the bulk. Indeed, molecules in the bulk are surrounded by neighbors all around, whereas surface molecules are attracted to molecules from the bulk, while they are unable to interact with the molecules of the adjacent gas. As a result, the net force acting on a surface molecule points toward the interior of the liquid, tending to drive it into the bulk. In other words, the surface area of a liquid tends to be minimal.

A liquid drop assumes a spherical shape (in the absence of other forces such as gravity or friction) because, for a given volume, the sphere is the geometrical object with the smallest surface area. Similarly, a soap bubble tends to maintain a spherical shape to minimize the liquid surface area. While doing so, it compresses the enclosed air, that is, the air pressure inside the bubble is slightly higher than the atmospheric pressure. To calculate it, one can apply Laplace's equation (Israelachvili, 2011), which expresses the difference in pressure needed to maintain a curved liquid surface in equilibrium. In the special case of a spherical surface, the pressure inside the sphere is higher than the outside pressure by $2\sigma/R$, where σ is the surface tension of the liquid and R is the radius of the sphere. Hence, the pressure inside a soap bubble is $4\sigma/R$ because the bubble consists of a thin film of soap solution, delimited by two spherical surface layers. (Here σ is the surface tension of the soap solution, about 3 times smaller than the surface tension of water.)

For a nonspherical, convex liquid surface, Laplace's equation is given by (Israelachvili, 2011).

$$\Delta p = \sigma\left(\frac{1}{R_1} + \frac{1}{R_2}\right) \tag{7.1}$$

Here σ is again the liquid's surface tension, whereas R_1 and R_2 are the so-called principal radii of curvature of the surface. To illustrate them, consider a very small portion of that surface and imagine fitting to it the lateral part of a properly sized egg (touching the surface from within the liquid). At the best fit, R_1 is the radius of the egg's equatorial cross-section, whereas R_2 is the radius of the circle that fits best onto the equatorial portion of the egg's meridional cross-section. More precisely, in a given point of a convex surface of a liquid, we consider two mutually perpendicular planes whose intersection is normal to the surface. Within each plane, we draw a circle that fits best to the intersection between the surface and the plane. We rotate the pair of planes around their intersection until one of the radii becomes minimal. Then, the radii of the two circles R_1 and R_2 are, by definition, the principal radii of curvature of the surface at the given point.

According to DAH, cells seek to establish as many strong bonds with their neighbors as possible. Cells on the tissue surface lack interactions on their membrane portion exposed to the cell culture medium. Therefore, they tend to dive into the bulk, surrounding themselves with other cells. Hence, cell–cell cohesion gives rise to a tissue surface tension—the free energy of the unit area of the tissue surface bathed by cell culture medium. Also, cohesive and adhesive interactions contribute to the interfacial tension, defined as the free energy of the unit area of the interface between two different tissues or between a tissue and a substrate.

To measure tissue surface tension, Foty et al. devised a parallel-plate compression apparatus (Foty et al., 1994). A tissue spheroid was compressed between the plates by raising the lower plate, while a microbalance maintained the upper plate in position and recorded the force exerted by the donut-shaped spheroid onto the plates until equilibrium was reached, within about an hour, depending on the tissue type. During the experiment, the spheroid was immersed in a cell culture medium, and the temperature of the system was maintained at 37°C (see (Foty et al., 1994, 1996) for technical details).

To compute the tissue surface tension, one can write the equilibrium of the forces acting on the upper compression plate (Fig. 7.1).

In its attempt to minimize its surface area by regaining the spherical shape, the squeezed spheroid pushes the plate upward by the force Δp multiplied by the area of contact between the plate and the spheroid πR_3^2. A microbalance maintains the plate in position by exerting a downward force, F, of the same magnitude. Hence, one option for calculating the surface tension is the equation

$$F = \Delta p \pi R_3^2 \tag{7.2}$$

Here, Δp results from Laplace's equation, where R_1 and R_2 denote the principal radii of curvature of the spheroid's surface at the equator. The spheroid is viewed as a liquid droplet in equilibrium. Gravity is negligible in the context of a submillimetric tissue spheroid whose weight is practically counterbalanced by buoyancy. Therefore, the pressure is the same everywhere within the spheroid, so Δp can be computed at any point of the surface.

If the angle θ measured in cell culture medium between the spheroid's surface and the compression plates is nonzero (because the cells stick to the plate), the above equation should also take into account the downward force exerted by the surface layer. This is the vertical component of the force $\sigma 2\pi R_3$, which acts along the

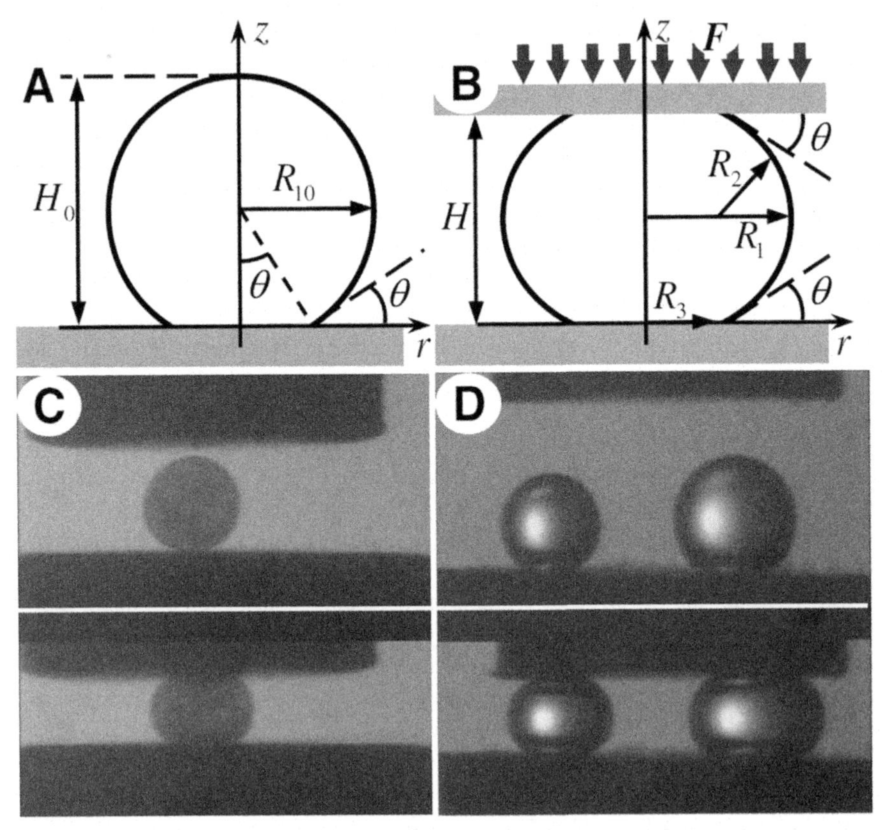

FIGURE 7.1

Parallel-plate compression of liquid droplets and tissue spheroids. Schematic of the meridional cross-section of a droplet before (A) and after (B) compression. A spheroid of chicken embryonic cardiac cushion tissue (C) and a pair of water droplets in olive oil (D) before and after compression (top and bottom, respectively).

From Norotte, C., Marga, F., Neagu, A., Kosztin, I., & Forgacs, G. (2008). Experimental evaluation of apparent tissue surface tension based on the exact solution of the Laplace equation. EPL (Europhysics Letters), 81*(4), 46003. https://doi.org/10.1209/0295-5075/81/46003.*

line of contact between the spheroid's surface and the plate. In this case, the expression of mechanical equilibrium becomes:

$$F = \Delta p \pi R_3^2 - \sigma 2\pi R_3 \sin\theta \tag{7.3}$$

Another option for computing the surface tension of the tissue spheroid is to write the mechanical equilibrium at the spheroid's equatorial cross-section. The top half of the compressed spheroid is in equilibrium under the action of three forces: the upward force $\Delta p \pi R_1^2$ due to the Laplace pressure, the downward force

$\sigma 2\pi R_1$ exerted by the surface layer, and the downward force F exerted by the balance. Their equilibrium can be written as

$$F = \sigma 2\pi R_1(\alpha - 1) \tag{7.4}$$

where $\alpha = \Delta p R_1/(2\sigma)$ is a dimensionless geometric parameter. Taking into account Laplace's equation, it can be rewritten as $\alpha = 0.5(1 + R_1/R_2)$, which indicates that $\alpha > 1$; the stronger is the compression, the larger is α.

As expected on the basis of the liquid analogy, tissue surface tension was found to be independent on the spheroid's diameter and on the degree of compression. The results, however, varied with culture duration, increasing for about a day, remaining relatively constant for the next few days, and increasing further afterward (Foty et al., 1996). Tissue surface tension is the term used for the plateau value; the others are referred to as "apparent tissue surface tension." The more recent literature, however, often calls "apparent" any surface tension that refers to a multicellular system to differentiate it from the surface tension of a true liquid (Mgharbel et al., 2009; Norotte et al., 2008).

After aggregate preparation, the constituent cells need to restore their adhesion apparatus and to squeeze out the intercellular liquid medium, gradually increasing the area of contact between neighbors. These phenomena explain the increase in spheroid cohesivity during the first 1–2 days in culture. Once these processes are completed, the plateau is reached, leading to a well-defined surface tension that is characteristic to a certain cell population. Later, as cells start to deposit their own extracellular matrix, the aggregate becomes less compressible. Moreover, the apparent surface tension of such aggregates depends on the extent of compression, as expected from an elastic solid.

Actually, living tissues are viscoelastic materials: when a sudden strain is applied to them, and it is maintained constant afterward, a stress relaxation is observed (the stress decreases, approaching an equilibrium value); when they are subject to a sudden stress, and it is maintained constant afterward, they continue to deform—a phenomenon called creep (Fung, 2013). On shorter time scales, of the order of minutes, viscoelastic features are observed also in cell aggregates and embryonic tissue spheroids (Forgacs et al., 1998).

An accurate methodology of tissue tensiometry relies on the exact solution of Laplace's equation (Norotte et al., 2008). While R_1 and H can be measured precisely from the side view of the system, R_2 is harder to assess. Fitting a circle on the spheroid's meridional cross-section (Fig. 7.1B) provides a good approximation for R_2. Nevertheless, such a procedure leads to a systematic error in surface tension assessment. Instead, using the exact Laplace profile (ELP) method, one computes the parameter α by numerically solving the equation

$$\frac{H}{2R_1} = \int_{\beta(\alpha,\theta)}^{1} \left[\left(\frac{x}{\alpha x^2 + 1 - \alpha}\right)^2 - 1\right]^{-1/2} dx \tag{7.5}$$

where $\beta(\alpha,\theta) = \left[\sin\theta + \sqrt{\sin^2\theta + 4\alpha(\alpha-1)}\right]/(2\alpha)$. Then, surface tension is given by Eq. (7.4).

Fig. 7.2 shows an accurate digital model of a compressed tissue spheroid in 3D view (panel A) and in meridional cross-section (panel B). This 3D model was built using the Rhinoceros 3D software (Robert McNeel & Associates, WA, USA) on the

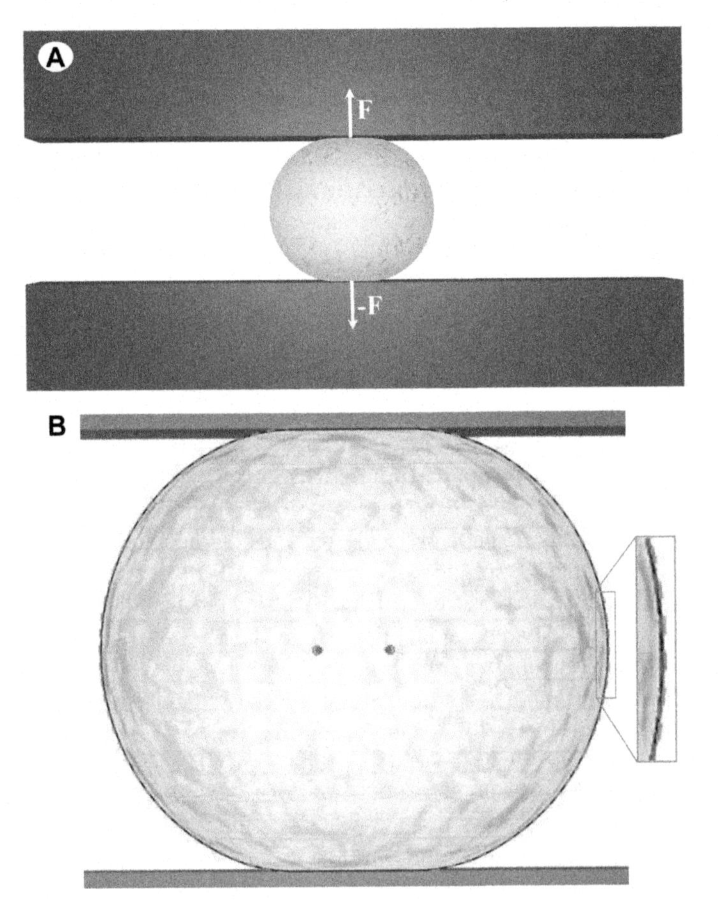

FIGURE 7.2

The digital model of a tissue spheroid compressed between a pair of nonadherent, parallel plates. The model is shown in 3D view (A) and in cross-section (B). The latter is taken in the vertical plane of symmetry of the model spheroid and is magnified threefold. In panel (B), the black solid line depicts the lateral profile obtained from the exact solution of the Laplace equation, the green dashed line represents the circular arc that provides the best fit of the lateral profile, whereas the red dashed-dotted line depicts the parabola that offers the best fit of the central portion of the lateral profile, equal in height to 25% of the spheroid's height. The centers of curvature of the circular and parabolic fit are represented by a green circle and a red star, respectively. For interpretation of the references to color in this figure legend, please refer online version of this title.

basis of the precise lateral profile obtained by solving the Laplace equation (Norotte et al., 2008). The texture of the rendering is meant to suggest the granular nature of the spheroid; indeed, it is composed of 10^3–10^4 cells, so the success of the continuum fluid dynamics description of cell aggregate biomechanics is somewhat surprising. Except for the rendering, the digital model is in accord with the continuum description (i.e., the tissue spheroid is treated as a homogeneous, incompressible viscous fluid with perfectly smooth surface layer).

The lateral profile of the compressed aggregate resembles a donut. Fitting it with a circular arc (green dashed line in Fig. 7.2B) results in a slight underestimation of R_2—the radius of curvature of the spheroid's surface measured in the meridional plane, in the vicinity of the equator (see also Fig. 7.1). An even rougher estimate would be obtained by supposing that the lateral profile of the compressed aggregate is a semicircle: $R_2 \cong H/2$. Such an approximation would lead to a more pronounced underestimation of R_2 (not shown in Fig. 7.2).

The ELP method enables one to analyze several compressions of spheroids of the same type, even if they were performed in different experiments. To this end, one fits the data points $(\alpha - 1,\ F/(2\pi R_1))$ with a straight line that passes through the origin. The slope of this line is the surface tension of the given tissue type. Moreover, if several spheroids are compressed simultaneously, a similar analysis applies. Consider the case of two spheroids, as shown in Fig. 1D: one obtains R_{1a} and α_a for the first and R_{1b} and α_b for the second. The force recorded by the balance is given by $F = \sigma 2\pi[R_{1a}(\alpha_a - 1) + R_{1b}(\alpha_b - 1)] = \sigma 2\pi(R_{1a} + R_{1b})(\overline{\alpha} - 1)$, where $\overline{\alpha}$ is the weighted average of α_a and α_b, given by $\overline{\alpha} = (R_{1a}\alpha_a + R_{1b}\alpha_b)/(R_{1a} + R_{1b})$. In other words, the analysis is similar to the case of a single aggregate, but R_1 should be replaced by the sum of the equatorial radii of all the spheroids compressed at the same time, and α should be replaced by $\overline{\alpha}$.

The ELP method, however, requires knowledge of the contact angle. The compression plates are coated with poly(HEMA), to which cells adhere poorly, ensuring that $\theta = 0°$ is a good approximation (Foty et al., 1994). Even if $\theta > 0°$ due to local impurities, the corresponding errors are small as long as $\theta \leq 20°$ (Norotte et al., 2008).

Another option for an accurate tissue surface tension measurement is the local polynomial fit (LPF) method. It requires fitting the lateral profile of the compressed droplet, $r(z)$, with a second order polynomial, $r = az^2 + bz + c$, and computing $R_2 = 1/(2a)$ (Mgharbel et al., 2009). More precisely, the fit is limited to a narrow window, centered on the equatorial plane of the aggregate. Ideally, this height should be small to furnish R_2, the local radius of curvature of the aggregate, measured in the meridional plane, next to the equator. The height of the fitting window, however, has a lower limit imposed by the image resolution and/or spheroid surface roughness. In practice, it is chosen between $H/2$ and $H/4$. The latter option provided an excellent fit to the central portion of the lateral profile, as shown in Fig. 7.2B. (Compare the red dashed-dotted line with the black solid line that depicts the exact profile according to the solution of Laplace's equation, also shown in a twofold magnification on the right side of Fig. 7.2B). The center of curvature obtained with the LPF method is depicted as a red star in Fig. 7.2B, whereas the one obtained from the circular arc fit is represented as a green circle; the two markers are not precisely concentric, indicating that the circular arc fit gave a smaller value for R_2 than the LPF method.

The polynomial fit enables one to calculate also R_1 and the surface tension can be computed using Eq. (7.4). The careful analysis of Mgharbel et al. (2009) demonstrates that LPF is as precise as ELP in estimating the principal radii of curvature of a compressed tissue spheroid, leading to similar values of the tissue surface tension. The LPF method is especially recommended when the contact angle is poorly known or when the compressed aggregate is asymmetric with respect to its equatorial plane.

3. Tissue viscosity and the fusion of tissue spheroids

3.1 Tissue viscosity

Tissue viscosity is a dynamic quantity, defined in the framework of the liquid analogy. It is a measure of the internal friction between the constituent cells as they move past one another. The kinetics of spontaneous cellular rearrangements in reaggregated embryonic tissue spheroids can be interpreted in terms of the hydrodynamics of highly viscous liquids. For example, the fusion of contiguous tissue spheroids is akin to the rounding of a jagged tissue fragment, a process driven by surface tension and resisted by viscosity (Gordon et al., 1972).

Three independent experiments led to estimates of the viscosity of certain embryonic tissues in the range of 0.4×10^8 to 1.5×10^{10} cP (Gordon et al., 1972). One of them is the breaking of a multicellular cylinder into a set of equally spaced clusters, another concerns the time course of the rounding of a tissue ellipsoid, whereas the third deals with the fusion of a pair of tissue spheroids.

Known in continuum hydrodynamics as the Plateau-Rayleigh instability. The surface-tension-driven breakup of a viscous liquid thread into a set of droplets has been described by Lord Rayleigh in 1892 (see (Eggers & Villermaux, 2008) for a review). The spacing between the emergent droplets is an indicator of the fluid's viscosity.

In the course of cell sorting, islets of 1 cell type segregate and fuse, forming a core of the more cohesive cell type, surrounded by the less cohesive cell population. While investigating the kinetics of cell sorting, Gordon et al. observed the time evolution of an oblate ellipsoid (shaped like a rugby ball) made of the more cohesive cells embedded in the less cohesive cell population (Gordon et al., 1972). They analyzed timelapse images to estimate the viscosity of the tissue ellipsoid using the formula

$$r - 1 = exp\left[-t\frac{\sigma}{\eta_t}\left(\frac{4\pi}{3V}\right)^{1/3}\frac{40(1+\eta_m/\eta_t)}{(2+3\eta_m/\eta_t)(19+16\eta_m/\eta_t)}\right] \tag{7.6}$$

where r is the ratio of the major (longer) axis and the minor (shorter) axis of the ellipsoid, V is the volume of the ellipsoid, σ is the medium-tissue interfacial tension (tissue surface tension if the ellipsoid is immersed in cell culture medium), η_m and η_t are the coefficients of viscosity of the medium and tissue, respectively. Eq. (7.6) describes the shape relaxation of an oblate ellipsoid of a highly viscous fluid immersed in another fluid of arbitrary viscosity. It was derived by J. A. Wheeler, being first published in (Harvey & Shapiro, 1941). For calculating η_t in the context of the ellipsoid of more cohesive cells incorporated in a shell of less cohesive cells, Gordon et al. assumed that the two cell populations have similar coefficients of viscosity (i.e., $\eta_m/\eta_t \cong 1$).

3.2 The fusion of volume-conserving tissue spheroid doublets

The most widely investigated phenomenon of in vitro morphogenesis is the fusion of tissue spheroids. It has attracted much attention recently because tissue spheroids are being used as building blocks of tissue-engineered constructs (Mironov et al., 2009). Assembled via 3D bioprinting, the desired construct emerges spontaneously as the adjacent tissue spheroids fuse (Moldovan et al., 2017). Computer simulations of postprinting structure formation rely on calibration experiments that monitor the fusion of tissue spheroid pairs (Flenner et al., 2012; McCune et al., 2014).

In the remainder of this section, we present a mathematical model that describes the time evolution of two identical tissue spheroids placed next to each other. Besides being practical, this model is also important for fundamental research because, to date, it provided the most accurate assessments of tissue viscosity (Shafiee et al., 2017; Stirbat et al., 2013).

Let us consider a pair of contiguous tissue spheroids of radius R_0 each. They are incubated in physiological conditions, immersed in cell culture medium, in a hanging drop, or on a nonadherent substrate (such as an agarose layer). Under these conditions, they fuse into a single spheroid, thereby minimizing the total area of the tissue–medium interface. Fusion commences as the cells next to the point of contact break up certain bonds with cells from the same spheroid and establish bonds with cells from the opposite spheroid. According to experiments conducted on a variety of cell types, the doublet of spheroids has the shape of two spherical caps of radius R that touch each other along a disk of radius $r = R\cdot\sin\theta$, where θ is the fusion angle (Fig. 7.3) (Flenner et al., 2012).

If the system's volume remains constant during fusion, the radius of the spherical caps depends on the fusion angle according to the formula

$$R = R_0 \cdot 2^{2/3}(1+c)^{-2/3}(2-c)^{-1/3} \tag{7.7}$$

where $c = \cos\theta$.

Following Frenkel (1945), we infer the time evolution of the system from energy balance, assuming that the surface energy released in unit time is equal to the energy dissipated in unit time due to internal friction. This assumption is justified in the case

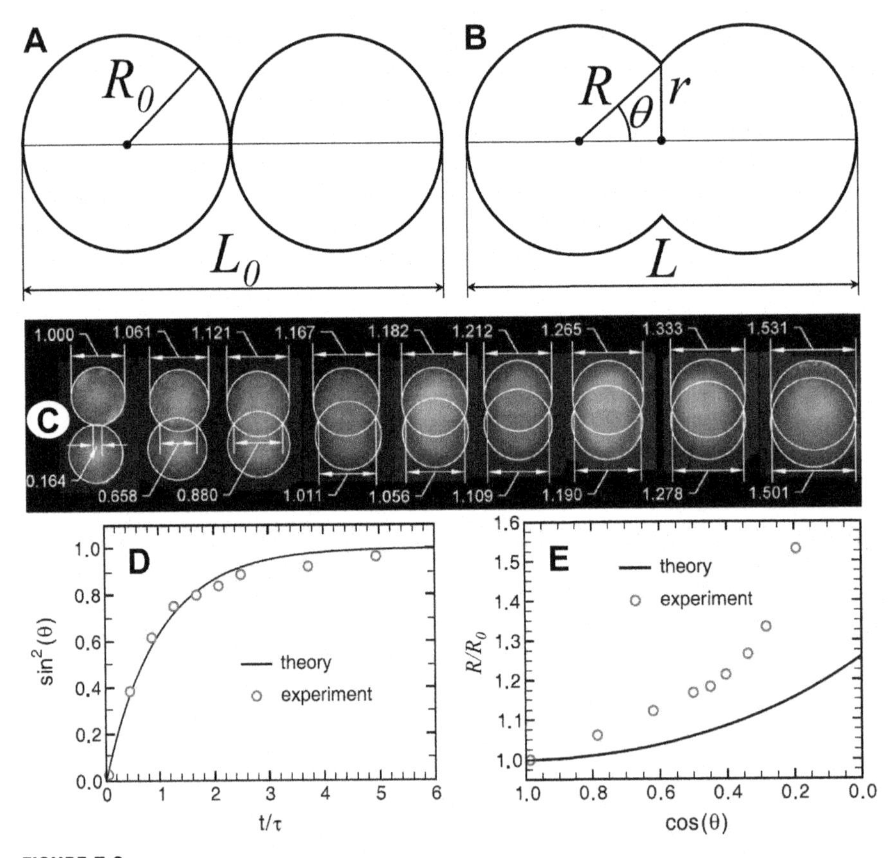

FIGURE 7.3

The geometry of two identical tissue spheroids that coalesce. Schematic diagram of the initial state (A) and intermediate state (B) of a pair of identical tissue spheroids placed next to each other. Stereomicroscopy snapshots (C) of two aggregates of Chinese hamster ovary (CHO) cells, taken at various stages of fusion. Theoretical fit (D) of the experimental data of panel (C) (circles) using Eq. (7.15) (line); here time is expressed in units of the characteristic fusion time, of about 70 h (Kosztin et al., 2012). The plot of the relative change in the radius of the spherical caps versus the cosine of the fusion angle (E) shows that the experimental points (circles) deviate from the theoretical plot given by Eq. (7.7) (line).

Panels (C-E) were reproduced from Kosztin, I., Vunjak-Novakovic, G., & Forgacs, G. (2012). Reviews of Modern Physics, 84*(4), 1791–1805. https://doi.org/10.1103/RevModPhys.84.1791, p. 1798, with permission from the American Physical Society.*

of an extremely viscous fluid because it evolves slowly, such that its kinetic energy is negligible at any instant of time. The equation of the energy balance reads:

$$\sigma\dot{S} \cong -2\eta\int\left(\frac{\partial v_i}{\partial x_j}\right)^2 dV \tag{7.8}$$

where σ and η are the surface tension and viscosity of the multicellular system, respectively. In Eq. (7.8), the indices i and j run from 1 to 3, denoting the spatial coordinates x, y, and z, respectively. Repeated indices are summed over, and dot denotes time derivative; that is, $\dot{S}$ denotes the rate of change in the surface area of the tissue, $S = 4\pi R^2(1+c)$. From volume conservation (Eq. 7.7), one can compute

$$\dot{S} = 4\pi R^2\frac{c}{2-c}\dot{c} \tag{7.9}$$

As fusion proceeds, θ increases from 0 to $\pi/2$ radians, so $c = \cos\theta$ decreases monotonously from 1 to 0.

In his analysis of the coalescence of highly viscous fluid droplets, Frenkel made a further assumption, considering a homogeneous extensional flow field, with the axial strain rate given by the approximate formula

$$u = \frac{\partial v_1}{\partial x_1} \cong \frac{1}{R}\frac{d}{dt}(Rc) = \frac{2}{(1+c)(2-c)}\dot{c} \tag{7.10}$$

Frenkel's expression of the strain rate tensor, however, did not obey volume conservation, $\partial v_i/\partial x_i = 0$, as noted by Eshelby (1949) who also proposed a correction. The elements of the corrected strain rate tensor are as follows: $\partial v_1/\partial x_1 = u$, $\partial v_2/\partial x_2 = \partial v_3/\partial x_3 = -u/2$, and $\partial v_i/\partial x_j = 0$ for $i \neq j$. Therefore, $(\partial v_i/\partial x_j)^2 = 3u^2/2$, and the energy balance equation (Eq. 7.8) becomes:

$$\dot{c} = -\frac{\sigma R^2}{8\eta R_0^3}c(1+c)^2(2-c) \tag{7.11}$$

From Eq. (7.7), one finds that $(1+c)^2(2-c) = 4(R_0/R)^3$, so Eq. (7.11) can be rewritten as follows:

$$\dot{c} = -\frac{1}{2\tau}\frac{R_0}{R}c \tag{7.12}$$

where $\tau = \eta R_0/\sigma$ is the characteristic fusion time. The ratio σ/η, known as visco-capillary velocity, has typical values of the order of 0.1 μm/min (Stirbat et al., 2013). Since the radii of tissue spheroids employed in biomedical research and bioengineering are typically in the range of 50–200 μm, τ is of the order of 10 h.

Finally, inserting the expression of R (Eq. 7.7) into Eq. (7.12), we obtain

$$\dot{c} = -\frac{1}{\tau}2^{-5/3}c(1+c)^{2/3}(2-c)^{1/3} \tag{7.13}$$

Eq. (7.13) is an ordinary differential equation satisfied by the function $c = c(t) = \cos\theta(t)$. It can be solved numerically with the initial condition

$c(t_0) = 1$, enabling one to describe the geometry of the system versus time. Here, t_0 denotes the instant of time when fusion commences, measured from the moment of the incubation of the contiguous pair of spheroids. This moment of the initial tack was assessed in experiments conducted on primary human chondrocytes, as well as on the MCF-7 human breast cancer cell line (Susienka et al., 2016). For both cell types, t_0 was less than 30 min in about 60% of the investigated cell aggregate doublets; all doublets started to fuse within 5 h.

From Eq. (7.13), one can recover the differential equation satisfied by the fusion angle, $\theta = \theta(t)$. It was first derived by Pokluda et al. (1997), being used also for analyzing the fusion of cell aggregates (Flenner et al., 2012; Kosztin et al., 2012). From the point of view of numerical analysis, however, Eq. (7.13) is preferable because it is free of singularities, whereas the differential equation satisfied by $\theta(t)$ is singular near zero (Pokluda et al., 1997).

As the two droplets fuse, R increases steadily from R_0 to its equilibrium value, $2^{1/3}R_0 \cong 1.26R_0$. Therefore, it seems reasonable to approximate $R/R_0 \cong 1$ in Eq. (7.12) (Flenner et al., 2012). The resulting differential equation can be solved analytically, with the result

$$c(t) \cong e^{-\frac{t-t_0}{2\tau}} \tag{7.14}$$

Consequently, the area of the contact disk divided by the area of the meridional cross-section of the spherical caps is given by Kosztin et al. (2012):

$$\left(\frac{r}{R}\right)^2 = \sin^2\theta \cong 1 - e^{-\frac{t-t_0}{\tau}} \tag{7.15}$$

Eq. (7.15) was used to fit the experimental data depicted by circular markers in Fig. 7.3D.

In the early stages of fusion, $R \cong R_0$ and $(t - t_0)/\tau$ is much smaller than 1. Retaining only the first two terms of the Taylor expansion of the exponential function $e^x = 1 + x + x^2/2! + x^3/3! + \cdots$, one recovers (the corrected version of) Frenkel's equation

$$r^2 = \frac{\sigma}{\eta}R_0(t - t_0) \tag{7.16}$$

This formula indicates that the slope of the straight line that fits r^2 vs. $R_0 \cdot t$ is the visco-capillary velocity, $v_p = \sigma/\eta$. The work by Stirbat et al. (2013) employs Eq. (7.16) to investigate the impact of various drugs on the fusion kinetics of cell aggregates.

Based on Eqs. (7.7 and 7.14), one can derive simple analytic expressions for the geometric quantities employed in the analysis of experimental data. In a large-scale study of tissue spheroid doublet formation and fusion (Susienka et al., 2016), the authors recorded time-lapse images of a statistically significant number of doublets (every 3 h, for 24 h) and developed an ImageJ (US National Institutes of Health, Bethesda, MD) macro for automated image analysis. Their results included the time dependence of the normalized doublet width R/R_0, and the normalized end-to-end doublet length, L/L_0. The former can be expressed inserting Eq. (7.14) into Eq. (7.7), whereas the latter is given by the following expression:

$$\frac{L}{L_0} = \frac{R}{R_0}\frac{1+c}{2} = 2^{-1/3}(1+c)^{1/3}(2-c)^{-1/3} \tag{7.17}$$

The continuum hydrodynamics model points out that the doublet length to doublet width ratio has a remarkably simple expression,

$$\frac{L}{2R} = 1 + c \tag{7.18}$$

Since it can be inferred from automatic image analysis, the length to width ratio might play an important role in the quantitative study of the kinetics of tissue spheroid fusion.

The analytic expressions given by Eqs. (7.14 to 7.18) are appealing because they are simple. The question arises, however, whether they are precise enough to capture the essential features of the kinetics of tissue spheroid fusion. This question is addressed in Fig. 7.4 by plotting various geometric quantities computed from the numerical solution of Eq. (7.13) (solid lines) and the approximate expressions based on Eq. (7.14) (dashed lines).

In the early stages of fusion, for $t - t_0 \leq \tau$, the dashed lines from Fig. 7.4 coincide precisely with the solid lines, indicating that, in this time window, the analytic expression of Eq. (7.14) describes the time course of $\cos\theta$ accurately. Later on, the approximation becomes less precise, but the deviations are relatively small, of the order of 1%. If the experimental errors exceed these deviations, the analytic expressions derived from Eq. (7.14) are suitable for data analysis; otherwise, more accurate numerical methods are needed (see, e.g., (McCune et al., 2014)).

3.3 Describing deviations from volume conservation

It is important to keep in mind that the whole argument presented so far relies on the assumption that tissue spheroids behave as extremely viscous fluid droplets whose volume does not change in time. In multicellular systems, however, volume conservation is the exception rather than the rule. Indeed, visual inspection of the stereomicroscopy snapshots from Fig. 7.3C leads to the conclusion that the volume of the CHO cell spheroid doublet increases due to cell proliferation. Quantitative image analysis confirms this conclusion, indicating that the normalized doublet width R/R_0 increases faster than predicted by Eq. (7.7) on the basis of volume conservation (Fig. 7.3E). Remarkably, the time dependence of $\sin^2\theta$ is well described by Eq. (7.15) even if the system's volume changes (Fig. 7.3D).

Cell division is not the sole reason for violations of volume conservation in systems built from tissue spheroids. Cell aggregate compaction causes the opposite effect, a decrease in the system's volume as adjacent cells tighten cohesive bonds with their neighbors and squeeze out the extracellular fluid trapped between cells in loosely packed spheroids. In spheroids assembled from certain cell types, such as

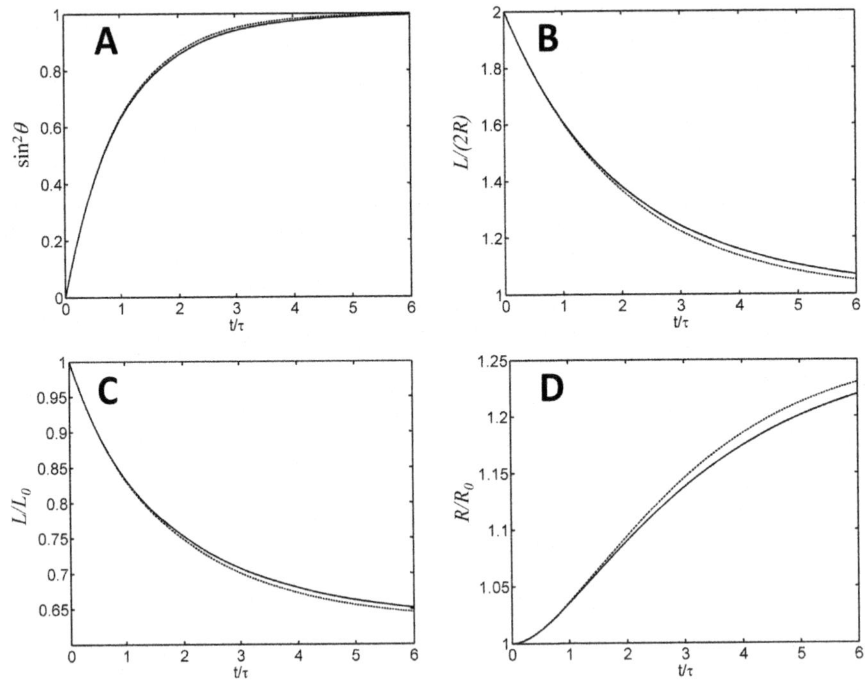

FIGURE 7.4

Characterizing the geometry of fusing tissue spheroid doublets. Plots of geometric quantities that characterize the time course of tissue spheroid fusion: the area of the contact disk, normalized by the area of the meridional cross-section of the fusing spheroids (A); the length-to-width ratio (B); the doublet length, normalized by its initial value (C); and the doublet width, normalized by its initial value (D). Solid lines plot the quantities derived from the numerical solution of Eq. (7.13), whereas dashed lines plot those obtained from the approximate solution given by Eq. (7.14). Time is expressed in units of the time constant of fusion.

human skin fibroblasts (McCune et al., 2014) or human chondrocytes (Susienka et al., 2016), compaction is the dominant effect. It has been described quantitatively by an exponential function (McCune et al., 2014)

$$a(t) = a_\infty + (1 - a_\infty)e^{-\lambda(t-t_0)} \tag{7.19}$$

where $a(t) = (V(t)/V(t_0))^{1/3}$ andλ is the rate of volume relaxation. This formula describes the change in the linear size of the system as the packing of the constituent cells is gradually optimized as a result of their mutual interaction. Here$a_\infty = \lim_{t\to\infty} a(t)$ is the relative change in the linear size of the system during the entire process of compaction.

Tissue spheroid compaction is independent of fusion but interferes with it. When individual aggregates are incubated in physiological conditions, their radius evolves according to the formula $a(t)R_0$. When a doublet of contiguous aggregates is incubated, compaction and fusion proceed simultaneously, and the radius of the meridional cross section evolves according to Eq. (7.7) augmented with an extra factor $a(t)$ on the right side. Therefore,

$$R/R_0 = 2^{2/3}(1+c)^{-2/3}(2-c)^{-1/3}a(t) \tag{7.20}$$

Similarly, the normalized doublet length can be expressed as follows:

$$L/L_0 = 2^{-1/3}(1+c)^{1/3}(2-c)^{-1/3}a(t) \tag{7.21}$$

Consequently, the entire derivation of the corresponding differential equation needs to take into account this factor. It turns out, however, that the numerical solution of the corresponding equation is very well approximated by that of the volume conserving equation, but with a slightly smaller, effective fusion time $\tau_{eff} = \tau(1+a_\infty)/2$ (McCune et al., 2014). Therefore, the simplest option for the quantitative analysis of the fusion of tissue spheroids that undergo volume relaxation is to fit experimental data using Eq. (7.15 or 7.18), thereby obtaining the (effective) characteristic fusion time. Then, one inserts Eqs. (7.14 and 7.19) into Eq. (7.20 or 7.21) and fits the corresponding data to find the parameters a_∞ andλ.

Fig. 7.5 represents the results of such an analysis performed on experimental data on human skin fibroblasts, reported in Fig. 7.4 of McCune et al. (2014).

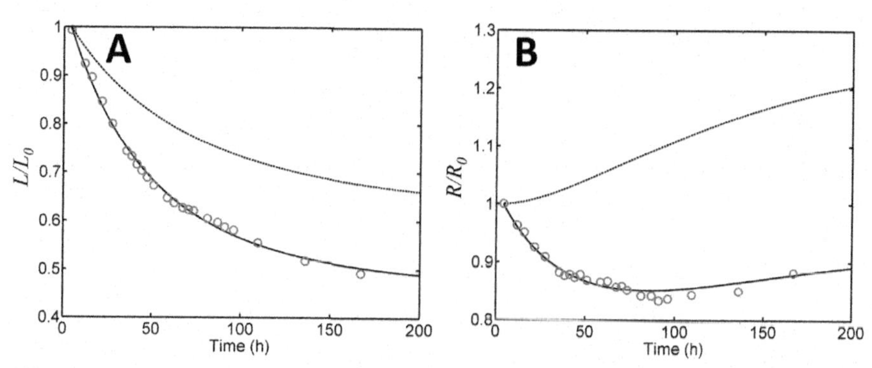

FIGURE 7.5

Shape evolution of a pair of spheroids that undergo compaction while fusing. Quantitative analysis of the human skin fibroblast data recorded by McCune et al. (2014). First, Eq. (7.15) was used to fit the corresponding experimental data (not shown). Then, the normalized doublet length data was fitted using Eq. (7.21) (A—solid line). Finally, the parameters inferred from these fits were inserted into Eq. (7.20) (B—solid line). Dashed lines plot the theoretical curves that disregard cell spheroid compaction (Eq. (7.17) for (A) and Eq. (7.7) for (B)).

To find the instant of time when fusion started, $t_0 = 4$ h, and the characteristic fusion time, $\tau = 42.4$ h, we used Eq. (7.15) and the fit function from MATLAB 7.13 (The MathWorks, Inc., Natick, MA, USA) to fit the experimental data reported by McCune et al. (2014) on the fusion of cell aggregate pairs composed of human skin fibroblasts. Using these parameters, we fitted the data regarding L/L_0 vs. time (Fig. 7.5A, circles) using Eq. (7.21) to obtain $a_\infty = 0.737$ and $\lambda = 0.0219\ h^{-1}$; the corresponding theoretical values are plotted as a solid line in Fig. 7.5A. The model parameters inferred this way provide an excellent agreement between the data concerning R/R_0 vs. time (Fig. 7.5B, circles) and the theoretical plot given by Eq. (7.20) (Fig. 7.5B, solid line). If cell spheroid compaction is neglected (i.e., using the theoretical Eqs. (7.17 and 7.7), one obtains the dashed curves from Fig. 7.5A and B, respectively.

Fitting the same data with the theory derived from the assumption of volume relaxation according to Eq. (7.19), (McCune et al., 2014) obtained a fusion time of 48 h. The value of τ derived from our analysis based on Eq. (7.15) (42.4 h) is close to the effective fusion time. $\tau_{eff} = (48\ h)(1 + a_\infty)/2 = 41.7\ h$.

The above procedure of data analysis is useful also in the case of rapidly dividing cells that boost the system's volume during fusion. This is the case for CHO cells (Kosztin et al., 2012) and for MCF-7 human breast cancer cells (Susienka et al., 2016).

Cell population kinetics can be described, in a first approximation, by supposing that all the cells of a spheroid are cycling at the same rate, r_p. Then, the number of daughter cells generated in a short time interval dt, is given by $dN = r_p N\, dt$, where N is the number of cells at the instant of time$t$. This differential equation can be integrated taking into account the initial condition $N(t_0) = N_0$, and the result is an exponential growth of the cell population:

$$N(t) = N_0 e^{r_p (t - t_0)} \tag{7.22}$$

It is convenient to characterize the growth rate in terms of the cell population doubling time T_D, defined as the time interval during which the number of cells increases twofold; that is, $N(t_0 + T_D) = 2\,N_0$. From Eq. (7.22), one obtains $r_p T_D = \ln(2)$. If all the cells would cycle at the same pace, needing the time T_c to complete a full cell cycle, and no cell would die, then$T_D = T_c$. Such a kinetics has been observed experimentally in Ehrlich ascites tumors in mice. For tumors comprising of up to 10^6 cells, the growth was consistent with all cells being proliferative, and the tumor doubling time and mean cell cycle time were roughly equal to 19 h. As the tumors increased in size, the growth curves fell below the exponential and saturated at a maximum size of about 10^9 cells per tumor (Lamerton & Steel, 1968).

Tissue spheroids employed in bioprinting are typically composed of 10^3 - 10^5 cells and do not present a necrotic core (Kosztin et al., 2012; Susienka et al., 2016). Therefore, in the case of spheroids made of rapidly growing cell types, the assumption of exponential growth is reasonable.

For a single spheroid of initial radius R_0, in the absence of compaction, $V/V_0 = N/N_0 = e^{r_p(t-t_0)}$; consequently, the time dependence of the aggregate radius is given by $R(t) = R_0\, e^{r_p(t-t_0)/3}$. If the spheroid also suffers compaction, the number of cells per unit volume $n = N/V$, increases according to $n = n_0(a(t))^{-3}$, where $a(t)$ is given by Eq. (7.19). In this case $V/V_0 = (n_0/n)(N/N_0) = (a(t))^3 e^{r_p(t-t_0)}$, so the aggregate radius evolves according to the formula $R(t) = R_0\, a(t)\, e^{r_p(t-t_0)/3}$.

The volume of a spheroid doublet might change during fusion because of spheroid compaction, cell proliferation, or both. The initial number of cells can be expressed as $N_0 = (2/3)\, n_0 4\pi R_0^3$, whereas the number of cells at a later time t, can be written as $N = (2/3)\, n\, \pi R^3(1+c)^2(2-c)$. (Here $(1/3)\pi R^3(1+c)^2(2-c)$ is the volume of one spherical cap.) Dividing N by N_0 and proceeding as in the previous paragraph, one obtains

$$R/R_0 = 2^{2/3}(1+c)^{-2/3}(2-c)^{-1/3}a(t)e^{r_p(t-t_0)/3} \tag{7.23}$$

A similar generalization is valid also for of the expression of L/L_0 (Eq. 7.21).

To analyze the geometry of fusing CHO aggregate pairs (Kosztin et al., 2012), we relied on the fit of $\sin^2\theta$ vs. time to determine the time constant of fusion, $\tau = 70$ h, and the moment of the initial tack, $t_0 = 0.1$ h. Then, we fitted the experimental data regarding L/L_0 using Eq. (7.21) with $a(t)$ substituted by $e^{r_p(t-t_0)/3}$ (i.e., by taking into account cell division and neglecting cell aggregate compaction). As a result, we obtained the rate of cell proliferation, $r_p = 2.936 \cdot 10^{-3}\ \text{h}^{-1}$, which corresponds to a cell population doubling time of 236 h. Fig. 7.6A represents experimental data as

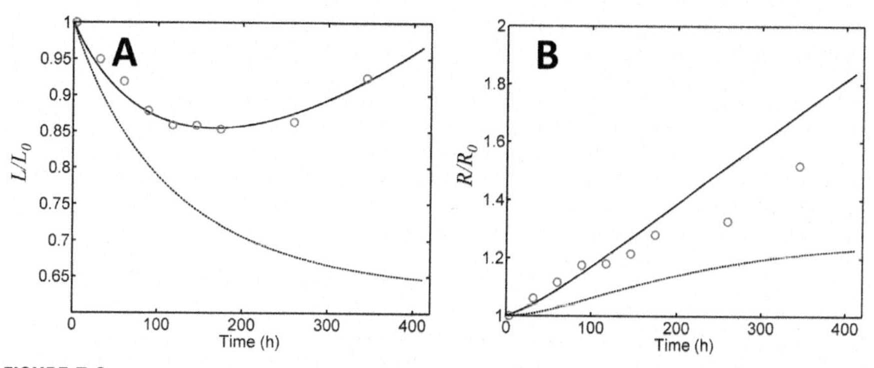

FIGURE 7.6

The time course of the geometry of fusing spheroid doublets whose volume increases due to cell proliferation. Analyzing the experimental data reported by Kosztin et al. (2012), we fitted the normalized doublet length versus time (A) to obtain the cell proliferation rate. The solid line represents the theoretical curve that takes into account cell proliferation, whereas the dashed line results from the volume conserving theory (Eq. 7.17). The normalized doublet width versus time (B) is plotted according to Eq. (7.23) (solid line) and Eq. (7.7) (dashed line).

circular markers, the theoretical fit curve as a solid line, and the volume-conserving theoretical curve as a dashed line.

Finally, in Fig. 7.6B we plotted the normalized doublet width versus time according to Eq. (7.23) with $a(t) = 1$, and other model parameters taken from the previous fits.

The approach outlined here can be employed also in the case of spheroids made of cell populations whose growth is not exponential. To this end, one needs to replace $e^{r_p(t-t_0)/3}$ with a different function$p(t)$, which describes the evolution of the linear size of the system due to cell division and/or cell death. In fact, most cell types deviate from exponential population growth (Lamerton & Steel, 1968). Even tumors display a more complex kinetics because they are heterogeneous, consisting of several cell populations of different genotype and phenotype, known as clones. Most cells of a tumor belong to just a few clones. To shed light on clonal dominance, several models have been devised (Palm et al., 2018). Some of them consider that dominant clones originate from cancer stem cells that are able to self-renew indefinitely. A different model considers that all cells divide at various division rates that are heritable and subject to mutations. Analytic models of the population kinetics of cell types employed in 3D bioprinting will provide valuable insights into postprinting tissue construct maturation. It is important to note, however, that Eq. 7.23 is an empirical extension of the volume-conserving Frenkel model. A more rigorous theory of the coalescence of growing multicellular spheroids has been developed recently (Dechristé et al., 2018).

This section provided simple views on multicellular self-assembly. Most of them are based on exponential functions. Although closer scrutiny often reveals discrepancies between these simple models and experimental data, they enable one to identify the time scales of the corresponding processes, establishing a link between experiments and computer simulations. As a result, one can tell the experimental time that corresponds to the time unit of the simulations.

References

Dechristé, G., Fehrenbach, J., Griseti, E., Lobjois, V., & Poignard, C. (2018). Viscoelastic modeling of the fusion of multicellular tumor spheroids in growth phase. *Journal of Theoretical Biology, 454*, 102–109. https://doi.org/10.1016/j.jtbi.2018.05.005

Eggers, J., & Villermaux, E. (2008). Physics of liquid jets. *Reports on Progress in Physics, 71*, 036601. https://doi.org/10.1088/0034-4885/71/3/036601

Eshelby, J. D. (1949). Discussion in: A. J. Shaler "seminar on the kinetics of sintering. *Metallurgical Transactions, 185*, 806.

Flenner, E., Janosi, L., Barz, B., Neagu, A., Forgacs, G., & Kosztin, I. (2012). Kinetic Monte Carlo and cellular particle dynamics simulations of multicellular systems. *Physical Review, 85*, 031907. http://link.aps.org/doi/10.1103/PhysRevE.85.031907.

Forgacs, G., Foty, R. A., Shafrir, Y., & Steinberg, M. S. (1998). Viscoelastic properties of living embryonic tissues: A quantitative study. *Biophysical Journal, 74*, 2227–2234. https://doi.org/10.1016/S0006-3495(98)77932-9

Foty, R. A., Forgacs, G., Pfleger, C. M., & Steinberg, M. S. (1994). Liquid properties of embryonic-tissues—measurement of interfacial-tensions. *Physical Review Letters, 72*, 2298–2301. https://journals.aps.org/prl/abstract/10.1103/PhysRevLett.72.2298.

Foty, R. A., Pfleger, C. M., Forgacs, G., & Steinberg, M. S. (1996). Surface tensions of embryonic tissues predict their mutual envelopment behavior. *Development, 122*, 1611–1620. https://www.ncbi.nlm.nih.gov/pubmed/8625847.

Frenkel, J. (1945). Viscous flow of crystalline bodies under the action of surface tension. *Journal of Physics, 9*, 385–391.

Fung, Y. C. (2013). *Biomechanics: Mechanical properties of living tissues* (2nd ed.). New York: Springer https://books.google.ro/books?id=yx3aBwAAQBAJ.

Gordon, R., Goel, N. S., Steinberg, M. S., & Wiseman, L. L. (1972). A rheological mechanism sufficient to explain the kinetics of cell sorting. *Journal of Theoretical Biology, 37*, 43–73. https://doi.org/10.1016/0022-5193(72)90114-2

Gumbiner, B. M. (1996). Cell adhesion: The molecular basis of tissue architecture and morphogenesis. *Cell, 84*, 345–357. http://www.sciencedirect.com/science/article/pii/S0092867400812799.

Harvey, E. N., & Shapiro, H. (1941). The recovery period (relaxation) of marine eggs after deformation. *Journal of Cellular and Comparative Physiology, 17*(2), 135–144. https://doi.org/10.1002/jcp.1030170202

Ingber, D. E., & Levin, M. (2007). What lies at the interface of regenerative medicine and developmental biology? *Development, 134*, 2541. http://dev.biologists.org/content/134/14/2541.abstract.

Israelachvili, J. N. (2011). *Intermolecular and surface forces: Revised* (3rd ed.). Elsevier Science https://books.google.ro/books?id=vgyBJbtNOcoC.

Kosztin, I., Vunjak-Novakovic, G., & Forgacs, G. (2012). Colloquium: Modeling the dynamics of multicellular systems: Application to tissue engineering. *Reviews of Modern Physics, 84*, 1791–1805. http://link.aps.org/doi/10.1103/RevModPhys.84.1791.

Lamerton, L. F., & Steel, G. G. (1968). Cell population kinetics in normal and malignant tissues. *Progress in Biophysics and Molecular Biology, 18*, 245–283. https://doi.org/10.1016/0079-6107(68)90026-6

McCune, M., Shafiee, A., Forgacs, G., & Kosztin, I. (2014). Predictive modeling of post bioprinting structure formation. *Soft Matter, 10*, 1790–1800. https://doi.org/10.1039/C3SM52806E

Mgharbel, A., Delanoe-Ayari, H., & Rieu, J.-P. (2009). Measuring accurately liquid and tissue surface tension with a compression plate tensiometer. *HFSP Journal, 3*, 213–221. http://www.tandfonline.com/doi/abs/10.2976/1.3116822.

Mironov, V., Visconti, R. P., Kasyanov, V., Forgacs, G., Drake, C. J., & Markwald, R. R. (2009). Organ printing: Tissue spheroids as building blocks. *Biomaterials, 30*, 2164–2174. http://www.sciencedirect.com/science/article/pii/S0142961209000052.

Moldovan, N. I., Hibino, N., & Nakayama, K. (2017). Principles of the Kenzan method for robotic cell spheroid-based three-dimensional bioprinting. *Tissue Engineering Part B Reviews, 23*, 237–244. https://doi.org/10.1089/ten.teb.2016.0322

Norotte, C., Marga, F., Neagu, A., Kosztin, I., & Forgacs, G. (2008). Experimental evaluation of apparent tissue surface tension based on the exact solution of the Laplace equation. *EPL, 81*, 46003. http://stacks.iop.org/0295-5075/81/i=4/a=46003.

Palm, M. M., Elemans, M., & Beltman, J. B. (2018). Heritable tumor cell division rate heterogeneity induces clonal dominance. *PLoS Computational Biology, 14*(2), e1005954. https://doi.org/10.1371/journal.pcbi.1005954

Pokluda, O., Bellehumeur, C. T., & Vlachopoulos, J. (1997). Modification of Frenkel's model for sintering. *AIChE Journal, 43*, 3253–3256. https://doi.org/10.1002/aic.690431213

Shafiee, A., Norotte, C., & Ghadiri, E. (2017). Cellular bioink surface tension: A tunable biophysical parameter for faster maturation of bioprinted tissue. *Bioprinting, 8*, 13–21. https://doi.org/10.1016/j.bprint.2017.10.001

Steinberg, M. S. (1963). Reconstruction of tissues by dissociated cells. Some morphogenetic tissue movements and the sorting out of embryonic cells may have a common explanation. *Science, 141*, 401–408. http://www.ncbi.nlm.nih.gov/entrez/query.fcgi?cmd=Retrieve&db=PubMed&dopt=Citation&list_uids=13983728.

Steinberg, Malcolm S. (1970). Does differential adhesion govern self-assembly processes in histogenesis? Equilibrium configurations and the emergence of a hierarchy among populations of embryonic cells. *Journal of Experimental Zoology, 173*, 395–433. https://doi.org/10.1002/jez.1401730406

Steinberg, M. S. (1996). Adhesion in development: An historical overview. *Developmental Biology, 180*, 377–388. https://www.sciencedirect.com/science/article/pii/S0012160696903127?via%3Dihub.

Stirbat, T. V., Mgharbel, A., Bodennec, S., Ferri, K., Mertani, H. C., Rieu, J.-P., & Delanoë-Ayari, H. (2013). Fine tuning of tissues' viscosity and surface tension through contractility suggests a new role for α-catenin. *PLoS One, 8*, e52554. https://doi.org/10.1371/journal.pone.0052554

Susienka, M. J., Wilks, B. T., & Morgan, J. R. (2016). Quantifying the kinetics and morphological changes of the fusion of spheroid building blocks. *Biofabrication, 8*, 045003. https://doi.org/10.1088/1758-5090/8/4/045003

Zhu, B., Chappuis-Flament, S., Wong, E., Jensen, I. E., Gumbiner, B. M., & Leckband, D. (2003). Functional analysis of the structural basis of homophilic cadherin adhesion. *Biophysical Journal, 84*, 4033–4042. https://doi.org/10.1016/s0006-3495(03)75129-7

CHAPTER

Postprinting evolution of 3D-bioprinted tissue constructs

8

Bioprinting is widely used in several fields of biomedical research including tissue engineering, regenerative medicine, and drug discovery. This relatively young, still teenager technology has evolved tremendously owing to the interdisciplinary research efforts of biologists, physicians, chemists, physicists, and engineers. The trial and error strategy of early bioprinting research has gradually turned into a quantitative, reproducible one. Along the way, bioprinting has benefitted from progress in materials science, cell biology, and affordable printer manufacturing, as well as from predictive mathematical modeling. As a result, the quality of the final product could be assured, opening the way for the high-throughput production of human tissue constructs. In the context of bioprinting, mathematical modeling sought (i) to optimize printing parameters by taking into account the needs of live cells and the physicochemical properties of the biomaterials delivered by the printer (see Chapter 6) and (ii) to characterize the spontaneous evolution of the bioprinted construct. The latter is the focus of this chapter.

Postprinting rearrangements of cells make bioprinting closely related to 4D printing because in both technologies, the final product emerges as a result of postprinting evolution. Therefore, part of the bioprinting community embraced the idea that bioprinting is actually a 4D process. Other researchers argued that the essential hallmark of 4D printing is the stimulus sensitivity of the 3D-printed construct. Therefore, to be deemed a product of 4D bioprinting, the multicellular structure built by the printer should evolve in a preprogrammed fashion under the action of a predetermined stimulus. The next chapter will outline the present status of the debate related to the definition of 4D bioprinting. This debate, however, did not hamper the exciting developments we witness in this field.

Multicellular self-assembly, presented in detail in Chapter 7, is one source of uncertainty pertaining to the final state of the bioprinted product. On the other hand, it opens up new opportunities for artificial tissue patterning by giving rise to structures whose complexity and feature size lie far beyond the printer's resolution. Once the desired final product is obtained, however, further evolution is undesirable. Computational modeling proved to be useful in this respect, too, by pointing out long-lived, metastable configurations of the tissue construct as a function of its geometry and cellular composition.

Computational models of postprinting structure formation will be presented at a level of technicality needed to understand the primary literature. The text will guide

Towards 4D Bioprinting. https://doi.org/10.1016/B978-0-12-818653-4.00009-7

the reader in selecting the optimal computational model for addressing a given problem under the limitations imposed by available hardware. By reading this chapter, hopefully, the reader interested in writing her/his own simulation programs will be able to decipher the original research articles discussed here.

1. Monte Carlo models of single-cell resolution

Bioprinted tissue constructs are typically made of millions of cells. Hence, their remodeling mainly involves cell movement as opposed to shape changes of individual cells. This assumption lies at the basis of Monte Carlo models of single-cell resolution that aim to describe postprinting evolution of multicellular structures (Jakab et al., 2004; Robu et al., 2012).

For computational efficiency, most models represent the tissue construct on a lattice, associating each lattice site to a cell or to a cell-sized volume element of the extracellular medium—extracellular matrix (ECM), cell culture medium, or a hydrogel soaked thereof. While cells are individual, motile entities of tissues, and, as such, it seems reasonable to represent them as a point-like particle able to move around on a lattice, fragments of the extracellular medium are different. Native ECM is a hydrated network of protein fibers synthesized by certain cell types (mainly fibroblasts). A hydrogel is, likewise, a set of cross-linked polymer filaments (synthetic or natural) bathed by an aqueous solution. A cell culture medium is an aqueous solution of small molecules and proteins able to sustain cell growth when kept at 37 °C in a humidified atmosphere augmented with 5% CO_2. Despite the inherent differences between them, a common feature of these three media is that cells can displace them as they move relative to their neighbors. Fibers of the extracellular matrix, as well as those of certain hydrogels, offer binding motifs for cell–substrate adhesion molecules, collectively called integrins—membrane proteins anchored to the cytoskeleton. Cells exert forces onto the adjacent fibers, move along them, and when their motion is hampered by fibers, they can depolymerize them with the help of a family of enzymes known as matrix metalloproteinases (MMPs). In conclusion, cells are capable of occupying a spot previously occupied by medium. This observation justifies the representation of the extracellular medium as a set of point-like particles bonded to one another, the bond strength being a measure of the ability of cells to displace volume elements of the given medium.

To represent a biological system, different types of entities (cells and/or media) are placed on a cubic or hexagonal lattice, in a geometric arrangement as close as possible to the real system. Lattice site occupancy is given by a number, σ, which takes one of the values 1,2, ..., T. The system's total energy of adhesion is expressed in terms of works of adhesion, $\varepsilon_{\sigma\sigma'}$, defined as the mechanical work needed to break the bond between two neighbors of types σ and σ' (Israelachvili, 2011). Neglecting a physically irrelevant constant term (Neagu et al., 2006), the total energy of adhesion can be rewritten as follows:

$$E = \sum_{\substack{\sigma,\sigma'=1 \\ \sigma<\sigma'}}^{T} \gamma_{\sigma\sigma'} N_{\sigma\sigma'} \tag{8.1}$$

were $N_{\sigma\sigma'}$ stands for the total number of bonds between particles of type σ and σ' (σ different from σ') and $\gamma_{\sigma\sigma'=0.5\cdot(\varepsilon_{\sigma\sigma}+\varepsilon_{\sigma'\sigma'})-\varepsilon_{\sigma\sigma'}}$ is the interfacial tension. The latter is defined as the energy of the unit area of the interface between the medium of type σ and another one, of type σ'. (More precisely, $\gamma_{\sigma\sigma'}$ is the interfacial tension only if the unit area is taken to be the cell membrane area divided by the number of bonds between a cell and its neighbors—see next paragraph. Otherwise, $\gamma_{\sigma\sigma'}$ is the interfacial tension divided by the area that corresponds to a single bond.)

In the lattice representation, to make the model as isotropic as possible, we assume to have similar interactions between nearest, next-nearest, and second-nearest neighbors. That is, we assume that each particle interacts with $n_{\ n} = 26$ neighbors, called the vicinity of that particle (see Fig. 8.1, right side).

Fig. 8.1 illustrates a model tissue using the visual molecular dynamics (VMD) software (Humphrey et al., 1996). In this schematic representation of a cell aggregate, individual cells are depicted as green spheres, whereas volume elements of the embedding medium are represented by silver spheres on the magnified view of a small portion of the aggregate-medium interface; medium particles are not shown in the image of the full aggregate, but they are included in the model.

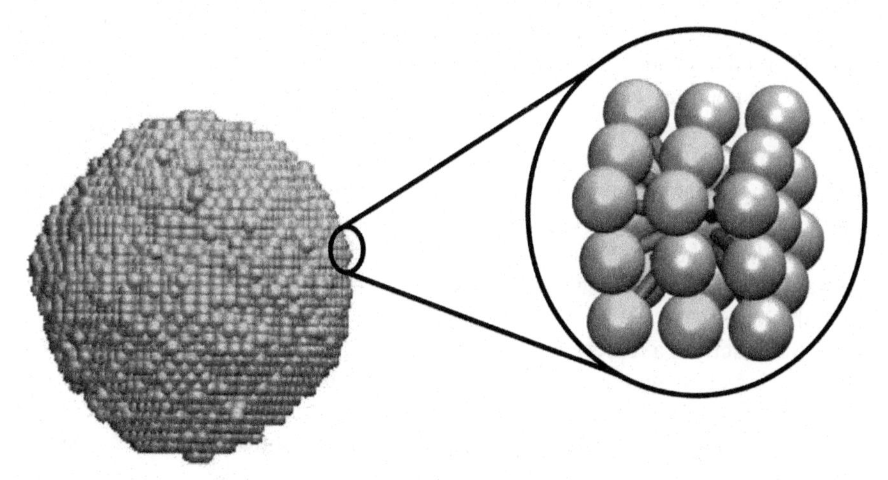

FIGURE 8.1 Lattice Representation of a Model Tissue at Single-cell Resolution.

A cell aggregate is depicted as a compact cluster of cells (green spheres), and the embedding medium is not shown (left). In the detailed view of a small portion of the aggregate surface (right), the medium particles are also displayed alongside the bonds formed between the central cell and its 26 neighbors: cell—cell bonds are visualized as red cylinders connecting green spheres, whereas cell—medium bonds are shown as blue cylinders connecting one green sphere with an adjacent medium particle. For interpretation of the references to color in this figure legend, please refer online version of this title.

From Neagu, A. (2017). Role of computer simulation to predict the outcome of 3D bioprinting. Journal of 3D Printing in Medicine, 1, *103–121. https://doi.org/10.2217/3dp-2016-0008. Copyright 2017 Future Science Group.*

1.1 Metropolis Monte Carlo simulations

Monte Carlo models describe the system's evolution via stochastic algorithms. For this purpose, one popular option is the Metropolis Monte Carlo (MMC) algorithm (Amar, 2006; Metropolis et al., 1953) modified to limit trial moves to biologically reasonable ones (Neagu et al., 2005, 2006; Robu & Stoicu-Tivadar, 2016).

Simulated cell movement is a succession of elementary moves that consist of swapping two neighbors of different types. If a swap causes no increase in the adhesive energy, it is readily accepted; otherwise, it might be accepted, albeit with a probability of less than one, $P = \exp\left(-\frac{\Delta E}{E_T}\right)$. Here, E_T is a quantitative measure of cell motility; it is the biological analog of the energy of thermal fluctuations from statistical physics, $k_B T$, where k_B is Boltzmann's constant and T is the absolute temperature (Beysens et al., 2000). A convenient option is to express works of adhesion and interfacial tensions in units of E_T.

The MMC algorithm is in accord with Steinberg's differential adhesion hypothesis because it has the tendency to minimize the system's total energy of adhesion.

An MMC simulation consists of a number of Monte Carlo steps (MCS). One MCS is defined as the set of operations that provide all interfacial cells with the chance to move once, in random order.

MMC simulations demonstrated that the evolution of a bioprinted construct results from the interplay of cell—cell cohesion and cell-matrix adhesion (Jakab et al., 2004). Simulations were in agreement with a variety of experiments. For example, when spherical aggregates were deposited in collagen gel contiguously, in a precise circular arrangement, tissue fusion gave rise to a multicellular torus, which shrunk at a rate modulated by collagen concentration in the embedding gel. As suggested by the model, the higher the collagen concentration (that is, the stronger the cell—gel interaction, and, therefore, the lower the cell—gel interfacial tension), the faster shrunk the torus.

Fig. 8.2 describes a representative simulation of the shape changes suffered by a bioprinted ring of 16 aggregates of 123 cells each. Conformations given by the postprinting rearrangements of the constituent cells are shown along with the plot of the energy of cohesion versus the elapsed MCS. These plots demonstrate that the MMC algorithm brings about a steady decrease in energy.

Long-lived, metastable conformations, such as a torus, correspond to a plateau of the plot of energy versus MCS (Fig. 8.2A). Nevertheless, instabilities can lead to the rupture of the ring. The resulting circular arc breaks down repeatedly, giving rise to three spheroids (Fig. 8.2B); these survived an additional ½ million MCS, while the energy remained roughly constant (results not shown), indicating that a new metastable state was reached (Jakab et al., 2004).

Steinberg's theory (Steinberg, 2007) states that any simply connected multicellular construct evolves toward a spherical shape because the sphere is the geometrical object of the smallest area for a given volume (see Chapter 7). How can one fabricate tissues of different geometry? MMC simulations indicate that, as they

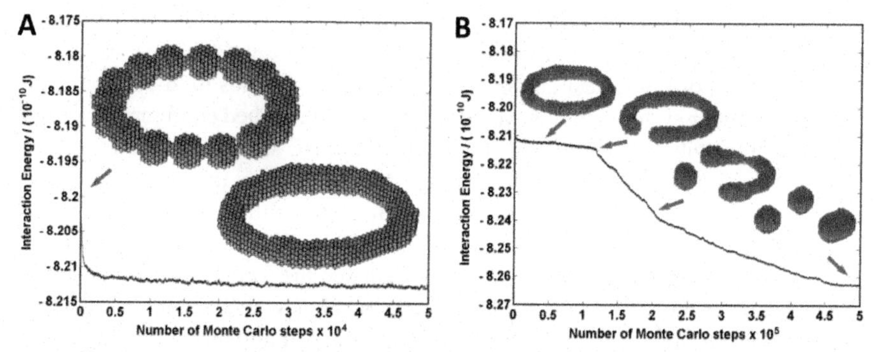

FIGURE 8.2 Postprinting Shape Changes of a Bioprinted Ring and the Associated Change in the Total Energy of Adhesion.

(A) A simulation of 50,000 MCS of cellular rearrangements in a chain of cell aggregates placed contiguously along a circle shows that the energy of adhesion decreases fast as the adjacent aggregates fuse. Once the torus is formed, in about 5000 MCS, the energy remains almost constant. (B) A longer simulation, of half a million MCS, demonstrates that the ring might eventually break down, opening up new opportunities for ample relocations of the constituent cells, which are associated with a higher rate of energy drop.

From Jakab, K., Neagu, A., Mironov, V., Markwald, R. R., & Forgacs, G. (2004). Engineering biological structures of prescribed shape using self-assembling multicellular systems. Proceedings of the National Academy of Sciences of the United States of America, 101, *2864–2869. Copyright (2004) National Academy of Sciences.*

evolve toward equilibrium, multicellular systems can be trapped in long-lived, metastable configurations. The simulations reproduced the experimentally observed shape changes of toroidal, tubular, and sheet-like structures assembled from multicellular spheroids delivered by a 3D bioprinter (Neagu et al., 2005).

Experiments conducted on various cell types demonstrated that, in about a week, the bioprinted tissue constructs become sturdy enough to be transferred into bioreactors that offer biomimetic conditions. Ideally, bioreactors promote extracellular matrix synthesis by facilitating mass transfer and assuring mechanical conditioning (such as a pulsatile flow in the case of tubular constructs akin to blood vessels) (Norotte et al., 2009).

The fusion of adjacent aggregates not only created the desired shape, but assured conditions for intercellular interaction and communication that replicated the in vivo conditions and, thereby, enabled the maturation of the tissue construct, making it functional. For example, a bioprinted multicellular patch of cardiac myocytes emerged within 70 h of cell spheroid fusion; one day later, the patch started to beat in synchrony (Jakab et al., 2008). Hence, if bioprinting is well planned, postprinting evolution gives rise to optimal shape and function.

Often computer simulations paved the way to successful experiments. For example, experiments aimed at fabricating branched tubular structures from multicellular spheroids were performed in parallel with computer simulations. The fusion of adjacent aggregates and the sorting out of the constituent cell populations were well-known from previous experiments and simulations. One concern remained, however, that bifurcations might undergo an unfavorable remodeling as cells seek the configuration of the lowest energy of adhesion. The viscous fluid-like behavior of the multicellular system might have destabilized the branched configuration, sealing off one of the tubes. To investigate this problem, MMC simulations were run on a model system–a wide tube with a side branch (Fig. 8.3A), and the results were surprising: even in the absence of a window between the main tube and the side branch in the printed structure (Fig. 8.3B), a suitable hierarchy of interfacial tensions has led to a spontaneous piercing of the wall at the branch and the formation of an endothelial cell layer on the surface of the tube wall. In other words, multicellular self-assembly opened up the entry of the side branch and maintained it open over a long simulation run. These findings suggest that a vascular tree is a remarkably stable construct. Bioprinting experiments indicated the same: tubes assembled from aggregates of human skin fibroblasts (Fig. 8.3C) turned into a sturdy branching tube within about a week of static culture (Fig. 8.3D) (Norotte et al., 2009).

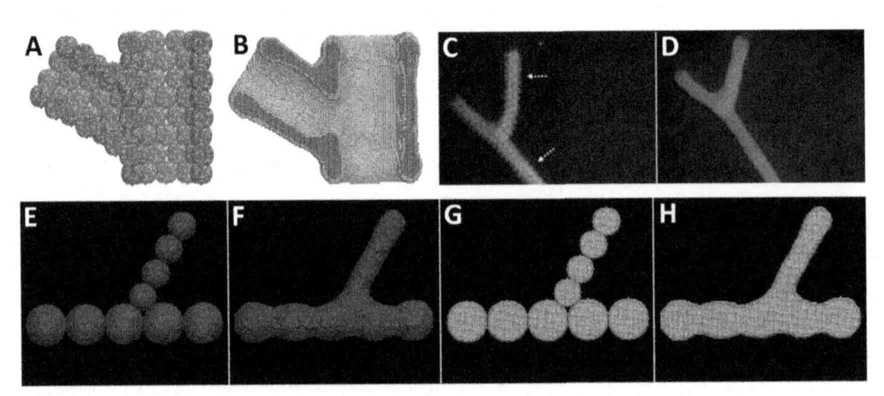

FIGURE 8.3 Representative Metropolis Monte Carlo (MMC) Simulations.

(A) The initial state of a computational model representing a branching tube made of multicellular spheroids composed of a random mixture of two types of cells—smooth muscle cells (red) and endothelial cells (pink). (B) The final state of the MMC simulation. (C) A bioprinted branching tube made of aggregates of human skin fibroblasts. (D) The configuration of the branching tube after 6 days in culture (Norotte et al., 2009). (E) A computational model of a branching tube made of composite aggregates—a gel core wrapped in endothelial cells. (F) The result of an MMC simulation for a biologically relevant set of model parameters. (G), (H) The axial cross-sections of the initial and final construct, respectively. For interpretation of the references to color in this figure legend, please refer online version of this title.

Panels A, B, E-H, reproduced with permission from Neagu, A. (2017). Role of computer simulation to predict the outcome of 3D bioprinting. Journal of 3D Printing in Medicine, 1, *103–121. https://doi.org/10.2217/3dp-2016-0008. Copyright (2017) Future Science Group. Panels C, D, adapted with permission from Norotte, C., Marga, F. S., Niklason, L. E., & Forgacs, G. (2009). Scaffold-free vascular tissue engineering using bioprinting.* Biomaterials, 30, *5910–5917. http://www.sciencedirect.com/science/article/pii/S0142961209006401 Copyright (2009), with permission from Elsevier.*

The synergy between in silico investigations and experimental efforts was apparent in a study aimed at building arteriole-like tubes. The wall thickness of such a tube is of the order of tens of microns, smaller than the resolution of an extrusion bioprinter. To test a working hypothesis, the experimental team suggested simulating the self-assembly of contiguously placed composite spheroids made of a gel droplet covered by cells. The bottom row of pictures from Fig. 8.3 represents such a construct in perspective (panel E) and in axial cross-section (panel G). MMC simulations indicated that a lumen emerges spontaneously for a certain hierarchy of interfacial tensions between cells, hydrogel, and medium (panels F and H). This prediction was confirmed experimentally in the case of uniluminal vascular spheroids (Fleming et al., 2010), demonstrating a practical way to build branched blood vessels of less than 1 mm in diameter. Chapter 10 will show alternative solutions to this problem, based on 4D biofabrication.

In numerous examples (Robu et al., 2012), the simulated evolution recapitulated the experimentally observed sequence of events, suggesting that the number of elapsed Monte Carlo steps would be proportional to the duration of the simulated process. It is important to note, however, that the MMC algorithm does not describe time evolution because it does not rely on a dynamical hierarchy of transition probabilities. Indeed, all the transitions that cause no increase in energy are accepted with the same probability, $P = 1$.

Another drawback of the MMC algorithm is that for a system close to equilibrium, or in a metastable state, most trial moves are rejected because acceptance probabilities are low.

The next section presents a rejection-free computational algorithm that also involves the time variable. Initially, called the n-fold way algorithm (Bortz et al., 1975), it became widely known as the Kinetic Monte Carlo (KMC) algorithm because it describes the system's evolution in terms of transition rates (Amar, 2006).

1.2 Kinetic Monte Carlo simulations

In a KMC simulation, rates are assigned to all the transitions the system is capable of. It is assumed that the transition rates are time-independent, and they only depend on the initial state and final state.

When the KMC algorithm is used to describe multicellular self-assembly in the framework of a lattice model of single-cell resolution, transition rates are associated with all possible swaps of the constituent cells with adjacent entities of different types (cells, ECM, or cell culture medium) (Flenner et al., 2012). The formula of such a transition rate is given by (Amar, 2006):

$$r = w_0 \exp\left(-\frac{E_b}{E_T}\right) \tag{8.2}$$

where w_0 is the frequency of the attempts to cross the energy barrier of height E_b between the initial state and the final state is the biological counterpart of the energy of thermal fluctuations from statistical physics (Beysens et al., 2000).

The formula of the transition rates is made specific by the detailed expression of the energy barrier, E_b. To describe the rearrangements of cells in accord with the differential adhesion hypothesis (Steinberg, 2007), Sun and Wang proposed the formula $E_b = \Delta E_{\text{int}}$, the change in the interfacial energy that would be caused by the transition (Sun & Wang, 2013).

A KMC simulation is a succession of steps. One step consists of the following set of operations: (1) identify all the transitions the system is capable of (N is their number); (2) compute their rates, $r_1, r_2, \ldots, r_N$, the partial sums of rates, $R_n = \sum_{i=1}^{n} r_i$ for $n = 1, 2, \ldots, N-1$, and the gross sum of rates, $R = R_N = \sum_{i=1}^{N} r_i$; (3) draw a random number, u, uniformly distributed between 0 and 1; (4) execute transition m that satisfies the inequality $R_{m-1} < uR < R_m$ (in the example of Fig. 8.4, transition four is selected to be performed next); (5) draw another uniformly distributed random number, u', situated between 0 and 1 and increment the time variable by $\Delta t = -R^{-1} \ln(u')$ (Amar, 2006).

To make the simulation computationally effective, at the beginning of the next step one needs to spot the transitions that might have been affected by the previous step. This amounts to identifying all possible transitions within the neighborhood of the pair of particles swapped right before, and computing their rates; the other rates can be retrieved from the set computed during the previous step.

A simulation runs until the time variable reaches a predefined value.

According to the works by Sun et al. KMC simulations reproduced the sequence of events observed in MMC simulations and in the corresponding validation experiments on bioprinted rings and patches (Jakab et al., 2004; Neagu et al., 2005). Importantly, the simulated events could be followed in time, making it possible to use KMC simulations to predict the duration of future experiments. Also, the mixing of cells from different spheroids was less intense in KMC simulations than in their MMC counterparts (Sun et al., 2014; Sun & Wang, 2013). In this respect, the KMC algorithm provides better agreement with experimental findings (Jakab et al., 2008) than the MMC algorithm.

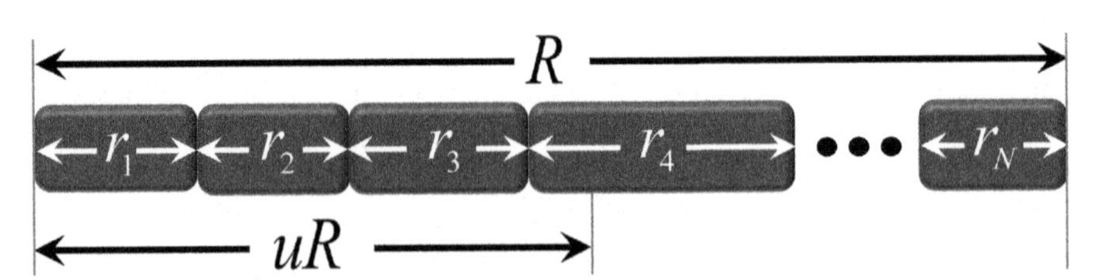

FIGURE 8.4 Selection of the Next Transition in a KMC Simulation.

Schematic representation of the rate selection process. The larger is the rate of a given transition, the higher is the probability to carry it out.

2. Particle dynamics models

As their name suggests, particle dynamics models rely on Newton's equations of classical mechanics to predict the time evolution of the building blocks of a multicellular system. They have the advantage of being off-lattice models able to describe cellular shape changes and movement in a friction-dominated milieu. They can also mimic traction forces actively exerted by cells on adjacent structures. Moreover, particle dynamics models can take into account cell growth and proliferation. Versatility is one of the most important strengths of particle dynamics models. Their most notable drawback is the computational load posed by simulations based on such models. This section will present examples of particle dynamics models and their applications, with special focus laid on biological phenomena involved in the spontaneous evolution of bioprinted tissue constructs.

The first particle dynamics model proposed for simulating multicellular rearrangements is the so-called subcellular element model (ScEM) (Newman, 2005). The ScEM sub-divides the cell into a number of subcellular elements and describes their dynamics in terms of potential energies of pairwise interactions. Both intracellular and intercellular interactions are described by a generalized Morse potential, $U(r) = U_1 \exp(-r/\xi_1) - U_2 \exp(-r/\xi_2)$.

The integrity of a biological cell is maintained by the cytoskeleton, a highly dynamic web of interconnected protein filaments—actin filaments, microtubules, and intermediate filaments (Alberts, 2008). Cell shape and motility are mainly governed by the self-assembly of actin filaments. Viewing the cell as being made of mobile subcellular elements amounts to modeling the rearrangements of volume elements of the cytoskeleton. The potential energy of intracellular interactions describes the forces transmitted along the filaments that connect neighboring elements.

Interactions between cells are mediated by two classes of membrane proteins: molecules that mediate cell–cell adhesion (e.g., cadherins) and molecules that connect cells with the extracellular matrix (integrins) (Alberts, 2008). Both intracellular and intercellular interactions are local; therefore, the potential energies employed to describe these interactions should rapidly decay with distance. This requirement can be met by a variety of potentials that involve two energy scales (to describe the repulsion that keeps elements apart and the short-range attraction that keeps them together) and two length scales (to characterize the typical separation of a pair of elements and the range of the attractive interaction that acts as their separation increases) (Newman, 2005).

Just like the ScEM model, the cellular particle dynamics (CPD) model views cells as being made of interacting point particles—cellular particles (CPs) (Kosztin et al., 2012). An aggregate of such model cells is represented on the left side of Fig. 8.5, whereas the right side shows a pair of cells made of CPs held together by intracellular and intercellular interactions. Fig. 8.5 was generated using VMD (Humphrey et al., 1996).

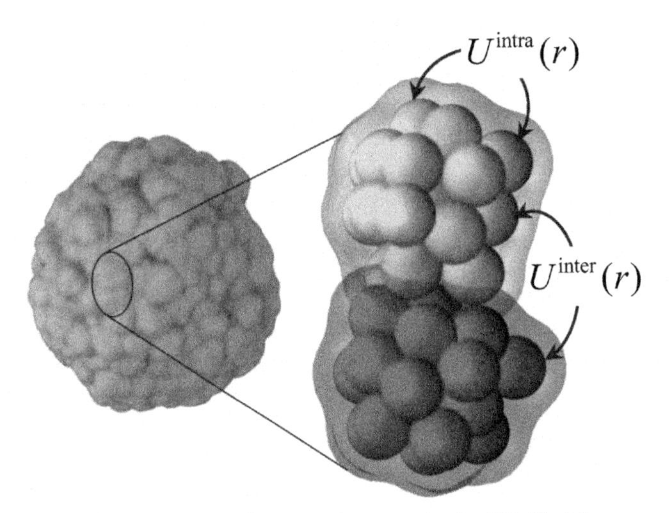

FIGURE 8.5 Representation of a Multicellular Structure in the CPD Model.

A cell aggregate is shown on the left and a magnified view of a pair of cells is represented on the right; in the latter, distinct colors are used merely to distinguish CPs from different cells. U^{intra} is the potential energy of interaction of a pair of CPs of the same cell, whereas U^{inter} is the potential energy of interaction of two CPs that belong to different cells.

Reprinted with permission from Kosztin, I., Vunjak-Novakovic, G., & Forgacs, G. (2012). Colloquium: Modeling the dynamics of multicellular systems: Application to tissue engineering. Reviews of Modern Physics, 84, *1791–1805. https://doi.org/10.1103/RevModPhys.84.1791 Copyright (2012) by the American Physical Society.*

In the CPD model, pairwise interactions between CPs of the same cell are described by the intracellular potential energy (Fig. 8.5, top right), composed of two terms (Kosztin et al., 2012):

$$U^{intra}(r) = U_{LJ}\left(r;\ \varepsilon^{intra},\ \sigma^{intra}\right) + \frac{k}{2}(r-\xi)^2\theta(r-\xi) \tag{8.3}$$

where r is the separation of the pair of interacting CPs that belong to the same cell. This intracellular potential energy accounts for the shape and mechanical properties of individual cells. The second term of $U^{\text{intra}}(r)$ is a harmonic confining potential, expressed in terms of the elastic constant, k, the characteristic length, ξ, which specifies the cell size. In this formula, θ denotes the Heaviside step function: $\theta(x) = 0$ if $x < 0$, $\theta(x) = 1/2$ if $x = 0$, and $\theta(x) = 1$ if $x > 0$. The confining potential assures that a cell that moves and suffers shape changes under the action of its neighborhood does not lose its CPs along the way.

In Eq. (8.3), U_{LJ} denotes the Lennard-Jones potential,

$$U_{LJ}(r;\ \varepsilon,\ \sigma) = 4\varepsilon\left[\left(\frac{\sigma}{r}\right)^{12} - \left(\frac{\sigma}{r}\right)^{6}\right] \tag{8.4}$$

written as a function of the distance between the interacting particles, r, the distance σ at which the Lennard-Jones potential becomes zero, and the binding energy, ε. The latter is the depth of the potential well; i.e., it is a positive number equal to the mechanical work needed to separate the pair of particles whose interaction is described by Eq. (8.4).

When $r < \sigma$, the Lennard-Jones potential is positive with a markedly negative slope, corresponding to a strong force of repulsion. Hence, to a good approximation, CPs can be imagined as rigid spheres of diameter σ^{intra}, denoted hereafter by σ (for simplicity).

Consider an isolated cell made of N_{CP} cellular particles. The CPD model describes cell adhesion in terms of a short-range potential energy of pairwise interaction between cellular particles that belong to different cells, given by:

$$U^{\,inter}(r) = \begin{cases} U_{LJ}\left(r+\delta;\ \varepsilon^{inter},\ \sigma^{inter}\right)\ if\ r < r_c \\ U_{LJ}\left(r_c+\delta;\ \varepsilon^{inter},\ \sigma^{inter}\right)\ if\ r \geq r_c \end{cases} \tag{8.5}$$

where r_c is a cutoff distance. Beyond this distance, the force of intercellular interaction vanishes. In other words, 2 cells interact only via those CPs whose separation is smaller than r_c. Typical values of this cutoff distance are about, 1.5σ, assuring that the adhesive interaction between contiguous cells only involves CPs located at their surface (Kosztin et al., 2012).

The equilibrium separation between CPs from different cells can be rendered the same as in the case of CPs of the same cell (Kosztin et al., 2012). To this end, the Lennard-Jones potential given by Eq. (8.4) is shifted by a distance, $\delta = \sqrt[6]{2}\left(\sigma^{inter} - \sigma^{intra}\right)$ in the expression of the potential energy of intercellular interactions (Eq. 8.5). Thus, both $U^{intra}(r)$ and $U^{intra}(r)$ have the minimum at the same separation, $r_0 = 2^{1/6}\sigma$. As the separation becomes larger than r_0, the repulsion between CPs turns into a mild attraction.

In the CPD model, the time evolution of a biological system made of N_C cells is simulated by solving the overdamped Langevin equations of motion:

$$\mu\dot{\mathbf{r}}_{\alpha_n} = -\sum_{m=1}^{N_{CP}} m \neq n \nabla_{\alpha_n} U^{\,intra}\left(\left|\mathbf{r}_{\alpha_n} - \mathbf{r}_{\alpha_m}\right|\right) - \sum_{\beta=1}^{N_C} \beta \neq \alpha \sum_{m=1}^{N_{CP}} \nabla_{\alpha_n} U^{\,inter}\left(\left|\mathbf{r}_{\alpha_n} - \mathbf{r}_{\beta_m}\right|\right) + \boldsymbol{f}_{\alpha_n}(t), \tag{8.6}$$

where μ is the friction coefficient. In Eq. (8.6), dot denotes the time derivative, Greek indices label cells ($\alpha, \beta = 1, 2, \ldots, N_C$), and latin indices specify CPs of a given cell ($m, n = 1, 2, \ldots, N_{CP}$).

The last term on the right side of Eq. (8.6) is a random force that mimics the impact of fluctuations taking place in the biological system in question—thermal fluctuations as well as nonequilibrium fluctuations caused by the continuous remodeling of the cytoskeleton (Alberts, 2008). This random force is a Gaussian white noise with zero mean and variance given by

$$\langle f_{\alpha_n,i}(t)\, f_{\beta_m,j}(0)\rangle = 2D\mu^2\delta(t)\delta_{\alpha\beta}\delta_{nm}\delta_{ij} \tag{8.7}$$

where i and j refer to components of the force vector f_{α_n} ($i,j = 1,2,3$ denote the x, y, and z components, respectively), δ_{ij} is the Kronecker delta function ($\delta_{ij} = 1$ if $i = j$ and $\delta_{ij} = 0$ otherwise), and is the Dirac delta function ($\delta(t) \to \infty$ if $t = 0$ and $\delta(t) = 0$ otherwise). In Eq. (8.7), D stands for the self-diffusion coefficient, and μ is the friction coefficient; according to the fluctuation-dissipation theorem, their product is equal to the energy of thermal fluctuations, ($\mu D = E_T$). In the context of a multicellular system, E_T is referred to by the term "biological fluctuation energy" and is associated with biochemical processes responsible for the dynamics of the cytoskeleton (Beysens et al., 2000).

To express model parameters, it is useful to introduce units of measurement: the unit of length is σ, the unit of time is σ^2/D, and the unit of energy is E_T. The other model parameters are dimensionless numbers expressed in these units. Biologically relevant model parameters have been identified by validating CPD simulations against experiments (Kosztin et al., 2012; McCune et al., 2014; Shafiee et al., 2014, 2015), leading to the following values: $D = 1$, $\mu = 1$, $\sigma^{intra} = 1$, $\xi = 10^{1/3}$ for a model with 10 CPs per cell, $\varepsilon^{intra} = 1.48$, $\sigma^{inter} = 5$, $\delta = 2^{1/6}\left(\sigma^{inter} - 1\right) = 4.5$, and $\varepsilon^{inter} = 40$.

Using the Langevin dynamics integrator from the Large-scale Atomic/Molecular Massively Parallel Simulator (LAMMPS) software package (a freeware), CPD simulations have been performed for systems made of thousands of cells (Flenner et al., 2012; McCune et al., 2014; Shafiee et al., 2014, 2015). Typical central processing unit (CPU) times for these simulations were of the order of days to weeks on clusters of hundreds of processors.

Although quite compute-intensive, CPD simulations are faster and less expensive than the corresponding laboratory experiments. For example, running on a cluster of 24 CPUs, a simulation of the fusion of a pair of multicellular cylinders (of 1535 cells each, 10 CPs per cell) evolved about 3000 CPD time units per day, an equivalent of 66 h of experimental time. Hence, such an in silico experiment is almost 3 times faster than a real, wet lab experiment. Since computational resources become cheaper every day, it will become more and more advantageous to rely on predictive computer simulations to test experimental strategies.

3. Phase field models

When a cluster of cells resides in a cell culture medium, or in a hydrogel soaked thereby, the system can be regarded as being composed of two distinct phases. Both may be treated as incompressible fluids provided that the phenomena of interest occur on the timescale of days—as usual in tissue engineering.

If we zoom into the interface between the cohesive cell population and the embedding medium, the discrete nature of the multicellular system becomes apparent and the reshaping of the interface is accomplished by cell movement. While such a microscopic view is appealing because it gives the impression of being able to dive into the biological system, it comes with prohibitive computational costs when the system at hand is realistic, comprising millions, or even billions of cells. In this section, we are going to show that in such cases it is worth zooming out and viewing the cell population as a continuum.

Phase field theory represents a binary mixture in terms of a phase variable, ϕ.

The evolution of the phase field is characterized by the Cahn-Hilliard equation (Cahn, 1959; Cahn & Hilliard, 1958, 1959):

$$\frac{\partial \phi}{\partial t} + \nabla \cdot (\boldsymbol{u}\phi) = \nabla \cdot (\lambda \nabla \mu), \tag{8.8}$$

where $\boldsymbol{u}$ is the average velocity of the fluid system, λ is the mobility, and μ is the chemical potential. The mobility may be a constant model parameter or a function of ϕ.

Let us consider the specific example of a biological system made of a homotypic multicellular construct embedded in a hydrogel. For such a system, the total free energy can be given by Yang et al. (2012):

$$F = \frac{k_B T}{2}\left[\int_\Omega \left(\gamma_1 |\nabla\phi|^2 + \gamma_2 \phi^2 (1-\phi)^2\right) d^3\boldsymbol{r} + \int_\Omega \left(\int_\Omega \phi(\boldsymbol{r},t) U_{LJ} (|\boldsymbol{r}-\boldsymbol{r}'|; \varepsilon, \sigma)\phi(\boldsymbol{r}',t) d^3\boldsymbol{r}'\right) d^3\boldsymbol{r}\right] \tag{8.9}$$

where k_B is Boltzmann's constant, T is the absolute temperature, whereas γ_1, γ_2, ε, and σ are model parameters.

The interfacial tension, defined as the interfacial energy per unit area, is proportional to $\sqrt{\gamma_1 \gamma_2}$ (Chen, 2002).

Although it does not attempt to account for the motion of individual cells, phase field modeling requires considerable CPU time. The Cahn-Hilliard equation is often coupled with the equations of continuum hydrodynamics, and the corresponding system of partial differential equations is solved numerically. When the finite difference method is applied for this purpose, the lattice spacing needs to be fine enough to avoid numerical instabilities—it should be smaller than 1/4 of the interface thickness (Qin & Bhadeshia, 2010).

4. Lattice Boltzmann models

While working on the kinetic theory of gases, in 1872, Ludwig Boltzmann devised a transport equation to describe heat flow in a dilute gas in the presence of a temperature gradient (Huang, 1987).

Instead of tracking each and every constituent particle (gas molecule), Boltzmann's theory focused on the evolution of the single particle distribution function, $f(\boldsymbol{r},\boldsymbol{v},t)$, defined in such a way that $dN = f(\boldsymbol{r}, \boldsymbol{v}, t)\, d^3\boldsymbol{r}\, d^3\boldsymbol{v}$ is the number of particles which, at time t, are located within the volume element $d^3\boldsymbol{r}$ about the position vector $\boldsymbol{r}$ and whose velocities lie within the velocity-space element $d^3\boldsymbol{v}$ about $\boldsymbol{v}$. The volume elements from this definition are not strictly infinitesimal; instead, they are large enough to contain a huge number of particles and small enough to be considered point-like in comparison with macroscopic dimensions (Huang, 1987).

For a system made of a single type of particles of mass m moving under the action of an external force field, $\boldsymbol{F}(\boldsymbol{r}, t)$, the single particle distribution function evolves according to the Boltzmann equation

$$\left[\frac{\partial}{\partial t}+\boldsymbol{v}\cdot\nabla+\frac{\boldsymbol{F}(\boldsymbol{r},t)}{m}\cdot\nabla_{\boldsymbol{v}}\right]f(\boldsymbol{r},\ \boldsymbol{v},\ t)=\left(\frac{\partial f}{\partial t}\right)_{collisions} \tag{8.10}$$

where the right-hand side is the collision term.

In the Bhatnagar-Gross-Krook (BGK) approximation, the collision term is linearized; i.e., it is written as $-(f-f^{eq})/\tau$, where τ is the relaxation time and $f^{eq}(\boldsymbol{r},\ \boldsymbol{v},\ t)$ is the equilibrium distribution function (He & Luo, 1997).

A lattice Boltzmann (LB) model is formulated by the discretization of the position vectors, $\boldsymbol{r}$, and velocities, $\boldsymbol{v}$, of the particles that make up the system. In other words, the distribution functions are defined only in the nodes of a lattice from a D-dimensional space ($D=1$, 2, or 3), and particles can have only a finite number of velocity vectors, $\boldsymbol{e}_i$, ($i=0, 1, \ldots, Q-1$). Thus, the distribution function is replaced by a set of Q distribution functions, $f_i=f(\boldsymbol{r},\ \boldsymbol{e}_i,\ t)$, and the Boltzmann equation, Eq. (8.10), is replaced by a system of Q equations, which, in the BGK approximation, read:

$$\frac{\partial f_i}{\partial t}+\boldsymbol{e}_i\cdot\nabla f_i-\frac{\boldsymbol{F}(\boldsymbol{r},t)}{m\chi c^2}[\boldsymbol{e}_i-\boldsymbol{u}(\boldsymbol{r},t)]f_i^{eq}=-(f_i-f_i^{eq})/\tau \tag{8.11}$$

In Eq. (8.11), $\boldsymbol{u}(\boldsymbol{r}, t)$ is the barycentric velocity of the system, and f_i^{eq} are the equilibrium distribution functions, given by

$$f_i^{eq}=w_i n\left[1+\frac{(\boldsymbol{e}_i\boldsymbol{u})}{\chi c^2}+\frac{(\boldsymbol{e}_i\boldsymbol{u})^2}{2\chi^2c^4}-\frac{(\boldsymbol{u})^2}{2\chi c^2}\right] \tag{8.12}$$

where w_i are weight coefficients, χ is a dimensionless characteristic constant of the model, n is the particle number density, and c is the thermal velocity of the particles ($c=\sqrt{(k_BT/\chi m)}$, where k_B is Boltzmann's constant and T is the absolute temperature) (He & Luo, 1997).

Two widely used LB models are the D2Q9 and D3Q27 models, formulated on square grids; in their acronyms, the first number refers to spatial dimensions, whereas the second number specifies the number of discrete velocities considered in the model. Both of these models share the same characteristic constant, $\chi=1/3$, but they differ in their weight coefficients (He & Luo, 1997). Fig. 8.6 depicts the velocity vectors available for the constituent particles in the D2Q9 model (panel A) and in the D3Q27 model (panel B).

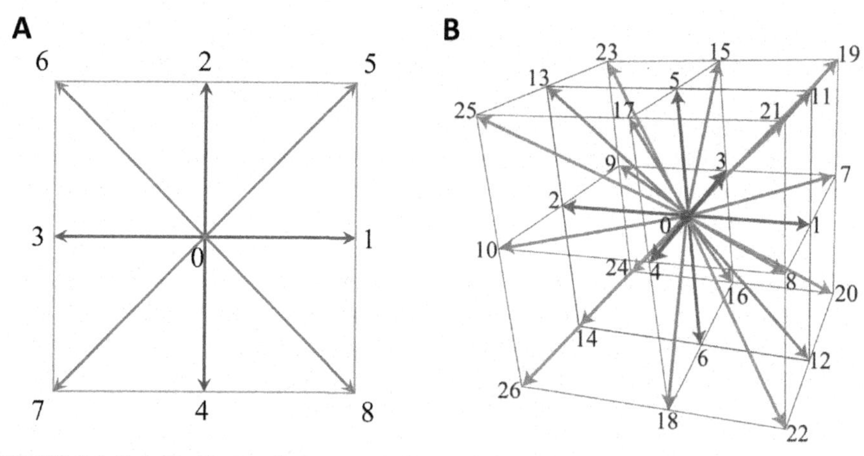

FIGURE 8.6 Velocity Vectors in Commonly Used LB Models.

(A) The set of discrete velocity vectors included in the D2Q9 model built on a two-dimensional square lattice. (B) The set of velocities considered in the D3Q27 model formulated on a three-dimensional cubic lattice.

Panel B is reproduced with permission from Suga, K., Kuwata, Y., Takashima, K., & Chikasue, R. (2015). A D3Q27 multiple-relaxation-time lattice Boltzmann method for turbulent flows. Computers and Mathematics with Applications, 69, *518–529. https://doi.org/10.1016/j.camwa.2015.01.010.*

In three spatial dimensions, LB models have been formulated on a cubic grid by only including the velocity vectors shown in red and blue in Fig. 8.6B (D3Q15), or those shown in red and green (D3Q19), but the errors were found minimal in the most complex, D3Q27 model (Suga et al., 2015).

When the system of interest is composed of several types of particles, a separate distribution function, $f^{\sigma}(\boldsymbol{r}, \boldsymbol{v}, t)$, is associated to each type, labeled by the index σ. For instance, to simulate the evolution of a homotypic cell population embedded in a biocompatible medium, such as a hydrogel soaked with cell culture medium, an LB model has been proposed to describe the system as a binary fluid made of two types of particles—cells ($\sigma = 1$) and volume elements of the embedding medium ($\sigma = 0$).

The rounding of irregular clusters of cells, as well as the segregation of cells randomly dispersed in the medium, giving rise to a multicellular spheroid, can be described by viewing the system as being made of two immiscible viscous fluids. While the fluid-like behavior of a multicellular construct on the timescale of days is well documented (see Chapter 7), the embedding hydrogel is not a fluid, but a viscoelastic medium. Despite this discrepancy between their mechanical properties, the cell–gel interface moves just like the surface of separation between two incompressible viscous fluids. More precisely, this statement is true only for hydrogels that can be remodeled by the cells with the help of MMPs (Rodríguez et al., 2010). As the cell cluster's shape evolves, the interfacial cells remodel the adjacent matrix, which otherwise would hamper tissue reshaping. In the binary fluid model, the viscosity of the fluid that represents the extracellular medium is a surrogate quantity that characterizes the ability of the cells to remodel the matrix. The viscosities of the two fluids are tuned in the LB model via the relaxation times of the two species—τ^0 for the hydrogel and τ^1 for the cell cluster, with the order of magnitude of 10^{-3} (Cristea et al., 2011; Cristea & Neagu, 2016).

In the LB model of a tissue construct placed in a medium (Cristea et al., 2011), the immiscibility of the two fluids has been described by considering that a particle of species σ is acted upon by the force $\boldsymbol{F}^{\sigma} = -\sum_{\sigma'} \omega^{\sigma\sigma'} \nabla X^{\sigma'}$, where $X^{\sigma}(\boldsymbol{r}, t)$ denotes the mole fraction of particle species σ, and $\omega^{\sigma\sigma'}$ are the interaction parameters, written as a function of a single model parameter, ω, as follows: $\omega^{01} = \omega^{10} = \omega$, and $\omega^{00} = \omega^{11} = 0$. To also account for the interfacial tension, a second term, $\kappa\nabla\left(\nabla^2 X^{\sigma}\right)$, needs to be included in the expression of the force (Cristea & Neagu, 2016).

The LB equations are solved numerically, using parallel computing. Due to the vast computational resources needed to tackle this problem, computational implementation is an essential feature of LB simulations. The Portable, Extensible Toolkit for Scientific Computation (PETSc) was developed at the Argonne National Laboratory, Argonne, IL, USA (https://www.anl.gov/mcs/petsc-portable-extensible-toolkit-for-scientific-computation) and has been used for decades to write LB simulation codes. PETSc is a set of data structures and routines for solving partial differential equations via parallel computing, which employs the Message Passing Interface (MPI) standard. PETSc codes distribute tasks to a cluster of CPUs. Although PETSc is still a valuable option, graphical processing units (GPUs) are increasingly popular in high-performance scientific computing. In 2007, NVIDIA released a general-purpose parallel computing platform and application programming interface, called Compute Unified Device Architecture (CUDA). Using CUDA, one can employ NVIDIA GPUs for general-purpose parallel computing. The CUDA Toolkit is composed of GPU-accelerated libraries, a compiler, development tools, and the CUDA runtime. It enables developers to program in C, C++, Fortran, Python, or MATLAB, and turn their program into a parallel computing code by using CUDA extensions. Since LB models are compute-intensive, but easy to parallelize, massively parallel GPU computing is widely used for LB simulations (Calore et al., 2016).

The D2Q9 model provided a good agreement with experimental findings on tubular tissue construct fabrication via 3D bioprinting of multicellular cylinders (Cristea & Neagu, 2016). The 2D model faithfully described the time course of the transversal cross-section of the construct. Indeed, the initial configuration, made of six multicellular cylinders placed in contact with one another in a hexagonal arrangement (Fig. 8.7A, panel 0), evolved due to the spontaneous coalescence of the cylinders. Eventually, the construct turned into a tube with a smooth wall composed of cohesive cells, as observed in experiments (Norotte et al., 2009). In the laboratory, the complete fusion took 3 days, whereas in the LB simulation it took 50,000 time steps (Fig. 8.7A, panel 5), suggesting that one simulation time step corresponds to an

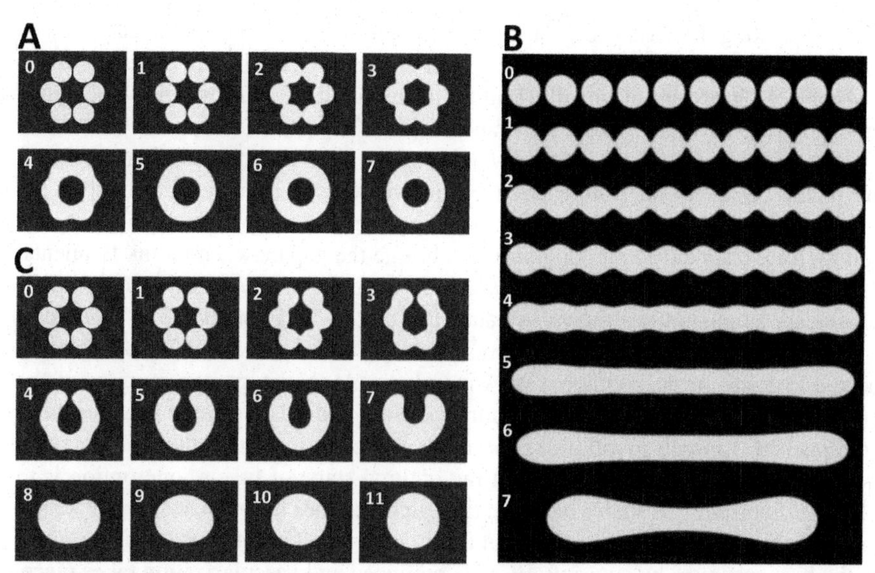

FIGURE 8.7 LB Simulations of the Evolution of Tissue Constructs Obtained by Multicellular Cylinder Bioprinting.

Snapshots of the simulated structures are numbered consecutively, with 0 corresponding to the initial configuration (the cross-section of the bioprinted construct in a plane oriented normally to the axes of the cylinders). (A) Configurations 1 to 7 of a tubular construct correspond to 5×10^3, 10^4, 1.5×10^4, 2.5×10^4, 5×10^4, 7.5×10^4, and 10^5 time steps, respectively. (B) Snapshots 1 to 7 of a bioprinted patch of cohesive cells obtained in 5×10^4, 6×10^4, 7×10^4, 10^5, 1.5×10^5, 2×10^5, and 5×10^5 time steps, respectively. (C) The evolution of an imprecisely printed tubular tissue construct: configurations 1 to 11 were obtained in 5×10^3, 10^4, 1.5×10^4, 2.5×10^4, 5×10^4, 7.5×10^4, 10^5, 1.5×10^5, 2×10^5, 2.5×10^5, and 5×10^5 time steps, respectively.

From Cristea, A., & Neagu, A. (2016). Shape changes of bioprinted tissue constructs simulated by the Lattice Boltzmann method. Computers in Biology and Medicine, 70, *80–87. https://doi.org/10.1016/j.compbiomed.2015.12.020.*

experimental time of $3 \times 24 \times 3600/50{,}000 = 5.184$ s. Hence, upon time step calibration, the LB modeling also enables one to estimate the duration of postprinting shape changes of tissue constructs obtained under similar conditions (using the same bioink and the same embedding medium).

The experimental validation of the LB model paved the way for in silico investigations of the expected outcome of various 3D printing processes. Among them, cell sheet fabrication was found feasible by the sidewise fusion of contiguous cylinders printed on a nonadherent surface, such as a layer of agarose gel (Fig. 8.7B). Also, the simulations suggested that a square grid of cylinders printed in an agarose groove would evolve into a multicellular block outfitted with a set of metastable longitudinal channels (not shown) (Cristea & Neagu, 2016). Moreover, the simulations could be used to test the impact of imprecise printing on the final product. In cell tube bioprinting, for instance, a misplaced cylinder is likely to miss the early stages of fusion (Fig. 8.7C, panel 2), and the resulting longitudinal slit will undergo an unfavorable progression, eventually leading to one multicellular cylinder 2.45 times larger in diameter than the bioink strand (Fig. 8.7C, panels 3–11).

According to the LB model, fusion takes place also when the adjacent cylinders are placed relatively far from each other, but in a precise arrangement. Despite the separation of about 40% of the cylinder radius, the random fluctuations involved in cell motility enable the cell populations to bridge the gap between bioink filaments, resulting in a uniform sheet. The large separation, however, slows down tissue fusion: the configuration shown in panel 1 of Fig. 8.7B is predicted to emerge in 3 days, whereas the relatively uniform sheet depicted in panel 5 of Fig. 8.7B is expected to result in six additional days in culture. At this stage, the margins of the patch are slightly thicker than the central portion, and this tendency is expected to continue; by 1 month in culture, it becomes twice as thick near the margins as in the center. To arrest this evolution at the optimal stage of fusion, one option is to plan the bioprinting in such a way that the patch assumes the right shape and constitution at the moment when it is needed for tissue repair. Once the patch is implanted in the host organism by suturing, ECM is produced and vascularization takes place, hampering its liquid-like behavior.

The set of LB simulation snapshots shown in Fig. 8.7C suggests that a bioprinted tube whose wall is incomplete might be repaired by depositing a cell cylinder over the longitudinal slit. The optimal time window for such an attempt lies between days 4 and 6, when the width of the slit is slightly smaller than the diameter of a bioink strand (Fig. 8.7C, panels 5 and 6). If the repair is successful, computational modeling can predict the size of the emerging tube; if the repair fails, it is advisable to discard the construct to avoid further experimental costs.

5. Conclusions and perspectives

A 3D bioprinter delivers live cells and supportive biomaterials in a precise, numerically controlled procedure that takes into account the needs of cells and the rheological properties of the materials. Regardless of the bioprinter's resolution, the outcome of the process remains unknown because the constituent cells will relocate relying on mechanisms known from developmental biology. Understanding postprinting evolution is essential for going beyond the trial-and-error approach in 3D bioprinting. Given the complexity of morphogenesis, computational modeling seems to be the most appropriate way for investigating this problem.

This chapter presented several computational models of postprinting structure formation. Although none of them was suitable for simulating the precise time evolution of a typical tissue construct made of millions or even billions of cells, most of them were able to predict the spontaneous reshaping of bioprinted multicellular structures of 10^3 to 10^5 cells.

The models proposed so far have one feature in common: they rely on the differential adhesion hypothesis to simulate cellular rearrangements within bioprinted tissue constructs. Otherwise, they differ in underlying assumptions, resolution, and computational load. Some of them are suitable to simulate time evolution, even in what concerns individual cell movement. Moreover, particle dynamics models also describe the mechanical properties of the simulated constructs. Further model developments will be needed to implement additional mechanisms involved in morphogenesis, such as chemical and mechanical signaling.

Model calibration has been devised relying on the theoretical analysis of dedicated experiments, such as the fusion of tissue spheroid doublets (see Chapter 7). Thereby, computational models gained predictive power, providing valuable hints for experimentalists. Despite the considerable CPU times needed for simulations, they still take less time than the corresponding laboratory experiments, and they are less expensive. Thus, computational models can be used to explore the outcomes of different bioprinting strategies in a fraction of the time and money needed for laboratory work.

Most importantly, computer simulations facilitate product quality assurance by enabling the user to optimize the digital model of the desired construct.

Provided that a trigger mechanism is discovered, the accurate prediction of bioprinted construct remodeling adds the fourth dimension to 3D bioprinting. Medical imaging, such as computed tomography or magnetic resonance imaging, creates a 3D digital model of the desired tissue construct (the target model, T). Computer simulations can be used to build a digital model of the structure that needs to be delivered by the printer (the model to be printed, P) in order to obtain the desired outcome as a result of postprinting shape changes. Finally, the user designs the path of the print head and sets the printing parameters to assure that the printer indeed creates a 3D object, O, whose geometry is identical to that of model P. Three-dimensional imaging can be employed to visualize object O and create its digital model, DO. Then, 3D image processing can quantify the overlap between DO and P. If they match (up to a certain error margin related to printing and subsequent imaging), the postprinting shape evolution, once triggered, will take place as programmed (predicted by the computer simulations). Quality control can be performed again by scanning the construct to obtain its digital model, DC, and comparing it with model T. Hence, the 3D-printed tissue construct, O, would be capable of preprogrammed reshaping under the action of a chemical or physical trigger. In other words, one could say that such a tissue construct was actually fabricated by 4D bioprinting. All we need is the trigger. Future investigations might shed light on methods for halting tissue fusion and triggering it on demand. One option is to control the properties of the embedding hydrogel to make it more or less permissive for cell movement, but biochemical methods might also be effective for slowing down or accelerating tissue fusion.

The questions that come to our mind naturally are "Do we need 4D bioprinting? Is it merely a fancy technology? What are its practical advantages over 3D bioprinting?". These questions are addressed in the remaining chapters of this book.

References

Alberts, B. (2008). *Molecular biology of the cell: Reference edition.* Garland Science. https://books.google.ro/books?id=iepqmRfP3ZoC.

Amar, J. G. (2006). The Monte Carlo method in science and engineering. *Computing in Science & Engineering, 8*, 9–19.

Beysens, D. A., Forgacs, G., & Glazier, J. A. (2000). Cell sorting is analogous to phase ordering in fluids. *Proceedings of the National Academy of Sciences, 97*, 9467–9471. https://doi.org/10.1073/pnas.97.17.9467

Bortz, A. B., Kalos, M. H., & Lebowitz, J. L. (1975). A new algorithm for Monte Carlo simulation of Ising spin systems. *Journal of Computational Physics, 17*, 10–18. https://doi.org/10.1016/0021-9991(75)90060-1

Cahn, J. W. (1959). Free energy of a nonuniform system. II. Thermodynamic basis. *The Journal of Chemical Physics, 30*, 1121–1124. https://doi.org/10.1063/1.1730145

Cahn, J. W., & Hilliard, J. E. (1958). Free energy of a nonuniform system. I. Interfacial free energy. *The Journal of Chemical Physics, 28*, 258–267. https://doi.org/10.1063/1.1744102

Cahn, J. W., & Hilliard, J. E. (1959). Free energy of a nonuniform system. III. Nucleation in a two-component incompressible fluid. *The Journal of Chemical Physics, 31*, 688–699. https://doi.org/10.1063/1.1730447

Calore, E., Gabbana, A., Kraus, J., Pellegrini, E., Schifano, S. F., & Tripiccione, R. (2016). Massively parallel lattice–Boltzmann codes on large GPU clusters. *Parallel Computing, 58*, 1–24. https://doi.org/10.1016/j.parco.2016.08.005

Chen, L.-Q. (2002). Phase-field models for microstructure evolution. *Annual Review of Materials Research, 32*, 113–140. https://doi.org/10.1146/annurev.matsci.32.112001.132041

Cristea, A., & Neagu, A. (2016). Shape changes of bioprinted tissue constructs simulated by the Lattice Boltzmann method. *Computers in Biology and Medicine, 70*, 80–87. https://doi.org/10.1016/j.compbiomed.2015.12.020

Cristea, A., Neagu, A., & Sofonea, V. (2011). Lattice Boltzmann simulations of the time evolution of living multicellular systems. *Biorheology, 48*, 185–197. https://doi.org/10.3233/BIR-2011-0595

Fleming, P. A., Argraves, W. S., Gentile, C., Neagu, A., Forgacs, G., & Drake, C. J. (2010). Fusion of uniluminal vascular spheroids: A model for assembly of blood vessels. *Developmental Dynamics, 239*, 398–406. https://doi.org/10.1002/dvdy.22161

Flenner, E., Janosi, L., Barz, B., Neagu, A., Forgacs, G., & Kosztin, I. (2012). Kinetic Monte Carlo and cellular particle dynamics simulations of multicellular systems. *Physical Review E, 85*, 031907. http://link.aps.org/doi/10.1103/PhysRevE.85.031907.

He, X., & Luo, L.-S. (1997). Theory of the lattice Boltzmann method: From the Boltzmann equation to the lattice Boltzmann equation. *Physical Review E, 56*, 6811–6817. https://doi.org/10.1103/PhysRevE.56.6811

Huang, K. (1987). *Statistical mechanics.* Wiley. https://books.google.ro/books?id=M8PvAAAAMAAJ.

Humphrey, W., Dalke, A., & Schulten, K. (1996). Vmd: Visual molecular dynamics. *Journal of Molecular Graphics, 14*, 33–38.

Israelachvili, J. N. (2011). *Intermolecular and surface forces: Revised* (3rd ed.). Elsevier Science https://books.google.ro/books?id=vgyBJbtNOcoC.

Jakab, K., Neagu, A., Mironov, V., Markwald, R. R., & Forgacs, G. (2004). Engineering biological structures of prescribed shape using self-assembling multicellular systems. *Proceedings of the National Academy of Sciences of the United States of America, 101*, 2864–2869.

Jakab, K., Norotte, C., Damon, B., Marga, F., Neagu, A., Besch-Williford, C. L., Kachurin, A., Church, K. H., Park, H., Mironov, V., Markwald, R., Vunjak-Novakovic, G., & Forgacs, G. (2008). Tissue engineering by self-assembly of cells printed into topologically defined structures. *Tissue Engineering Part A, 14*, 413–421. https://doi.org/10.1089/tea.2007.0173

Kosztin, I., Vunjak-Novakovic, G., & Forgacs, G. (2012). Colloquium: Modeling the dynamics of multicellular systems: Application to tissue engineering. *Reviews of Modern Physics, 84*, 1791–1805. https://doi.org/10.1103/RevModPhys.84.1791

McCune, M., Shafiee, A., Forgacs, G., & Kosztin, I. (2014). Predictive modeling of post bioprinting structure formation. *Soft Matter, 10*, 1790–1800. https://doi.org/10.1039/C3SM52806E

Metropolis, N., Rosenbluth, A. W., Rosenbluth, M. N., Teller, A. H., & Teller, E. (1953). Equation of state calculations by fast computing machines. *The Journal of Chemical Physics, 21*, 1087–1092. https://doi.org/10.1063/1.1699114

Neagu, A., Jakab, K., Jamison, R., & Forgacs, G. (2005). Role of physical mechanisms in biological self-organization. *Physical Review Letters, 95*, 178104.

Neagu, A., Kosztin, I., Jakab, K., Barz, B., Neagu, M., Jamison, R., & Forgacs, G. (2006). Computational modeling of tissue self-assembly. *Modern Physics Letters B, 20*, 1217–1231.

Newman, T. J. (2005). Modeling multicellular systems using subcellular elements. *Mathematical Biosciences and Engineering, 2*, 613–624.

Norotte, C., Marga, F. S., Niklason, L. E., & Forgacs, G. (2009). Scaffold-free vascular tissue engineering using bioprinting. *Biomaterials, 30*, 5910–5917. http://www.sciencedirect.com/science/article/pii/S0142961209006401.

Qin, R. S., & Bhadeshia, H. K. D. H. (2010). Phase field method. *Materials Science and Technology, 26*, 803–811. http://www.msm.cam.ac.uk/phase-trans/2010/PHreview.html.

Robu, A., Aldea, R., Munteanu, O., Neagu, M., Stoicu-Tivadar, L., & Neagu, A. (2012). Computer simulations of in vitro morphogenesis. *Biosystems, 109*, 430–443. http://www.sciencedirect.com/science/article/pii/S0303264712001049.

Robu, A., & Stoicu-Tivadar, L. (2016). Simmmc – an informatic application for modeling and simulating the evolution of multicellular systems in the vicinity of biomaterials. *Romanian Journal of Biophysics, 26*, 145–162. http://www.rjb.ro/articles/447/art01%20pag%20I.pdf.

Rodríguez, D., Morrison, C. J., & Overall, C. M. (2010). Matrix metalloproteinases: What do they not do? New substrates and biological roles identified by murine models and proteomics. *Matrix Metalloproteinases, 1803*(1), 39–54. https://doi.org/10.1016/j.bbamcr.2009.09.015

Shafiee, A., McCune, M., Forgacs, G., & Kosztin, I. (2015). Post-deposition bioink self-assembly: A quantitative study. *Biofabrication, 7*, 045005. http://stacks.iop.org/1758-5090/7/i=4/a=045005.

Shafiee, A., McCune, M., Kosztin, I., & Forgacs, G. (2014). Shape evolution of multicellular systems; application to tissue engineering. *Biophysical Journal, 106*, 618a. https://doi.org/10.1016/j.bpj.2013.11.3418

Steinberg, M. S. (2007). Differential adhesion in morphogenesis: A modern view. *Current Opinion in Genetics & Development, 17*, 281–286. http://www.sciencedirect.com/science/article/pii/S0959437X07001062.

Suga, K., Kuwata, Y., Takashima, K., & Chikasue, R. (2015). A D3Q27 multiple-relaxation-time lattice Boltzmann method for turbulent flows. *Computers & Mathematics with Applications, 69*, 518–529. https://doi.org/10.1016/j.camwa.2015.01.010

Sun, Y., & Wang, Q. (2013). Modeling and simulations of multicellular aggregate self-assembly in biofabrication using kinetic Monte Carlo methods. *Soft Matter, 9*, 2172–2186. https://doi.org/10.1039/C2SM27090K

Sun, Y., Yang, X., & Wang, Q. (2014). In-silico analysis on biofabricating vascular networks using kinetic Monte Carlo simulations. *Biofabrication, 6*, 015008. http://stacks.iop.org/1758-5090/6/i=1/a=015008.

Yang, X., Mironov, V., & Wang, Q. (2012). Modeling fusion of cellular aggregates in biofabrication using phase field theories. *Journal of Theoretical Biology, 303*, 110–118. https://doi.org/10.1016/j.jtbi.2012.03.003

CHAPTER

The definition of 4D bioprinting

9

Why dedicate a chapter to a definition? Although definitions are essential in any scientific discipline, most of them are neither engaging nor inspiring. Moreover, they are usually formulated in a single sentence. What is different here? If 4D printing is in its infancy, 4D bioprinting is a neonate: its principles and terms are under development, and its definition is still debated in the literature. This chapter aims at reporting the present status of this debate, hoping to engage the readers to explore the frontiers of this exciting field of biomedical research. Specific examples will be given to clarify the distinction between 3D and 4D bioprinting, thereby demonstrating that 4D bioprinting encompasses a diverse group of emerging technologies, which bear the promise to solve long-standing problems of tissue engineering and regenerative medicine.

1. Working definitions of 4D bioprinting

The formal definition of a term first assigns it to a class and then explains how it is different from other terms in the class. Since 4D printing is well defined (see Chapter 2), it is appealing to assign 4D bioprinting to the class "4D printing" and provide enough information to distinguish it from other forms of 4D printing. A simple definition along these lines could be the following: "4D bioprinting is the 4D printing of structures that contain live cells."

While the above definition is formally correct, it is unsatisfactory because it is not specific enough. For example, it does not state that the created structure should be made of biocompatible materials, ensuring cell viability during and after printing. Also, this definition lacks information concerning the stimulus-responsiveness of the printed structure. Does it rely on the inanimate fraction of the system (smart biomaterials) or also on the constituent cells? Including both seems to be a wise choice, thereby accounting for a wide variety of future developments.

What is most disturbing, this simple definition assumes that the printed structures are static (i.e., that they do not change unless stimulated—as they are supposed to be in 4D printing). The previous chapters of this book have demonstrated that a system comprising a population of viable cells is anything but static. How do we interpret the postprinting self-assembly of cells? Besides their rearrangement, cells in culture can differentiate and/or secrete extracellular matrix (ECM) components and signaling molecules. Provided that all these changes are known in advance, do we

Towards 4D Bioprinting. https://doi.org/10.1016/B978-0-12-818653-4.00011-5

consider them as hallmarks of 4D bioprinting (i.e., part of the anticipated, preprogrammed response)?

In the early days, the maturation of printed microtissues via cellular coating, multicellular self-organization, and ECM deposition was considered a signature of 4D bioprinting (Gao et al., 2016). This view, however, proved to be too inclusive, making it difficult to draw a limit between 3D bioprinting and 4D bioprinting. Indeed, from this perspective, a 3D-printed tissue construct that naturally evolves in a well-known, desirable way would be considered as being 4D printed, although no specific stimulus was applied to elicit the evolution. Then, the term "3D bioprinting" would merely refer to the additive manufacturing of tissue constructs whose evolution is yet unpredictable and, therefore, impossible to reproduce.

An et al. proposed a definition that encompassed the experimental approaches known at that time (An et al., 2016):

> *"4D bioprinting refers to groups of programmable self-assembly, self-folding or self-accommodating technologies which include three main defining or essential components: (i) man-made and not nature-made programmable design, (ii) 2D or 3D bioprinting process, and (iii) post-printing programmable evolving of bioprinted constructs which could be driven by cells or biomaterials and triggered by external signals."*

The above definition was inspired by the 4D bioprinting methodologies proposed until the publication of the article by An et al. (2016). In one of them, a cell population is printed on a smart substrate that folds under the action of the stimulus. The multicellular system follows the substrate, and, eventually, the desired shape (a tube) emerges by the spontaneous self-assembly of the cells. In this approach, the as-printed construct is artificially structured, and the substrate is programmed to curl upon stimulation (due to material properties and microarchitecture). Both the substrate and the cellular mass are delivered by a multimaterial 3D printer. In the absence of stimulation, the construct is relatively stable as a result of cell–substrate and cell–cell interactions. At least this is true on the time scale of minutes to hours. Nevertheless, it evolves within days because, for the construct to function (to fuse into a tube and detach from the substrate), the cohesion forces that keep the cells together need to be stronger than the adhesion forces that anchor the cells to the substrate. It is clear from this example that, in certain cases, cellular rearrangements work against the intended response of the 4D-printed construct. In about a week, the patch of cells would shrink and develop round edges (see Chapter 8), making it difficult to accomplish the preprogrammed task.

In the second approach, a biodegradable, 3D-printed medical device is implanted. In vivo, the device gradually breaks down and changes in size and shape as the underlying tissue grows. Here, the stimulus consists of the construct's deployment, and the anticipated response is driven by the concomitant action of chemical and physical factors: biodegradation and tissue growth. One might legitimately argue that this is actually an example of 4D printing (as opposed to bioprinting) because the additive manufacturing stage only involves biomaterials; cells attach to the construct upon implantation and bring about the preprogrammed response. Nevertheless, during its intended function, the construct is cellularized and responds as planned. It satisfies the three requirements stated in the definition of An et al.: (i) the construct has an artificial design, (ii) it is created by 3D printing, and (iii) when stimulated by environmental cues provided by the host organism, it evolves as planned.

In analogy with the application presented in the previous paragraph, a 3D-printed biodegradable scaffold could also be considered a 4D-bioprinted object. Although it does not violate the text of the definition, such an interpretation deviates from its spirit: the response of a 4D-(bio)printed object is expected to be predictable, reproducible, and purposeful. The biodegradation of a tissue engineering scaffold and the associated shape change are useful because they leave room for the growing tissue. Nevertheless, such a process is not strictly reproducible; the overall rate of degradation is known but its details are not. Two identical scaffolds implanted in the same host will not evolve in precisely the same way. Therefore, including 3D-printed scaffolds under the 4D printing umbrella is highly controversial (Pedro Morouço et al., 2020).

The third method developed at the debut of 4D printing is based on the computer-controlled deposition of cell-laden microdroplets in a specific pattern. Then, the cells are stimulated to undergo self-assembly, giving rise to a preenvisioned new pattern (An et al., 2016). The design of the construct is man-made (i), it is created by 2D printing (ii), and it undergoes an anticipated shape change as a result of stimulation (iii).

The definition of An et al. provides a generous framework for future developments, mainly because it refers to a family of technologies based on different principles.

Another important feature of this definition is the requirement of purposeful, man-made design in the spirit of programmable matter research. The geometry and material properties of the constituent voxels are akin to a computer code, assuring the intended response via autonomous conformational changes. The material programming can assure a variety of stepwise responses initiated by environmental changes of different types and/or intensities. Moreover, a reversible response could be programmed for certain applications, such as the deployment and removal of medical devices.

Printability is also an essential distinctive feature of 4D-bioprinted constructs. This feature is borrowed from 4D printing, which delimits itself within the larger research domain of programmable matter by strictly referring to objects that can be fabricated by additive manufacturing. Although such systems are usually fabricated via 3D printing, some 2D-printed objects might also assume a preprogrammed 3D conformation.

According to the definition of An et al. an object is considered 4D-bioprinted only if its evolution is not a spontaneous, natural process. Instead, it should be accomplished by forces exerted by biomaterials or cells but only as a response to a specific change in the environment (a stimulus). Hence, postprinting structure formation in bioprinted multicellular systems (such as the fusion of tissue spheroids into a construct of predefined shape) is not a hallmark of 4D bioprinting, unless the system can be maintained static, as delivered by the printer, and the fusion can be triggered by a stimulus (An et al., 2016). To this end, further research will be needed to understand the molecular details of tissue fusion and devise instruments to control it noninvasively. But is it worth the effort? The next chapter will argue that it is, indeed.

From the perspective of the current definition of 4D bioprinting, the essential feature that differentiates 3D bioprinting from 4D bioprinting is the ability of the printed structure to change over time in a predefined way if and only if an external stimulus is applied. In this respect, the use of a bioreactor to stimulate the maturation of a 3D-bioprinted construct might be considered a form of 4D bioprinting, provided that the evolution of the construct is known beforehand (An et al., 2016). Anticipated stimulus-responsiveness of a 3D-bioprinted object renders it 4D bioprinted.

The definition of An et al. has withstood the test of time: it accommodates the vast majority of 4D bioprinting approaches developed so far (Kirillova et al., 2017; Li et al., 2017; Morouço et al., 2017; Pedro Morouço et al., 2020). Most of them rely on transformations driven by smart materials. That is, the stimulus-responsiveness of the bioprinted object is assured by the physicochemical properties of its inanimate components.

In their inspirational review article, Li et al. analyzed a vast set of smart materials and envisioned 4D bioprinting as the additive manufacturing of shape-morphing tissue constructs from cell-laden, stimulus-responsive bioinks (Li et al., 2017). In a more recent work published by the same research group, this concept is illustrated as shown in Fig. 9.1 (Ashammakhi et al., 2018). Note that, in this illustration, cells are involved in all bioprinting technologies. This research group defined *4D bioprinting* as

> *"3D printing of cell-laden materials in which the printed structures would be able to respond to external stimulus due to stimuli-responsive bioinks or internal cell forces" (Ashammakhi et al., 2018).*

Their analysis suggests that future research will be needed to tailor the properties of existing smart biomaterials to turn them into bioinks. To this end, they should become biocompatible, have appropriate rheological parameters to ensure printability, possess physical or chemical mechanisms to assure the stability of the bioprinted structures, and, most importantly, display shape-changing capacity without affecting cell viability. That is, the cells need to survive the stimulus, as well as the ensuing rearrangement of the bioprinted construct. It is clearly hard but not impossible to satisfy all these requirements, as demonstrated by the work (Kirillova et al., 2017). These authors added methacrylate groups to alginate and hyaluronic acid to synthesize photo-crosslinkable hydrogels. They loaded them with commercially available mouse bone marrow stromal cells and created small-diameter tubes by 4D bioprinting. Cell viability was about 95% at the end of the entire biofabrication process and did not drop during 1 week in culture (Kirillova et al., 2017).

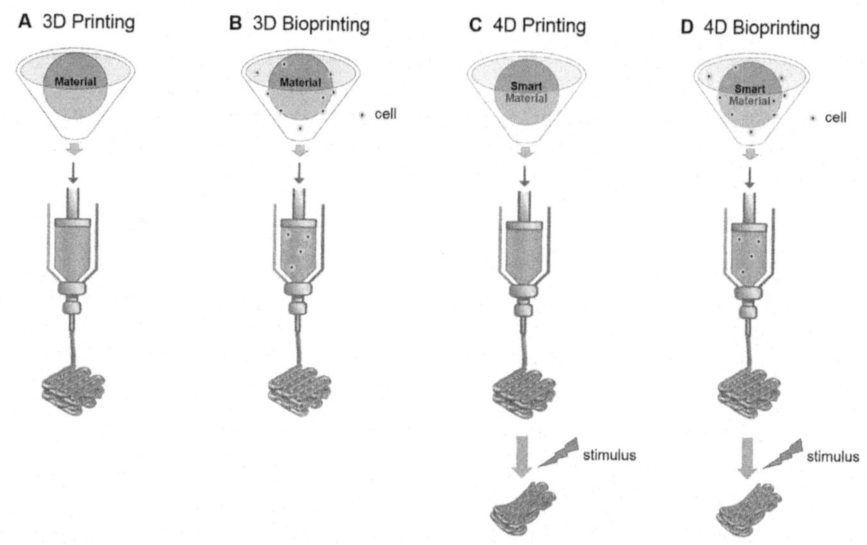

FIGURE 9.1 Comparative illustrations of 3D/4D printing and 3D/4D bioprinting

Schematic drawings of (A) 3D printing, (B) 3D bioprinting, (C) 4D printing, and (D) 4D bioprinting.

Reproduced with permission from Ashammakhi, N., Ahadian, S., Zengjie, F., Suthiwanich, K., Lorestani, F., Orive, G., Ostrovidov, S. & Khademhosseini, A. (2018). Advances and future perspectives in 4D bioprinting. Biotechnology Journal, *13, e1800148. https://doi.org/10.1002/biot.201800148. Copyright 2018 John Wiley & Sons, Inc.*

2. Potential refinements

Future developments in the additive manufacturing of stimulus-sensitive biomedical products might outgrow the current definition of 4D bioprinting. Extensions might become necessary to explicitly consider various biological responses that do not involve a change in tissue shape, size, or microarchitecture (Whitford, 2016). Since a stimulus-dependent shift in function is one option in 4D printing, it seems justified to consider such a possibility also in 4D bioprinting.

The spectrum of known cellular responses is wide, including changes in gene expression, signaling molecule synthesis, cell polarization, or multipotent cell differentiation.

Stimuli able to elicit such responses include physicochemical stress, the spatial pattern of nutrient and oxygen concentration, the vicinity of a basement membrane, immunological signals, differentiation signals, and epigenetic modulators.

Certain changes in cell physiology do not bring about tissue remodeling, but they might impact tissue function, such as contractility or secretory behavior. Nevertheless, nonstructural, biochemical responses often establish a change at the tissue level. For example, the in situ differentiation of a cell population within a tissue construct can lead to cell polarization and subsequent morphological changes.

Focusing merely on the shift in the organizational state of the tissue results in an oversimplified view of the process, which might miss the true nature of the underlying phenomena (Whitford, 2016).

Hence, well-understood and predictable nonstructural changes can be engineered into 4D-bioprinted constructs. To explicitly accommodate this option, Whitford proposed the following definition for 4D bioprinting (Whitford, 2016):

> *"The construction of individual multi-material printed objects from either living or preliminary substrates for medical or biotechnological applications when they are specifically engineered to respond in anticipated ways to user-demands or a changing environment by self-actuated morphogenesis, cellular differentiation, tissue patterning, defined biological characteristic alteration or functionality development."*

Although not stated in the definition of 4D bioprinting by An et al., according to the take-home message of their paper, a change in functionality is considered an acceptable response to the postprinting stimulation (An et al., 2016).

Certain active players in the field use the term *4D bioprinting* to refer to the *4D printing of an object destined for biomedical applications, both* in vitro *and* in vivo (Yang, Yeo, et al., 2019). In their acceptance, the examples presented in the last section of Chapter 2 are actually examples of 4D bioprinting.

A recent review of the applications of 4D bioprinting (Yang, Yeo, et al., 2019) assigns the term *4D bioprinting* to any *"4D printing process that has at least one of the following characteristics: (i) the printed objects can be used in biomedical engineering (e.g., as biomedical devices), (ii) the printed materials are biocompatible and suitable for being implanted into the human body, or (iii) the printed materials incorporate live cells"*. Moreover, it states that the postprinting evolution "can occur spontaneously without a stimulus," although it is "more often triggered by an external stimulus" (Yang, Yeo, et al., 2019).

Other research groups use the term *4D biofabrication* to refer to a wider set of technologies that exploit stimulus-sensitive materials in the context of biological systems (Ionov, 2018). This broader concept relaxes the requirement of printability but maintains the other criteria satisfied by 4D bioprinting in the view of An et al. (2016). Three main approaches have been proposed in 4D biofabrication. In the first, a biomaterial construct is fabricated, shape transformation is elicited, and the final structure is populated with cells. In the second, a biomaterial construct is fabricated, it is seeded with cells, and the shape transformation is accomplished by stimulation. In the third, the construct is fabricated by the simultaneous deposition of biomaterials and biological cells, and it is followed by a shape transformation. In all of them, the biomaterials must be cytocompatible, nontoxic, and cell-adhesive. In the first approach, represented in Fig. 9.2A, biomaterial deposition can involve high temperatures and organic solvents, and the materials can be sensitive to any physicochemical signal (e.g., temperature, light, metal ions, extreme pH). Nevertheless, just as in scaffold-based tissue engineering, the final structure is hard to be cell-seeded at the desired cell density and uniformity. In the second approach (Fig. 9.2B),

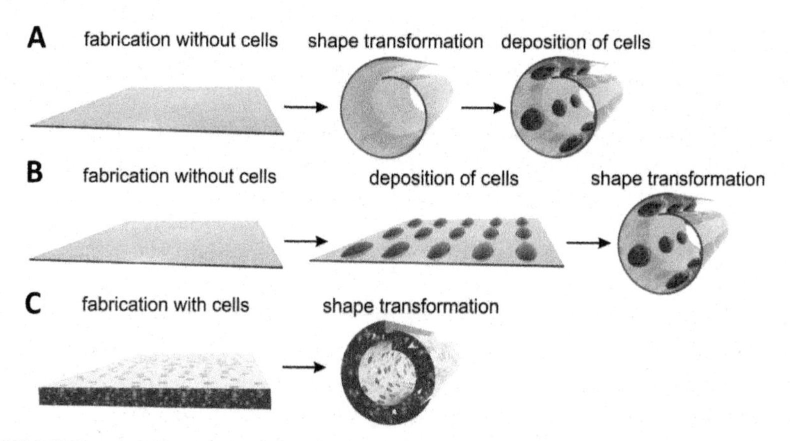

FIGURE 9.2 Approaches to 4D biofabrication.

Schematic drawings of currently explored methods of 4D biofabrication: (A) build a smart material structure, apply a stimulus to trigger the preprogrammed shape transition, and seed cells onto the final structure; (B) fabricate a smart material structure, deposit cells, and provoke shape transformation while keeping cells alive; and (C) create the initial structure from cell-laden, stimulus-responsive material, and apply a gentle stimulus to achieve shape transformation without affecting cell viability.

Reproduced with permission from Ionov, L. (2018). 4D Biofabrication: Materials, methods, and applications. Advanced Healthcare Materials, *7, 1800412. https://doi.org/10.1002/adhm.201800412. Copyright (2018) John Wiley and Sons.*

cells are deposited on top of the initial construct, which is specifically designed to facilitate this process, and the final construct emerges via the shape transition of the stimulus-responsive construct. The shape transformation, of course, should occur in cell-friendly conditions. In the third approach (Fig. 9.2C), the specific needs of cells are to be taken into account during fabrication and shape change. Albeit restrictive, the last two approaches are considered the most promising in the literature (Ionov, 2018) because they enable precise cell patterning prior to stimulation and can produce constructs with intricate microarchitecture as a result of shape transition.

It is apparent from the above examples that the concept of 4D bioprinting is quite dynamic. Just as a language evolves along with the people who speak it, the terminology of a research field evolves as new phenomena are discovered or new methods are devised.

This chapter is not meant to settle the debate on the definition of 4D bioprinting. (One cannot, and clearly should not, keep a language from evolving). Rather, it aimed at showing the variety of viewpoints concerning this concept and to illustrate the myriad of exciting avenues opened up by combining smart materials and biological systems in additive manufacturing.

The reader might rightfully ask "What is meant by 4D bioprinting in this book?" Here, 4D bioprinting refers to the set of technologies that rely on bioprinting to fabricate tissue constructs of man-made design that respond to a well-defined change in their environment by an anticipated change in shape, size, structure, function, or combination thereof.

The community of researchers interested in biomedical uses of smart materials will eventually reach a consensus on the definition of 4D bioprinting, and further developments will show whether it will become the dominant way of 4D biofabrication.

References

An, J., Chua, C. K., & Mironov, V. (2016). A perspective on 4D bioprinting. *International Journal of Bioprinting, 2*, 3–5. https://doi.org/10.18063/ijb.2016.01.003

Ashammakhi, N., Ahadian, S., Zengjie, F., Suthiwanich, K., Lorestani, F., Orive, G., Ostrovidov, S., & Khademhosseini, A. (2018). Advances and future perspectives in 4D bioprinting. *Biotechnology Journal, 13*. https://doi.org/10.1002/biot.201800148. e1800148.

Gao, B., Yang, Q., Zhao, X., Jin, G., Ma, Y., & Xu, F. (2016). 4D bioprinting for biomedical applications. *Trends in Biotechnology, 34*, 746–756. https://doi.org/10.1016/j.tibtech.2016.03.004

Ionov, L. (2018). 4D biofabrication: Materials, methods, and applications. *Advanced Healthcare Materials, 7*, 1800412. https://doi.org/10.1002/adhm.201800412

Kirillova, A., Maxson, R., Stoychev, G., Gomillion, C. T., & Ionov, L. (2017). 4D biofabrication using shape-morphing hydrogels. *Advanced Materials, 29*, 1703443. https://doi.org/10.1002/adma.201703443

Li, Y.-C., Zhang, Y. S., Akpek, A., Shin, S. R., & Khademhosseini, A. (2017). 4D bioprinting: The next-generation technology for biofabrication enabled by stimuli-responsive materials. *Biofabrication, 9*, 012001. http://stacks.iop.org/1758-5090/9/i=1/a=012001.

Morouço, Pedro, Azimi, B., Milazzo, M., Mokhtari, F., Fernandes, C., Reis, D., & Danti, S. (2020). Four-dimensional (Bio-)printing: A review on stimuli-responsive mechanisms and their biomedical suitability. *Applied Sciences, 10*. https://doi.org/10.3390/app10249143

Morouço, P., Lattanzi, W., & Alves, N. (2017). Four-dimensional bioprinting as a new era for tissue engineering and regenerative medicine. *Frontiers in Bioengineering and Biotechnology, 5*, 61. https://doi.org/10.3389/fbioe.2017.00061

Whitford, W. G. (2016). *Another perspective on 4D bioprinting*. https://cellculturedish.com/another-perspective-on-4d-bioprinting/.

Yang, Q., Gao, B., & Xu, F. (2019). Recent advances in 4D bioprinting. *Biotechnology Journal, 15*, 1900086. https://doi.org/10.1002/biot.201900086

Yang, G. H., Yeo, M., Koo, Y. W., & Kim, G. H. (2019). 4D bioprinting: Technological advances in biofabrication. *Macromolecular Bioscience, 19*, 1800441. https://doi.org/10.1002/mabi.201800441

CHAPTER

Applications of 4D bioprinting

10

The field of 4D bioprinting is very young but highly regarded because it opens up new horizons in biomedical engineering as well as in tissue engineering and regenerative medicine.

Why is the fourth dimension so appealing when it comes to fabricating biological tissue structures? There are several reasons for this. First, the ability of a bioprinted construct to transform its shape under a stimulus facilitates the fabrication process. The initial patterning of the system is done fast, with no need for auxiliary structures to support the construct during bioprinting. Then, an external signal is applied, and a complex shape emerges according to the predesigned material programming. Second, the postprinting evolution can give rise to microstructures whose feature size is far below the resolution of the bioprinter. Third, the initial shape can be designed to optimize cell survival, thereby extending the shelf-life of the bioprinted construct. When needed, shape transition can be initiated, and the final product can be transferred into a bioreactor or implanted in the host organism. Fourth, because cells sense and interact with their microenvironment, a 4D-bioprinted construct can dynamically regulate cell function (e.g., it can influence gene expression in specialized cells or drive differentiation in pluripotent stem cells) (Lee et al., 2021; Miao et al., 2020). Fifth, and, perhaps, most important, unlike the static structures created by 3D bioprinting, 4D-bioprinted tissue constructs might recapitulate the dynamics of certain native tissues, such as the rhythmic contractions of heart chambers, or the peristaltic movement of the esophagus or gut (periodic waves of contraction and relaxation of the tube wall that propel the contents of the tube). Also, 4D-bioprinted systems might mimic the response of blood vessel walls to biochemical signals (e.g., vasodilation caused by alcohol or nitric oxide or vasoconstriction caused by caffeine). All these dynamic behaviors of biological tissues are of utmost importance for their function and rely on their ability to respond to certain internal or external stimuli (Lee et al., 2021; Li et al., 2017).

This chapter illustrates the potential of 4D bioprinting by showing biomedical applications developed to date, whereas the next chapter will discuss the perspectives of the field. The separation of present-day results from envisioned developments is meant to convey a hype-free picture of 4D bioprinting as we know it today. It might well happen, however, that reality will beat our imagination.

Towards 4D Bioprinting. https://doi.org/10.1016/B978-0-12-818653-4.00010-3

1. Self-folding tubes

A planar geometry is a popular choice for the initial state of a 4D-printed/bioprinted structure. Straightforward fabrication is a strong incentive in this respect, but not the only one—as shown later in this subchapter. A 3D printer works in a layer-by-layer fashion, with excellent in-plane precision. Therefore, planar structures display reproducible shape transformation according to the initial material programming. Theoretical knowledge of the bending behavior of multilayered sheets and our fascination with origami are valuable for the design of the initial construct.

Solvent-induced autonomous folding of bioprinted cell-laden hydrogel sheets created hollow tubes for diverse biomedical applications (Kirillova et al., 2017). Tube diameters could be controlled by tuning the printing and postprinting parameters.

1.1 Vascular tissue constructs

Kirillova et al. modified two affordable, biocompatible hydrogels (alginate and hyaluronic acid) by reacting them with methacrylic anhydride. The resulting methacrylated alginate (AA-MA) and methacrylated hyaluronic acid (HA-MA) could be crosslinked with cell-friendly green light by using a photoinitiating system composed of Eosin Y in 1-vinyl-2-pyrrolidinone and triethanolamine. At usual concentrations, of 1%–3% by weight, aqueous solutions of these polymers displayed shear thinning, which favored printability by extrusion bioprinting at relatively low shear stress. That is, these bioinks ensure mild conditions for cells during both printing and photo-crosslinking.

The entire fabrication process is presented schematically in Fig. 10.1A–C. First, the bioprinter delivered rectangular (25 mm by 2 mm) patches of polymer solution onto a glass or polystyrene (PS) substrate (Fig. 10.1A). The solutions were previously loaded with the photoinitiating system—5 μL of 0.5% Eosin Y solution in 1-vinyl-2-pyrrolidinone and 5 μL of 2.0 M triethanolamine were mixed with 2 mL polymer solution. The printing took place in air. Then, the structures were crosslinked by exposing them, for 2 min, to green (530 nm) light generated by a powerful (100 W) light-emitting diode (LED) lamp. The resulting hydrogel patches were left to dry for 10 min (Fig. 10.1B). Finally, they were immersed into water, phosphate-buffered saline (PBS), or cell culture medium. Within seconds, they folded into tubes (Fig. 10.1C). This procedure is suitable for the large-scale production of cell-laden hydrogel tubes (Kirillova et al., 2017).

The discovery of Kirillova et al. is exciting because self-folding is usually observed in bilayers of materials that differ in their volume expansion coefficients. Here, by contrast, a single-component hydrogel film folds upon swelling, presumably because the extent of photocrosslinking is larger next to the light source; the crosslinking gradient results in a nonuniform swelling ratio (i.e., the swelling is less pronounced at the top than at the bottom). The differential swelling of hydrogels is the phenomenon underlying most of the current applications of 4D

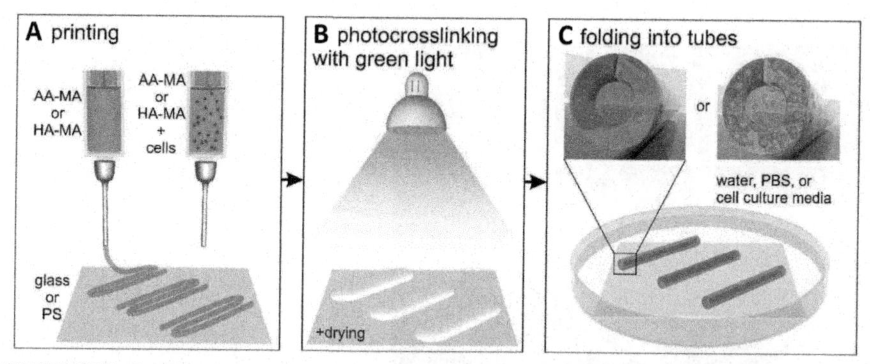

FIGURE 10.1 An Approach to Creating Hollow Tubes With Cell-Laden Hydrogel Walls.

Schematic representation of the 4D bioprinting process: (A) 3D printing of methacrylated alginate (AA-MA) or methacrylated hyaluronic acid (HA-MA) polymer solutions with or without cells in suspension; (B) achieving the sol-gel transition by crosslinking the printed polymer layer with 530 nm green light, followed by an optional drying process; and (C) autonomous folding of hydrogel patches into tubes as a result of immersion in aqueous media.

From Kirillova, A., Maxson, R., Stoychev, G., Gomillion, C. T. & Ionov, L. (2017). 4D Biofabrication using shape-morphing hydrogels. Advanced Materials, 29, *1703443. https://doi.org/10.1002/adma.201703443.*

bioprinting. Such systems are moisture-sensitive, so shape morphing is triggered by the immersion of the construct in an aqueous environment, such as the cell culture medium. This stimulus is cell-friendly and does not require specific equipment.

The self-folding tubes ranged from 20 to 250 μm in inner diameter. Their geometry was influenced by polymer type, the concentration of the polymer solution, the material of the substrate, the crosslinking time, and the soaking medium. Tubes made of HA-MA were larger in diameter than those made of AA-MA. For both polymers, a PS substrate determined larger tube diameters than a glass substrate, perhaps because PS is hydrophobic, whereas glass is hydrophilic. For AA-MA, the tube diameters were similar in deionized water and in PBS, whereas in cell culture medium, the tube diameter was larger, probably due to the small $CaCl_2$ content, which caused an additional crosslinking of the alginate, thereby diminishing the gradient created during photo-crosslinking. Remarkably, tubes emerged also after a brief photo-crosslinking, of merely 1s, albeit the tube diameters were larger and less consistent. Longer times, of ½ min to 2 min, resulted in smaller tube diameters with a narrower distribution, which is consistent with a decrease in the swelling grade (Kirillova et al., 2017).

Knowledge of the physicochemical factors that influence the geometry of self-folding tubes enables one to tailor them for specific applications. To date, this technique is the only one able to produce tubular tissue constructs with internal diameters as low as 20 μm, comparable with native arterioles and venules.

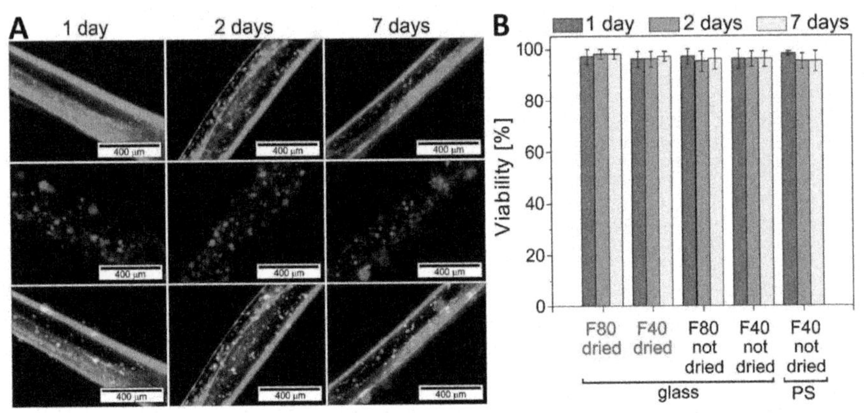

FIGURE 10.2 Cell Viability Within 4D-Bioprinted Methacrylated Alginate (AA-MA) Tubes.

(A) Fluorescence microscopy image of a typical tube from the "F80, dried, glass" sample after 1 day (left column), 2 days (middle column), and 1 week (right column) in culture. In each column, the top row consists of green fluorescence images showing live cells inside the tube wall; the middle row contains red fluorescence images showing all cells, whether live or dead; the bottom row displays the overlay of the green and red fluorescence images, showing live cells in orange and dead cells in red. (B) Bar plots representing the results of the cell viability analysis for five different samples after 1 day (green), 2 days (orange), and 7 days (yellow). The number following the capital letter "F" specifies the printing speed (mm/min.); the next line states whether the bioprinted hydrogel layer has been dried or not, and the bottom line specifies the substrate.

From Kirillova, A., Maxson, R., Stoychev, G., Gomillion, C. T. & Ionov, L. (2017). 4D Biofabrication using shape-morphing hydrogels. Advanced Materials, 29, *1703443. https://doi.org/10.1002/adma.201703443.*

Another important feature of the 4D bioprinting approach shown in Fig. 10.1 is the lack of harsh conditions that might affect cell survival. Indeed, Fig. 10.2 shows that a variety of tubes fabricated from cell-laden AA-MA hydrogels assured at least 95% cell viability, which did not decrease during 1 week in culture. These experiments have been conducted on murine bone marrow stromal cells (from the commercially available D1 cell line), but the cell viability assessments are likely to be generalizable because the entire procedure is cell-friendly.

In the experiments presented in Fig. 10.2, a 3% by weight AA-MA solution was prepared in PBS, and the photoinitiator was added along with 10^5 U/mL penicillin and 10^5 μg/mL streptomycin (to avoid bacterial contamination); then, mouse bone marrow cells were suspended in PBS and added to the mixture to reach a final cell density of half a million cells per mL. The cell-laden polymer solution was loaded into the printer cartridge, and the 4D printing was conducted as shown in Fig. 10.1.

To investigate the impact of printing and postprinting parameters on cell viability, five different sample types were produced in triplicate, each sample consisting of five tubes. In Fig. 10.2, the sample types are labeled by the experimental conditions, specifying the substrate (glass or PS), the printing speed (e.g., F80 stands for a printing speed of 80 mm/min), and whether the hydrogel sheet was dried before immersion or not. Nondried samples were regarded as controls in the cell viability analysis because they were the least likely to harm the incorporated cells. Nevertheless, they were unsatisfactory from the point of view of postprinting shape transition, displaying partial folding at best. Fortunately, it turned out that cell viability did not decrease during the drying process (see Fig. 10.2B). Thus, it has been demonstrated that 4D bioprinting is an effective way of fabricating biocompatible hydrogel tubes with live cells evenly distributed within their walls (Kirillova et al., 2017).

1.2 Nerve grafts

The idea of stimulus-responsive planar constructs served also in the 4D biofabrication of artificial nerve grafts (Apsite, Constante, et al., 2020). Self-folding bilayer mats were fabricated from uniaxially aligned polycaprolactone-poly(glycerol sebacate) (PCL-PGS) fibers and randomly aligned HA-MA fibers deposited sequentially by electrospinning. Immersion in aqueous media triggered the transformation of the rectangular PCL-PGS/HA-MA bilayer into a scroll-like, hollow tube. The outcome of the shape transformation could be controlled via the bilayer thickness, the ratio of the individual layer thicknesses, and the Ca^{2+} ion concentration in the aqueous medium. PCL is hydrophobic, whereas PGS is hydrophilic, with water contact angles of 77 degrees and 35 degrees, respectively. A blend of 75% PCL and 25% PGS by weight could be used to fabricate bead-free nanofibers of 0.6 ± 0.2 μm in diameter and displayed an acceptable hydrophobicity, with a water contact angle of 64 degrees. PGS acted as a plasticizer, lowering the storage modulus of fibers made from the blend threefold in comparison to pure PCL. Moreover, PGS made the fibers more degradable: while pure PGS fibers have lost less than 5% of their mass within 4 weeks in culture, fibers made from the blend containing 25% PGS have lost about 40% of their mass under the same conditions. For the entire bilayer, the mass loss was about 70% within 4 weeks of incubation, corresponding to the degradation of virtually all HA-MA fibers and the PGS fraction of the PCL-PGS fibers. Meanwhile, the construct remained tubular, but its diameter increased about 3 times. The PC-12 cell line is widely used in neurobiology as it exhibits certain features of mature dopaminergic neurons. PC-12 cells were seeded on HA-MA mats, pure PCL mats, PCL-PGS mats, and PCL-PGS/HA-MA bilayers. The HA-MA mats assured the weakest cell adhesion and the lowest cell proliferation during 1 week in culture. Cell proliferation was roughly the same on the other three types of mats. Despite the poor proliferation on the HA-MA mat, the bilayer offered an excellent substrate for PC-12 cells seeded on the aligned fibrous side of the bilayer. In cell culture medium augmented with 100 ng/mL of nerve growth factor (NGF), the cells attached to the PCL-PGS side of the bilayer were elongated along the fibers and started to form neurites. These results suggest that 4D-biofabricated PCL-PGS/HA-MA bilayer tubes could be used as nerve guide conduits for treating neural injuries (Apsite, Constante, et al., 2020).

Miao et al. fabricated a versatile nerve guide conduit by stereolithography (SL)-based 4D bioprinting of soybean oil epoxidized acrylate (SOEA), a photocrosslinkable resin of natural origin (Miao et al., 2018). Prior to photocuring, SOEA is a yellow, sticky liquid. It was poured into a glass Petri dish and exposed to 355 nm UV light generated by a fiber optic-coupled solid-state laser. The tip of the optical fiber was attached to a commercial 3D printer, which moved it at various speeds to create lines of crosslinked polymer, as designed. The faster the movement of the laser head, the thinner the polymer strands. Moreover, they are inhomogeneous. Part of the UV light beam is absorbed as it penetrates the resin layer; therefore, the surface proximal to the laser head will have the highest crosslink density, whereas the most distal surface will have the lowest crosslink density and includes pores filled with uncured, liquid resin. This structural inhomogeneity is a valuable asset because it makes the 3D-printed construct prone to postprinting shape morphing. It can be further amplified by adding photoabsorbers to the liquid resin to increase the light intensity gradient within the sample.

Despite the internal stress accumulated in the course of printing, the construct remained flat; it remained so even after being taken out of the vat. Nevertheless, as soon as it was immersed in ethanol, the structure curved within seconds. The less cured face became concave, presumably because ethanol dissolved the thick uncured resin from the pores, thereby breaking the stress—strain balance. In time, as ethanol diffused into the pores, soaking the strand, the bending became less pronounced (but did not disappear completely). The curvature increased again when the construct was transferred into an aqueous medium, such as phosphate-buffered saline, to make it suitable for cell seeding. This surprisingly favorable behavior can be explained as follows. When the construct is immersed in water, the ethanol gradually diffuses out from the pores, but it is not replaced by water because the polymerized SOEA is hydrophobic. Hence, the strand bends again.

To fabricate a self-folding nerve conduit, Miao et al. incorporated graphene nanoparticles into the liquid SOEA resin; these limited light penetration into the printing ink, thereby boosting the light intensity gradient. The radius of curvature of a 3D-printed rectangular construct could be tuned by varying the graphene content. The result was a scroll-like, tubular mesh, of about 1 mm in inner diameter, suitable to wrap the loose ends of an injured nerve. It was unclear, however, how to manipulate these delicate structures.

An elegant solution to this problem is based on the shape memory effect displayed by the SOEA polymer. Its glass transition temperature, $T_g = 20°C$, makes it suitable for thermomechanical programming followed by shape recovery at physiological temperatures. The self-folded tubular construct was unrolled at 37°C, the resulting rectangular mesh was cooled down to −18°C, thereby fixing this temporary shape. For deployment, the flat mesh was placed below the model of a damaged nerve; upon warming up to 37°C, the mesh recovered its permanent shape, gently wrapping the nerve model (Miao et al., 2018).

Although axon growth was not yet demonstrated along this multiple responsive nerve guide conduit, it was found suitable to host human mesenchymal stem cells and favored their neurogenic differentiation. The microarchitecture of the polymer mesh promoted cell alignment, and the presence of graphene in the SOEA polymer was associated with the upregulation of neurogenic gene expression (Miao et al., 2018).

1.3 Muscle tissue constructs

Skeletal muscle consists of bundles of aligned muscle fibers, which form as a result of the fusion of mononuclear myoblasts into myotubes. These are elongated, multinuclear cells, which undergo differentiation to become muscle fibers.

The importance of muscle tissue engineering is easy to recognize if one keeps in mind that muscles make up about 40% of our body mass and their regenerative capacity is limited. Self-healing works when the damage is light, but massive muscle loss due to trauma or illness can only be addressed by transplantation. Functional human skeletal muscle tissue constructs would be invaluable also in drug development. They could be used to test potential therapies for muscular disorders or to evaluate the myotoxicity of drug candidates. Statins, for example, are important cholesterol-lowering drugs, but they are detrimental to muscles (Armitage, 2007). One of them had even to be withdrawn from the market. It would be desirable to use engineered muscle tissues to assess beforehand whether a drug candidate affects or treats muscles.

Self-folding electrospun mats were found suitable also for muscle tissue engineering (Apsite, Uribe, et al., 2020). Myoblast alignment is vital for myogenesis. The mats fabricated for this application consisted of a layer of anisotropic AA-MA fibers covered by a layer of aligned PCL fibers. When mouse myoblasts from the C2C12 cell line were cultured on these mats, they aligned along the PCL fibers and differentiated into myotubes capable of contraction under electrical stimulation. This work by Apsite et al. demonstrates that such a 4D biofabrication approach leads to the formation of hollow, contractile tubular constructs that can be used as building blocks of larger, vascularized, functional muscle tissue constructs.

The last two applications described in this section deviate from the definition of 4D bioprinting in that the initial construct is not 3D printed. How did they end up here? The reason for discussing them in this chapter is that similar structures can be obtained by 4D bioprinting, too. Indeed, the same research group fabricated myotube-like constructs also by using a combination of 3D printing of AA-MA and melt-electrowriting of PCL fibers followed by shape-morphing (Constante et al., 2021). PCL has a low melting point, of about 65 °C, which makes it especially suitable for melt-electrowriting. Their approach is illustrated in Fig. 10.3.

A 5% AA-MA solution was mixed with a stock solution of the photoinitiator, IRG2959, to obtain a final concentration of 0.1% of the photoinitiator. A 3D printer was used to build a double-layered disk of 20 mm in diameter by placing threads of the above mixture at 0.5 mm from one another. The solution was extruded from a

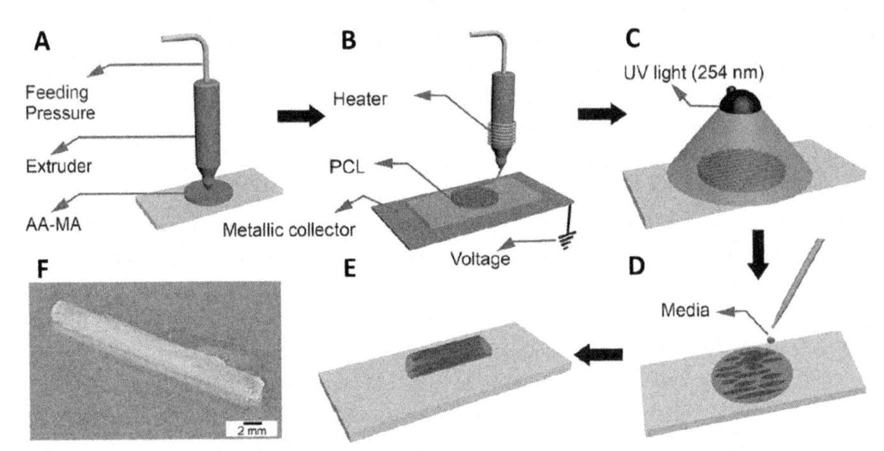

FIGURE 10.3 Schematic Representation of the 4D Bioprinting of Myotube-Like Tissue Constructs.

(A) 3D printing of a double-layered disk of methacrylated alginate (AA-MA) hydrogel; (B) melt-electrowriting of parallel polycaprolactone (PCL) fibers; (C) photo-crosslinking of the hydrogel; (D) cell seeding by pipetting a muscle cell suspension on top of the 3D printed scaffold; (E) shape-change triggered by immersion in the aqueous medium in which the cells were dispersed; (F) photograph of a typical tissue construct—a tubular structure with enclosed cells—obtained by the procedure shown in panels A–E.

Reprinted with permission from Constante, G., Apsite, I., Alkhamis, H., Dulle, M., Schwarzer, M., Caspari, A., Synytska, A., Salehi, S. & Ionov, L. (2021). 4D Biofabrication using a combination of 3D printing and melt-electrowriting of shape-morphing polymers. ACS Applied Materials & Interfaces, *13, 12767. https://doi.org/10.1021/acsami.0c18608. Copyright 2021 American Chemical Society.*

needle of 0.33 mm in inner diameter, at a pressure of 0.4 bar and movement rate of 5 mm/s; these parameters determined the printed layer thickness. Then, PCL of 45 kDa molecular weight was melted and heated up to 110°C for ensuring proper rheological properties, and PCL fibers were printed on top of the AA-MA disk, in a parallel arrangement, at 0.1 mm from each other. Typical diameters of fibers produced by melt-electrowriting were of the order of 10 μm, an order of magnitude larger than the diameters of fibers produced by electrospinning. The disk was exposed for 10–15 min to UV radiation of 240 nm wavelength. AA-MA/PCL bilayers have been produced also by first crosslinking the AA-MA and then depositing the PCL fibers. For cell seeding, a Teflon ring was placed on the margins of the crosslinked disk to prevent its folding as the cell suspension was pipetted over it. Finally, the ring was removed to allow the disk to fold Fig. 10.3E, turning into a scroll-shaped hydrogel structure with myoblasts uniformly distributed along its walls (Constante et al., 2021).

Had the tube been produced by 3D printing, or by wrapping a biomaterial sheet around a rod, the cell seeding process would have been less effective; because of gravity, the cells would have settled down on one side of the tube.

Uniform cell seeding of tubular scaffolds is only achieved in a perfusion bioreactor that rotates the tube around its longitudinal axis maintained in a horizontal plane. Instead, 4D printing facilitates uniform cell seeding prior to self-folding. Instead of pipetting a cell suspension over the 3D-printed disk (Fig. 10.3D), for even better control of cell patterning, one might as well deliver the cells using a 3D bioprinter.

The diameters of the self-folded tubes were influenced by the fabrication process. They depend on the photoinitiator concentration, on the duration of the photo-crosslinking of AA-MA, as well as on the ratio of PCL layer thickness to AA-MA layer thickness. Large photoinitiator concentrations and long crosslinking times both contributed to reducing the self-folded tube diameters. The presence of the PCL layer hampered the self-folding of the AA-MA layer. For example, the tube diameter was about 2 mm when the PCL layer had half the thickness of the AA-MA layer, and it was twice larger when the PCL layer was 1.8 times thicker than the AA-MA layer.

Moreover, the tube diameter could be fine-tuned by the Ca^{2+} concentration of the aqueous solution that caused the self-folding of the bilayer: an increase of $[Ca^{2+}]$ from 0 to 50 mM caused a twofold increase in the tube diameter (Constante et al., 2021). It turned out that further addition of calcium salts brings about the unfolding of the tubes, and the process can be reversed by the addition of ethylenediaminetetraacetic acid (EDTA)—a calcium chelator (Fig. 10.4A).

The bilayer tubes proved to be suitable to host muscle cells. When C2C12 mouse myoblasts were cultured on AA-MA films and on AA-MA/PCL bilayers, about 80% of the cells remained viable and their metabolic activity increased with time, at a bit higher rate on bilayers than on AA-MA films. In Fig. 10.4, cells are visualized by fluorescence microscopy; the actin cytoskeleton was stained with phalloidin Dylight 488 (green), while the nucleus was stained with DAPI (blue). Most importantly, about 65% of the cells seeded over the bilayer scaffolds were aligned with the PCL fibers. The fraction of aligned cells slightly decreased with time, dropping by about 10% during 1 week in culture (Fig. 10.4D, bottom plot), presumably because of cell clustering and weak adhesion to the alginate. By contrast, no cell alignment was observed on AA-MA films (Fig. 10.4D, top plot).

Yang et al. exploited the shape-morphing abilities of gelatin films to devise a 4D printing approach to muscle tissue engineering (Yang et al., 2021). First, they developed an electric field-assisted 3D bioprinting technique to fabricate cell-laden gelatin methacryloyl (GelMA) fibers. These were deposited onto gelatin films, whose swelling gave rise to cylindrical bundles. Fig. 10.5 represents the schematics of their methodology.

The technique proposed by Yang et al. relies on recent progress in skeletal muscle regeneration based on electric stimulation. Previous research demonstrated that exposure to electric fields has the potential to enhance myogenic precursor cell proliferation as well as myotube formation (Di Filippo et al., 2017). In tissue-engineered human myobundles, intermittent electrical stimulation tripled the

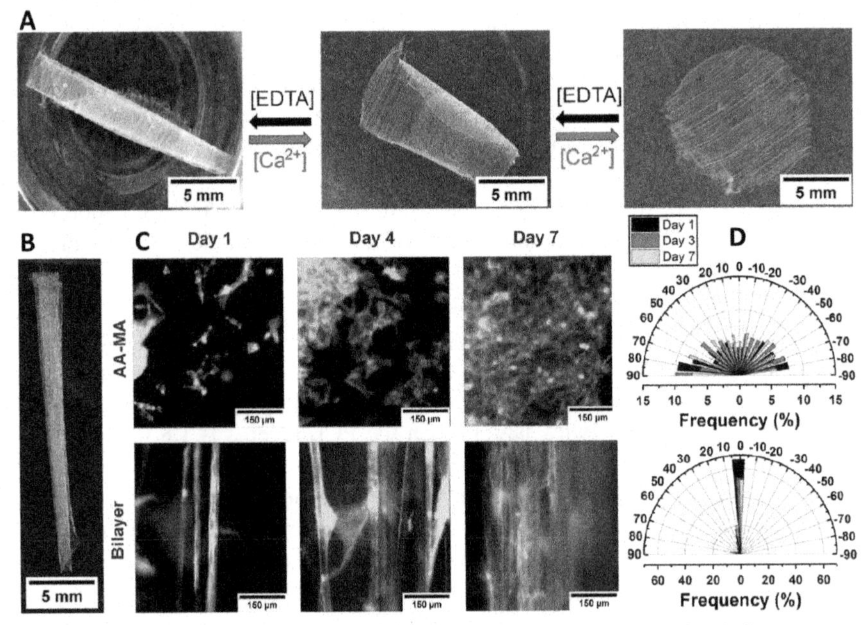

FIGURE 10.4 Evaluation of Scroll-Like Bilayer Scaffolds Created by 4D Bioprinting.

(A) Snapshots of the reversible unfolding of methacrylated alginate (AA-MA)/ polycaprolactone (PCL) bilayer mats under the action of Ca^{2+} ions; when calcium ions are sequestered by ethylenediaminetetraacetic acid, the construct refolds. (B) Self-folded tube incorporating mouse muscle cells stained green with calcein. (C) Fluorescence microscopy image of cells cultured on AA-MA films (top row) and on AA-MA/PCL bilayers (bottom row); actin filaments are green, nuclei are blue. (D) Radial plots showing the frequency distribution of the alignment of muscle cells on AA-MA sheets (top) and on AA-MA/PCL bilayers (bottom). For interpretation of the references to color in this figure legend, please refer online version of this title.

Adapted with permission from Constante, G., Apsite, I., Alkhamis, H., Dulle, M., Schwarzer, M., Caspari, A., Synytska, A., Salehi, S. & Ionov, L. (2021). 4D biofabrication using a combination of 3d printing and melt-electrowriting of shape-morphing polymers. ACS Applied Materials & Interfaces, 13, *12767–12776. https://doi.org/10.1021/acsami.0c18608. Copyright 2021 American Chemical Society.*

contractile force, increased myobundle size, as well as myotube diameter and length, and boosted glucose and fatty acid metabolism. Myotubes were larger when the stimulation frequency was 10 Hz than for 1 Hz, suggesting that the maturation of engineered muscle tissue is optimal when the electrical stimulation pattern mimics the in vivo muscle activity (Khodabukus et al., 2019).

The two-in-one bioprinting and electrical stimulation setup represented in Fig. 10.5B is able to extrude a bioink thread, while applying an electric field to induce the alignment and myogenic differentiation of the C2C12 cells embedded in the bioink. The 3D printer is outfitted with a Teflon tube of 350 μm in inner diameter connected to the tip of the extrusion nozzle and a cylindrical polydimethylsiloxane (PDMS) mold. The cell-laden bioink is exposed to the electric field applied

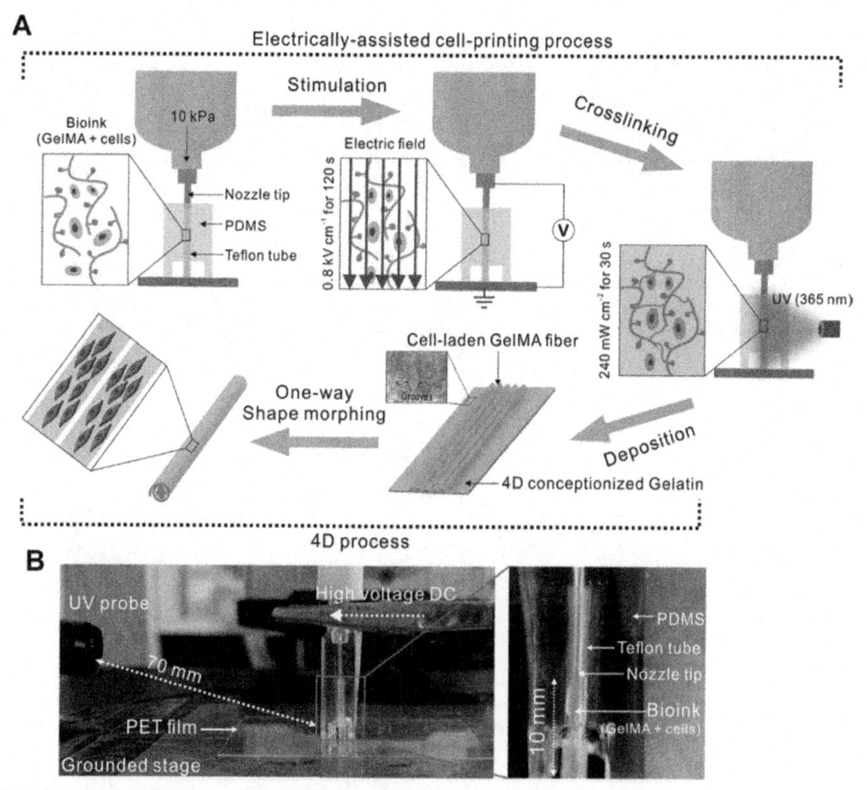

FIGURE 10.5 Illustration of the Physical Principle of Electrically Assisted Bioprinting.

(A) Schematic representation of electrically assisted 4D bioprinting, which eventually leads to the formation of self-assembled bundles of cell-laden gelatin methacryloyl (GelMA) fibers. (B) Photograph of the actual experimental setup.

From Yang, G. H., Kim, W., Kim, J. & Kim, G. (2021). A skeleton muscle model using GelMA-based cell-aligned bioink processed with an electric-field assisted 3D/4D bioprinting. Theranostics, *11, 48–63. https://doi.org/10.7150/thno.50794. Reprinted under the terms of the Creative Commons Attribution 4.0 International License (https://creativecommons.org/licenses/by/4.0/legalcode).*

between the tip of the nozzle and the grounded stage. After 2 minutes of electrical stimulation, UV radiation was applied to stabilize the structure, and a new dose of bioink was injected into the Teflon tube to expel the crosslinked hydrogel cylinder and deposit it onto the gelatin film. The hydrophobicity of Teflon facilitates the extrusion of the hydrogel cylinder.

The result of the process shown in Fig. 10.5A is a muscle-fiber-like tissue construct. Fig. 10.6A shows the fluorescence microscopy image of a GelMA fiber fragment obtained in the absence and presence of electrical stimulation (left and right image, respectively). Myosin heavy chain (MHC) staining (green) indicates that C2C12 cells have undergone myogenic differentiation in both conditions. When the

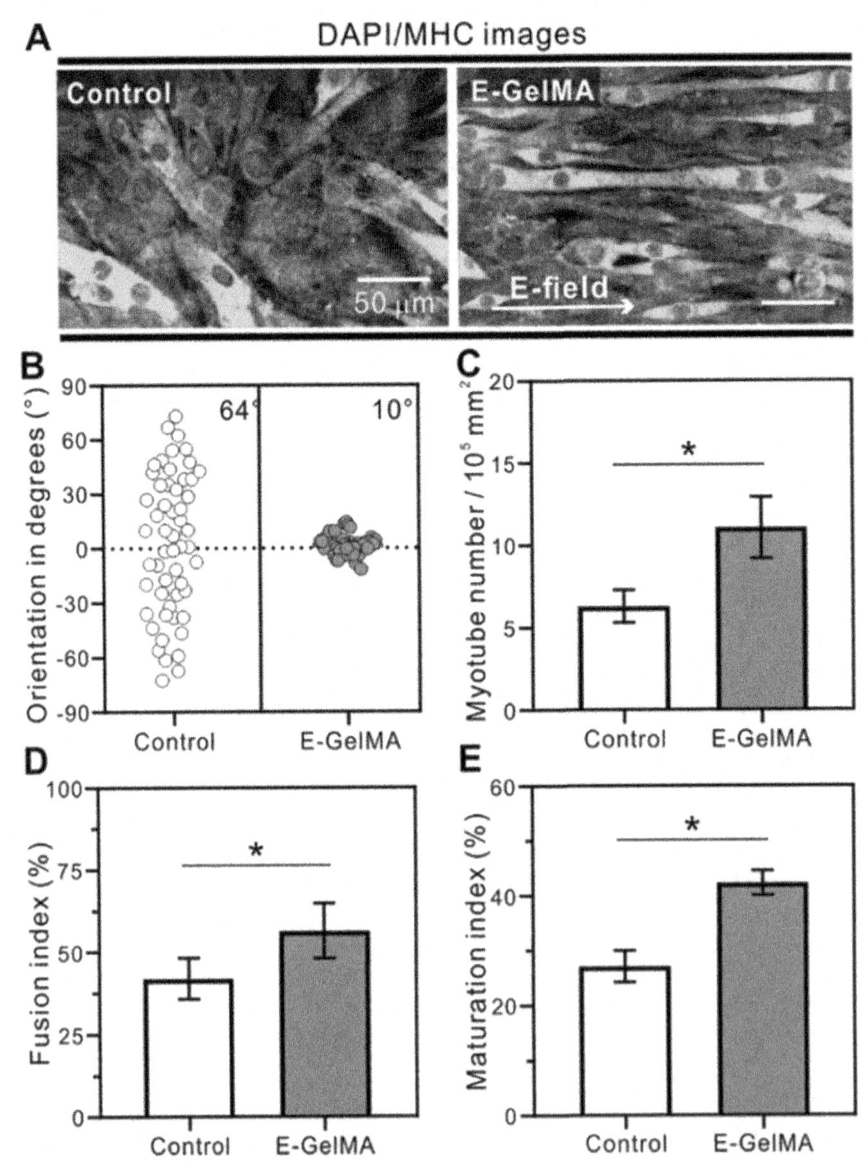

FIGURE 10.6 The Impact of Electrical Stimulation on the Microarchitecture of a Cell-Laden Gelatin Methacryloyl (GelMA) Fiber.

(A) Fluorescence microscopy image of a GelMA fiber fragment, obtained by myosin heavy chain (MHC) (green) and 4′,6-diamidino-2-phenylindole (DAPI) (blue) staining after 21 days in culture; the left (right) image represents a typical sample obtained in the absence (presence) of electrical stimulation. The plots shown in panels B–E describe (B) cell orientation, (C) myotube numbers, (D) fusion index (the percentage of nuclei electric field was applied, most cells were aligned parallel to the axis of symmetry (the direction of the applied electric field) (compare panels A and B of Fig. 10.6). Moreover, the number of myotubes was almost twice larger in GelMA fibers fabricated under electrical stimulation than in nonstimulated fibers Fig. 10.6C. The electric field also facilitated cell fusion and maturation (Fig. 10.6D and E, respectively).

To create skeletal muscle tissue constructs, cell-laden GelMA microfibers were deposited on top of a 100 μm-thick 3D printed gelatin film (Fig. 10.7A, center, before). Depending on the initial state, the shape morphing of the construct created bundles (Fig. 10.7A, center, After) akin to subunits of multipennate muscles, such as the deltoid, or parallel fusiform muscles, such as the biceps brachii (Fig. 10.7A, left). According to cross-sectional SEM images, the microfibers became tightly packed as they were wrapped in the gelatin film (Fig. 10.7A, right). Confocal microscopy images of constructs cultured for 3 weeks demonstrate the proper alignment of myotubes within the hydrogel bundles.

Cell-laden microfiber bundles mimic the structure of native skeletal muscle tissue. Therefore, they could be incorporated into muscle-on-a-chip devices for in vitro drug testing.

Electrically assisted bioprinting still faces challenges due to limited printing resolution. If they turn out to be surmountable, this technique will enable the biofabrication of vascularized muscle tissue constructs.

2. Shape morphing patches

Tissue damage is often localized, so one therapeutic option is to cover the weak spot with a customized patch. In an ideal scenario, the patch integrates into the native tissue, promotes regeneration, and restores function. Necessary conditions for such benefits include suitable geometry, mechanical properties, and composition. Again, it is tempting to start with a bioprinted flat sheet, of the appropriate shape and size, and rely on stimulus-responsiveness to bring about the desired 3D geometry. Here, we discuss biomedical applications based on this strategy.

2.1 Tissue-engineered trachea

A moisture-sensitive hydrogel sheet loaded with two kinds of cells was created by using a digital light processing (DLP) bioprinter and a photopolymerizable bioink,

located in polynucleated MHC-positive myotubes), and (E) maturation index (the percentage of myotubes that contain more than five nuclei). For interpretation of the references to color in this figure legend, please refer online version of this title.

From Yang, G. H., Kim, W., Kim, J. & Kim, G. (2021). A skeleton muscle model using GelMA-based cell-aligned bioink processed with an electric-field assisted 3D/4D bioprinting. Theranostics, *11, 48–63. https://doi.org/10.7150/thno.50794. Reprinted under the terms of the Creative Commons Attribution 4.0 International License (https://creativecommons.org/licenses/by/4.0/legalcode).*

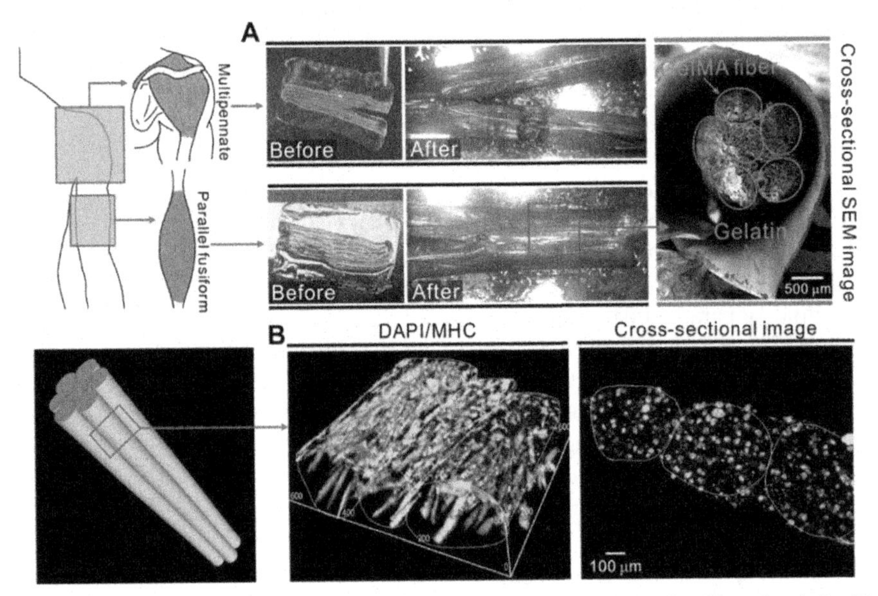

FIGURE 10.7 The Microarchitecture of 4D-Printed Bundles of GelMA Microfibers Loaded with Muscle Cells.

(A) Schematic representation of typical skeletal muscle structures (left); 3D-printed constructs of cell-laden hydrogel microfibers deposited on a gelatin film and the corresponding bundles formed by the shape change of the gelatin film immersed in cell culture medium (center); cross-sectional view of a 4D-bioprinted microfiber bundle captured by scanning electron microscopy (SEM) (right); (B) A set of closely packed microfibers drawn schematically (left); 3D reconstruction (center) and transversal cross-section (right) of a portion of three adjacent filaments visualized by confocal fluorescence microscopy; myosin heavy chain (MHC) antibody was used to stain myosin filaments from the C2C12 cells dispersed in the hydrogel (green) and DAPI staining was applied to cell nuclei (blue). For interpretation of the references to color in this figure legend, please refer online version of this title.

From Yang, G. H., Kim, W., Kim, J. & Kim, G. (2021). A skeleton muscle model using GelMA-based cell-aligned bioink processed with an electric-field assisted 3D/4D bioprinting. Theranostics, *11, 48–63. https://doi.org/10.7150/thno.50794. Reprinted under the terms of the Creative Commons Attribution 4.0 International License (https://creativecommons.org/licenses/by/4.0/legalcode).*

Sil-MA, synthesized from silk fibroin and glycidyl methacrylate (Kim et al., 2020). The DLP printer projects light on a thin layer of the photocurable solution (which can also contain cells in suspension), thereby crosslinking an entire layer at once (see Chapter 5). This technique presents several advantages over other bioprinting methods: (i) it is faster because it relies on the selective hardening of the bioink, as opposed to bioink dispensing; (ii) it is less harmful to cells because it is nozzle-free, so cells are not exposed to high shear stress; (iii) it ensures better mechanical properties of the bioprinted construct because successive layers adhere well to each other and buoyancy roughly cancels the weight force, preventing the deformation of the construct while being printed. The layer-by-layer photopolymerization leads to slightly anisotropic, but otherwise uniform hydrogel structures. Hence, the ability of DLP-printed bilayers to change their shape when immersed in aqueous media does not result from gradients of cross-link density but from differences between the two layers in composition or architecture.

Shape morphing is vital for creating complex tissue constructs via DLP. Based on planar projection, DLP is highly impractical for building contorted 3D structures made of multiple materials. A separate projection is needed for each material, and the switch from one material to the next consists in replacing the contents of the vat. In contrast, only one switch is required for printing a flat bilayer. Then, the autonomous shape transition gives rise to the final product.

To optimize their bioprinting methodology, Kim et al. conducted experiments on a variety of cell-free Sil-MA hydrogels. They dissolved Sil-MA powder in distilled water, phosphate-buffered saline, or phenol-free Dulbecco's Modified Eagle's Medium (DMEM) at Sil-MA concentrations of 15% or 30%, added a photoinitiator, lithium phenyl(2,4,6-trimethylbenzoyl) phosphinate (LAP), at a concentration of 0.2%, and stirred the mixture until fully dissolved. These solutions were used to print hydrogel sheets of different designs. Photocuring of each new layer, of 40 μm in thickness, was done for about 5 s at a UV light intensity of 3.5 mW/cm^2. They printed monolayers and bilayers and varied Sil-MA concentration as well as construct architecture. It turned out that the degree of curvature was favored by (i) the presence of embossed patterns in the top layer, (ii) a larger thickness of the patterned layer compared to that of the base layer, and (iii) a lower Sil-MA concentration in the patterned layer than in the base layer. The final configuration was affected also by the osmolar concentrations of the Sil-MA dissolvent and the soaking solvent. These experiments demonstrated that the outcome of 4D bioprinting can be tuned by the geometry and composition of the 3D-printed bilayer and its environment (Kim et al., 2020).

Finite element analysis was used to predict the shape changes of the 3D-printed structures. Using the Abaqus CAE software package, Kim et al. modeled the hydrogel as a porous material (of porosity 0.5, Poisson ratio 0.1, and Young modulus 1.22 MPa—measured experimentally) and simulated its deformation caused by moisture-driven swelling. Fig. 10.8 depicts the result of such a simulation (panel A) and its experimental counterpart (panel B).

In the computational model, volume expansion started from the surface, causing a gradual deformation of the hydrogel bilayer, in good agreement with experimental findings (compare panels A and B of Fig. 10.8).

Trachea tissue constructs were fabricated along the same lines, starting from a digital model of a thin bilayer: a 0.8 mm thick base layer covered by a second layer consisting of parallel, equidistant stripes (0.8 mm high and 1 mm wide) separated by 1 mm wide gaps. The same type of hydrogel was used for both layers, obtained from a 30% Sil-MA solution prepared in phenol-free DMEM, sterilized via pasteurization, and supplemented with 0.2% LAP passed through a 0.22 μm syringe filter. The bioink used for printing the base layer was loaded with tracheobronchiolar

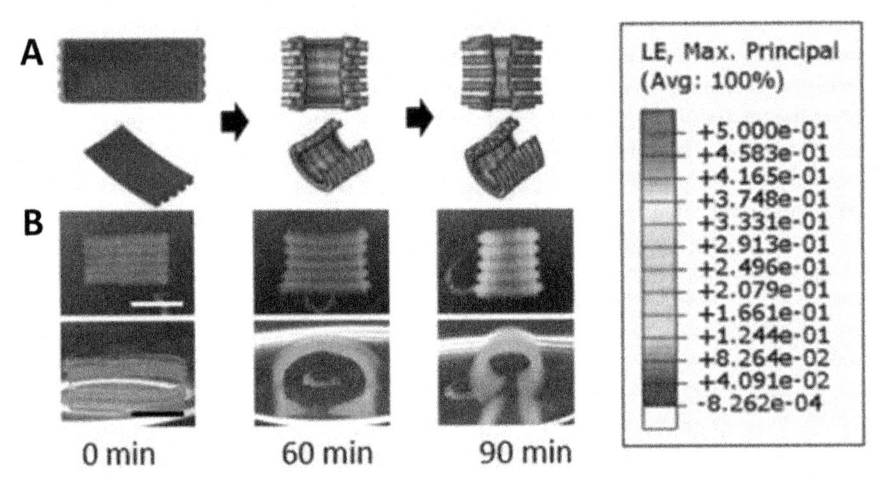

FIGURE 10.8 Shape Morphing Snapshots Predicted by Finite Element Analysis and Probed by Experiments.

(A) Finite element analysis of shape morphing caused by moisture-driven swelling; the color map depicted on the right indicates the pore pressure from low (blue, black in print) to high (red, dark gray in print); (B) experimental snapshots of a hydrogel bilayer immersed in distilled water shown in top view (top row) and perspective view (bottom row). Scale bar = 10 mm.

Reprinted from Kim, S. H., Seo, Y. B., Yeon, Y. K., Lee, Y. J., Park, H. S., Sultan, M. T., Lee, J. M., Lee, J. S., Lee, O. J., Hong, H., Lee, H., Ajiteru, O., Suh, Y. J., Song, S. H., Lee, K. H. & Park, C. H. (2020). 4D-bioprinted silk hydrogels for tissue engineering. Biomaterials, 260, *120281. https://doi.org/10.1016/j.biomaterials.2020.120281, with permission from Elsevier.*

stem cells (TBSCs), whereas the patterned layer contained chondrocytes to mimic the hyaline cartilage ring; the cell density was 10 million cells/mL in both layers. For human tissue constructs, chondrocytes were harvested from septal cartilage, and TBSCs were isolated from inferior turbinate tissue obtained from patients. For rabbit constructs, chondrocytes were isolated from the auricular cartilage of New Zealand White rabbits, and TBSCs were isolated from their inferior turbinate tissue. The bioprinted sheets were washed three times with phosphate-buffered saline to remove nonadherent cells, as well as residual monomers and LAP molecules.

Human trachea tissue constructs were cultured in vitro for 2 weeks. Their shape morphing was recorded every 10 h, and histological evaluation was carried out after 1 week and after 2 weeks in culture. The corresponding experimental protocol is represented schematically in Fig. 10.9A.

Fig. 10.9B shows that shape morphing occurs within 2 h, and the curved conformation remains stable for 2 weeks, presumably because of β-sheet formation in the Sil-MA hydrogel, which is known to increase its stiffness. The slightly increased degree of curvature in comparison to the initial bending was ascribed to construct remodeling due to cell proliferation. Indeed, the fluorescence microscopy images shown on the left side of Fig. 10.9C (arrows) indicate that both cell types managed to proliferate within the bioprinted construct, showing marked progress during the second week. Moreover, histological sections show that a cartilage-specific extracellular matrix has formed in the patterned layer of the construct (Fig. 10.9C, Saf-O, and MT). Safranin-O staining reveals the presence of a proteoglycan-rich extracellular matrix, whereas Masson's trichrome staining shows the collagen produced by the embedded cells. Tissue morphology is visualized by hematoxylin and eosin staining (HE).

Rabbit tissue constructs were cultured for 3 days and evaluated in vivo, in a rabbit tracheal damage model. To generate the damage, Kim et al. incised the neck skin vertically in the pretracheal region and resected a 1 cm long portion of the anterior tracheal wall. The defect extended about 210 degrees along the total circumference of the trachea. Then, the 4D-bioprinted construct was trimmed to become a bit larger than the defect; it was placed over the damage and sutured to the tracheal wall along the line of excision. The engineered trachea did not suffer any damage during the surgical procedure. The evolution of the repair site was monitored by endoscopy up to 8 weeks post surgery. During the first 4 weeks, the implant was exposed, but its internal surface became completely covered by epithelial mucosa by week 6, a process that caused mild stenosis, but the airways remained open during the entire follow-up period of 8 weeks. The histopathological analysis revealed that, at 8 weeks post surgery, the regenerated epithelium was similar to the ciliated columnar epithelium of the native trachea. Also, Safranin-O and Fast Green staining demonstrated tissue maturation, proving that the 4D-bioprinted construct is biocompatible and facilitates tissue regeneration.

2.2 Cardiac patches

Ischemic heart disease is the major cause of mortality, accounting for 16.2% of all deaths worldwide according to the Global Burden of Disease Study 2019 (Vos et al., 2020). In the case of myocardial infarction, when reduced blood supply (ischemia) causes irreversible damage to the heart muscle, the myocardium is unable to self-repair and suffers irreversible remodeling into scar tissue. This process leads to heart dilatation and failure, leaving the patient with few therapeutic options. The most effective one, heart transplantation, is severely limited by organ donor shortage. Cell-based regenerative therapies are promising alternatives because they directly address the loss of cardiac output due to cardiomyocyte loss or dysfunction. Despite considerable research efforts, fueled by the success of animal studies conducted 2 decades ago, clinical studies did not show statistically significant benefits of therapies based on injecting various cell types into the epicardial infarct zone (Normand et al., 2019).

To address this problem, tissue engineers sought to build functional cardiac tissue constructs in the laboratory. Besides providing mechanical support, cellularized cardiac patches are meant to restore the contractile function of the damaged myocardium. Patches made from hydrogels of natural origin were found to provide a favorable microenvironment for cell growth and differentiation. Since primary

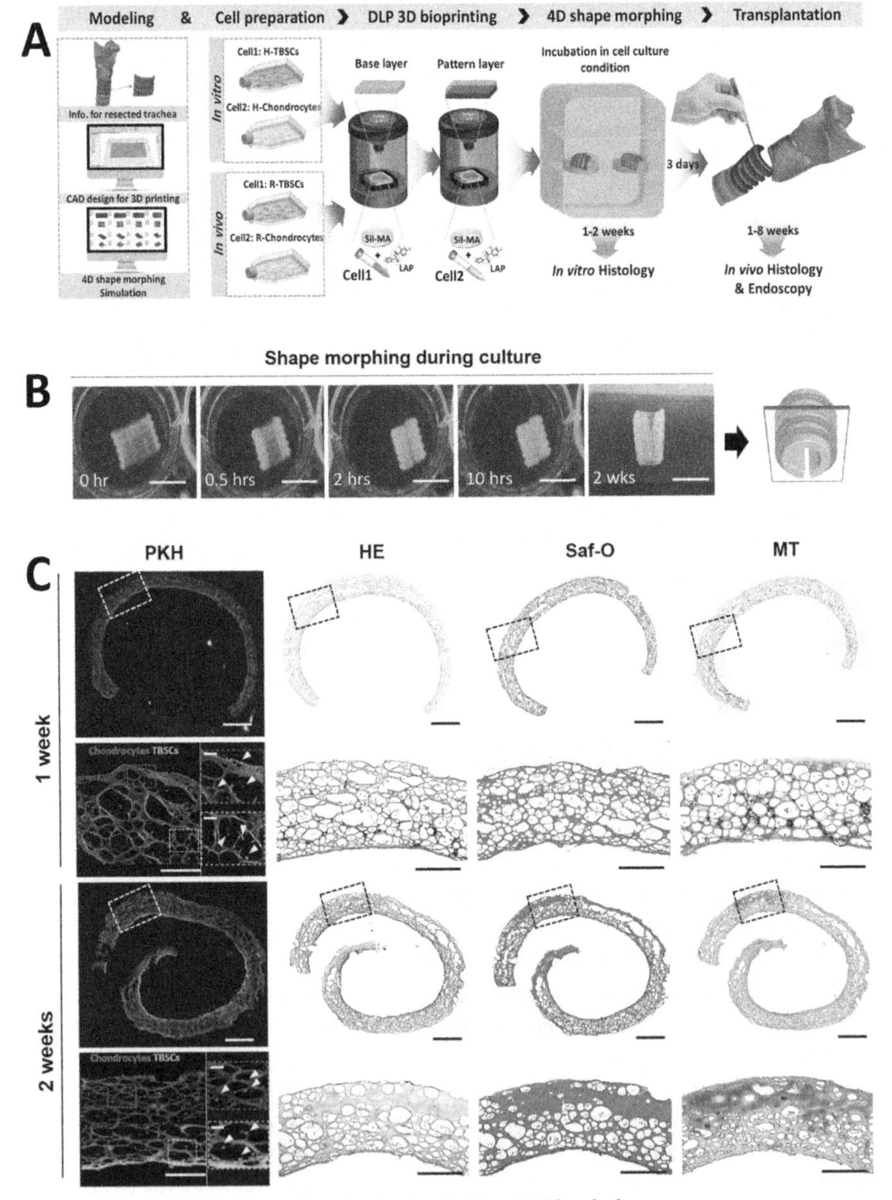

FIGURE 10.9 Tracheal Tissue Constructs Created by 4D Bioprinting.

(A) Schematics of the study design; (B) experimental snapshots of the shape morphing of a representative human trachea construct cultured in vitro; the target configuration is shown on the right—the TBSC-laden base layer is green and the chondrocyte-laden hydrogel stripes are red; scale bar = 10 mm; and (C) fluorescence microscopy and histological evaluation of human tracheal constructs after 1 week and 2 weeks in culture;

cardiomyocytes have a limited expansion capacity, in recent years researchers focused on cardiac patches populated with human induced pluripotent stem cell-derived cardiomyocytes (hiPSC-CMs). Nevertheless, clinical applications of hydrogel-based cardiac patches are hampered by their low mechanical strength and insufficiently streamlined manufacturing techniques (Cui et al., 2020).

The most prominent challenges of cardiac tissue engineering are (i) inadequate cardiomyocyte maturation, (ii) small patch surface area, (iii) insufficient patch thickness, limited by lack of vascularization, and (iv) lack of electromechanical integration between the implanted cardiac patch and the host myocardium (Shadrin et al., 2017). The "Cardiopatch" platform developed by Shadrin et al. successfully addressed the first two challenges by hydrogel-molding and free-floating dynamic culture of hiPSC-CM-laden cardiac patches of clinically relevant surface area (16 cm^2). After 3 weeks of in vitro culture, cardiopatches displayed spontaneous contractions at rates ranging from 30 to 60 beats per minute, while the constituent hiPSC-CMs were comparable in maturation with adult cardiomyocytes. The observed force-length relationships were in the physiological range, and the passive stiffness of the patches was similar to the diastolic cross-fiber stiffness of adult human ventricles. The in vivo evaluation of cardiopatches in mice and rats demonstrated the presence of vascularization and blood perfusion within 2 weeks of implantation. Although the patches remained functional, maintaining their electrical properties acquired in vitro, they were separated from the underlying epicardium by a thin noncardiac tissue layer and failed to develop functional coupling with the host myocardium (Shadrin et al., 2017).

Cui et al. proposed a 4D bioprinting approach to fabricate cardiac patches with physiological adaptability, populated with 3 cell types: hiPSC-CMs, human mesenchymal stromal cells (hMSCs), and human endothelial cells (hECs) (Cui et al., 2020). The association of these cell types is known to improve cardiac contractility because it favors angiogenesis and myogenesis via paracrine signaling. A hiPSC-CM:hEC:hMSC cell seeding ratio of 4:2:1 provided, at confluency, a ratio of cellular proportions of 30:40:30 (%), respectively, recapitulating the cellular composition of the human heart (25%–35% cardiomyocytes, 40%–45% endothelial cells, and supporting cells—fibroblasts, smooth muscle cells, mast cells, immune cells, etc.).

the scale bar corresponds to 1 mm in the top row of each week, and 100 μm in the magnified images shown on the second row of each week; abbreviations: PKH—fluorescence labeling by the lipophilic long-chain carbocyanine dyes PKH26 (red) for chondrocytes and PKH67 (green) for TBSCs, *HE*, hematoxylin & eosin staining; *MT*, Masson's trichrome staining; *Saf-O*, Safranin-O staining.

Reprinted from Kim, S. H., Seo, Y. B., Yeon, Y. K., Lee, Y. J., Park, H. S., Sultan, M. T., Lee, J. M., Lee, J. S., Lee, O. J., Hong, H., Lee, H., Ajiteru, O., Suh, Y. J., Song, S. H., Lee, K. H. & Park, C. H. (2020). 4D-bioprinted silk hydrogels for tissue engineering. Biomaterials, 260, *120281. https://doi.org/10.1016/j.biomaterials.2020.120281, with permission from Elsevier.*

Their cardiac patches were made of a photocurable bioink, an aqueous solution of 5% GelMA and 15% polyethylene glycol diacrylate (PEGDA), using laser-beam-scanning stereolithography. This technique provides precise control over the crosslinking degree of the inks by tuning the printing speed and laser beam intensity. During the fabrication of the multilayered construct, the photocuring of a new layer contributed to the additional crosslinking of previous layers. Thus, the bottom layer had the highest crosslinking density, and, therefore, the patches had a tendency to bend toward the most recently cured layer Fig. 10.10A. When the printing speed was 6 mm/s, the curvature generated by shape morphing matched the curvature of the left ventricular surface of the murine heart. The shape transition did not change the microstructure of the construct—200 μm-thick hydrogel filaments, with 45 degrees angles between adjacent layers, and 40% fill density (Fig. 10.10B). The stress–strain curve of such a patch was found in good agreement with the tensile properties of the myocardium under physiological strain (Cui et al., 2020).

Cardiac patches made of hiPSC-CMs dispersed in the bioink (10^6 cells/mL) did not exhibit spontaneous beating. Hydrogel-embedded cells were less active metabolically than those in conventional culture. Therefore, Cui et al. chose to create cellularized patches by seeding them with cells after the printing was completed. Within 3 days, patches seeded with hiPSC-CMs exhibited spontaneous contractions along the fibers. After 1 week, the hydrogel mesh was covered by an interconnected cell population, which started to beat in synchrony due to

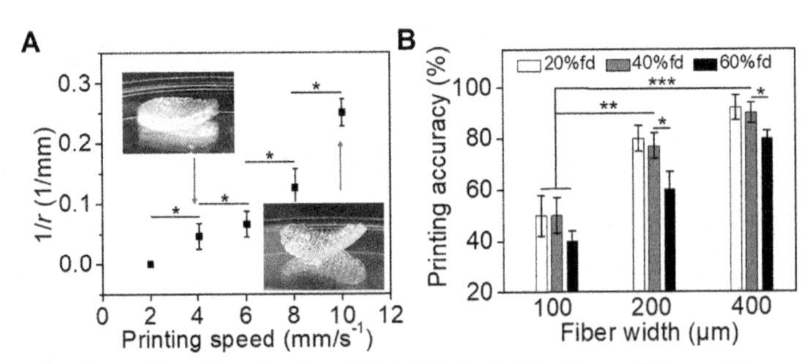

FIGURE 10.10 Shape Morphing Cardiac Patch Fabricated by Stereolithography.

(A) The self-bending of the patch can be controlled by choosing the right printing speed; (B) the printing accuracy plotted versus hydrogel fiber width for different values of the fill density (fd). In both panels, the plotted values represent the mean of at least six experiments, whereas error bars represent standard deviations. The statistical significance of differences between mean values was evaluated via one-way analysis of variance (ANOVA); *, **, and *** stand for $P < .05$, $P < .01$, and $P < .001$, respectively.

Reprinted with permission of AAAS from Cui, H., Liu, C., Esworthy, T., Huang, Y., Yu, Z., Zhou, X., San, H., Lee, S., Hann Sung, Y., Boehm, M., Mohiuddin, M., Fisher John, P. & Zhang Lijie, G. (2020). 4D physiologically adaptable cardiac patch: A 4-month in vivo study for the treatment of myocardial infarction. Science Advances, *6, eabb5067. https://doi.org/10.1126/sciadv.abb5067.*

electrophysiological coupling. Their contraction rate was about the same as the rate observed in hiPSC-CM monolayers. In patches seeded with a mixture of hiPSC-CM, hEC, and hMSC, cells adopted a longitudinal alignment along the fibers, proliferated (covering the patch uniformly within 1 week), and generated spontaneous contractions. The contractile activity of hiPSC-CMs, however, did not lead to in-plane contractions of the entire patch, probably because of the high mechanical resistance of the hydrogel mesh.

To provide biomimetic conditions during the in vitro culture, cardiac patches were subjected to dual mechanical stimulation consisting of (i) periodic radial compressions meant to mimic the mechanical loading encountered in vivo and (ii) fluid flow devised to mimic the shear stress caused by blood flow. This stimulation regime elicited, within 2 weeks, an increased expression of genes associated with cardiac maturation (α-actinin—a measure of sarcomere density; Connexin 43 (Cx43)—a marker associated with the formation of gap junctions; myosin light chain 2 (MYL2)—a marker of ventricular cardiomyocyte maturation), and angiogenesis (von Willebrand factor (vWf), and CD31). Also, the stimulated patches contained a higher number of longitudinally aligned vascular cells than their nonstimulated counterparts (Cui et al., 2020).

The cardiac patches were also tested in vivo, in a murine model of chronic myocardial infarction based on ischemia-reperfusion injury. In each animal (NSG immunodeficient mouse), a patch was placed over the infarcted portion of the heart (Fig. 10.11A and B). Although no fibrin glue was used, the patch adhered firmly to the epicardium, as shown in the optical images taken after 3 weeks of implantation (Fig. 10.11C). Histological analysis based on H&E staining reveals dense cell clusters at the interface between the epicardium and the implanted patch, providing firm engraftment (Fig. 10.11D), whereas fluorescent microscopy proved that the hiPSC-CMs remained viable (these cells were transfected to express the green fluorescent protein (GFP) to differentiate them from the cells of the host animal) (Fig. 10.11E). The immunofluorescence analysis of cardiac troponin I (cTnI) shows regions populated by cardiac myocytes, whereas the analysis of vWf expression reveals the presence of vascular cell clusters crossing the interface between the epicardium and the patch (Fig. 10.11F). After 10 weeks of implantation, the mean infarct size of mice treated with cardiac patches was less than half of the mean infarct size of untreated mice (Fig. 10.11G). Cardiac magnetic resonance imaging proved the existence of blood flow from the heart to the patch (Fig. 10.11H).

At week 10, both the patch and the underlying heart muscle are traversed by capillaries (Fig. 10.11I). Capillaries generated by the implanted cells are visualized in Fig. 10.11J via DAPI staining combined with the immunostaining of human CD31, also known as platelet endothelial cell adhesion molecule-1 (PECAM-1). The progressive increase in the extent of vascularization is shown in Fig. 10.11K, however, mainly resulted from host blood vessel ingrowth because the proportion of CD31+ cells did not increase from week 10 to 4 months (Cui et al., 2020).

A rich body of evidence suggests that cells survive in epicardially implanted patches and take advantage of dense neovascularization originating from the host

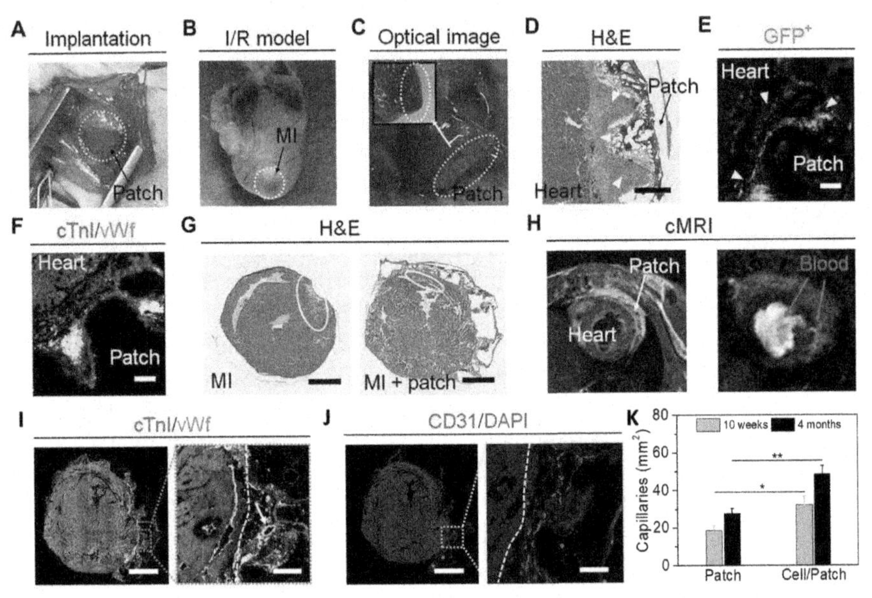

FIGURE 10.11 In vivo Evaluation of a 4D Physiologically Adaptable Cardiac Patch.

(A) Photograph of the patch taken right after implantation; (B) Picture of a murine heart taken after 4 months from the implantation of a patch covering the infarcted region; (C) Photograph taken after 3 weeks in vivo, showing that the patch adheres firmly to the ventricular wall (inset); (D) histological image of the cross-section of the ventricular wall covered by the cellularized patch (H&E staining), with yellow arrowheads pointing toward cell clusters of high density; scale bar, 400 μm; (E) Fluorescence microscopy image with yellow arrowheads indicating regions inhabited by GFP^{+} hiPSC-CMs; scale bar, 100 μm; (F) Immunostaining of the heart/patch interface at week three to show cardiac myocytes, which express cTnI (red), and endothelial cells, which express vWf (green); scale bar, 100 μm; (G) H&E staining of untreated (MI) and treated (MI + patch) hearts at week 10; yellow ellipses highlight infarcted regions; scale bars, 800 μm; (H) Cardiac magnetic resonance imaging (cMRI) of treated hearts at week 10, in spin echo mode (left) to show tissue structures and cine mode (right) to visualize blood (white); (I) Immunostaining of a section of the treated heart at week 10; cTnI (red), vWf (green), in low magnification (left—scale bar, 800 μm) and in close-up view (right—scale bar, 100 μm) of the epicardial surface (white dashed line) and the overlying patch exhibiting capillary vessels (white arrow); (J) Fluorescence microscopy image of a section of the treated heart at week 10; DAPI staining of cell nuclei (blue) and immunostaining of human-specific CD31 (red) showing capillaries generated by human endothelial cells (hECs); left—scale bar, 800 μm; right—scale bar, 100 μm; (K) the area covered by capillaries identified by immunostaining of vWf in infarcted hearts treated with cell-free constructs (Patch) or cellularized constructs (Cell/Patch); here * and ** denote $P < .05$ and $P < .01$, respectively, where the P-value refers to Student's t-test. For interpretation of the references to color in this figure legend, please refer online version of this title.

Reprinted with permission of AAAS from Cui, H., Liu, C., Esworthy, T., Huang, Y., Yu, Z., Zhou, X., San, H., Lee, S., Hann Sung, Y., Boehm, M., Mohiuddin, M., Fisher John, P. & Zhang Lijie, G. (2020). 4D physiologically adaptable cardiac patch: A 4-month in vivo study for the treatment of myocardial infarction. Science Advances, *6, eabb5067. https://doi.org/10.1126/sciadv.abb5067.*

as well as from the implanted hECs and hMSCs. The patches adhered to the epicardial surface over the entire implantation period and diminished the infarct sizes, presumably because the mechanical support provided by the patch reduced scar tissue formation. These features were attributed to the appropriate geometry of the patch combined with its capability of physiological stretching. The cellularization of the patch also contributed to its robust engraftment; indeed, acellular patches had a weak adhesion by month 4. The patch, however, did not improve cardiac function, probably because the patch was unable to perform in-plane contractions. Nevertheless, the lack of adverse effects and several benefits plead for further studies involving more complex printing techniques and in vivo experiments performed on larger animals.

Within 1 year, progress has been reported by the same research group. Wang et al. used DLP-printed molds to create stimulus-responsive mats with microgrooves on their surface and seeded them with the same cocktail of cells (hiPSC-CM:hEC: hMSC at a ratio of 4:2:1) (Wang et al., 2021). The molds were made of a photocrosslinkable ink composed of 30% polyethylene glycol diacrylate (PEGDA), 70% deionized water, and 1% Irgacure-819 (the latter was the photoinitiator). The PEGDA hydrogel molds were airdried for 1 day, while they lost 77% of their weight and shrunk by about 40% in linear size; as such, the pattern sizes decreased by the same percentage. Hence, the feature sizes of the dried hydrogel molds were beyond the printer's resolution. These molds were used to build substrates for cell seeding (curved mats with microgrooves), made of a shape memory polymer—bisphenol A diglycidyl ether, poly(propylene glycol) bis(2-aminopropyl) ether, and decylamine at a molar ratio of 10:3:4, ensuring a glass transition temperature, T_g, of about 45 °C. The mats were flattened at a high temperature $T > T_g$; they were maintained so while being cooled down to physiological temperature. This thermomechanical programming brought about the fixation of the flat shape, which enabled uniform cell seeding. In about a week in culture, the cells aligned along the microgrooves (except for hECs) and gave rise to synchronous contractions of the entire tissue construct—a hallmark of the electrophysiological coupling of hiPSC-CMs. Immunostaining revealed myocardial protein expression (cTnI and α-actinin), indicating that the triculture conditions were favorable for hiPSC-CM maturation. During the fabrication process, the shape memory polymer was doped with 15% graphene to make it capable of absorbing near-infrared radiation (NIR). On day 7, the construct was exposed to NIR, so the substrate was remotely heated above T_g, eliciting the recovery of its initial, curved shape. The cells were not affected by this process because NIR absorption raised the shape memory polymer's bulk temperature, while the cells were in contact with the cooler cell culture medium. The curved multicellular patch exhibited rhythmic, directional beating, which was slower but more uniform than the beats of hiPSC-CM monolayers formed on multiwell plates. Micropatterned, NIR-sensitive substrates might be used in the future for building functional cardiac constructs, with aligned myofibers, whose curvature can be precisely adjusted to specific recipient hearts (Wang et al., 2021).

References

Apsite, I., Constante, G., Dulle, M., Vogt, L., Caspari, A., Boccaccini, A. R., Synytska, A., Salehi, S., & Ionov, L. (2020). 4D Biofabrication of fibrous artificial nerve graft for neuron regeneration. *Biofabrication, 12*, 035027. https://doi.org/10.1088/1758-5090/ab94cf

Apsite, I., Uribe, J. M., Posada, A. F., Rosenfeldt, S., Salehi, S., & Ionov, L. (2020). 4D biofabrication of skeletal muscle microtissues. *Biofabrication, 12*, 015016. https://doi.org/10.1088/1758-5090/ab4cc4

Armitage, J. (2007). The safety of statins in clinical practice. *Lancet, 370*, 1781–1790. https://doi.org/10.1016/s0140-6736(07)60716-8

Constante, G., Apsite, I., Alkhamis, H., Dulle, M., Schwarzer, M., Caspari, A., Synytska, A., Salehi, S., & Ionov, L. (2021). 4D biofabrication using a combination of 3D printing and melt-electrowriting of shape-morphing polymers. *ACS Applied Materials & Interfaces, 13*, 12767–12776. https://doi.org/10.1021/acsami.0c18608

Cui, H., Liu, C., Esworthy, T., Huang, Y., Yu, Z., Zhou, X., San, H., Lee, S., Hann Sung, Y., Boehm, M., Mohiuddin, M., Fisher John, P., & Zhang Lijie, G. (2020). 4D physiologically adaptable cardiac patch: A 4-month in vivo study for the treatment of myocardial infarction. *Science Advances, 6*. https://doi.org/10.1126/sciadv.abb5067. eabb5067.

Di Filippo, E. S., Mancinelli, R., Marrone, M., Doria, C., Verratti, V., Toniolo, L., Dantas, J. L., & Fulle, S. (2017). Neuromuscular electrical stimulation improves skeletal muscle regeneration through satellite cell fusion with myofibers in healthy elderly subjects. *Journal of Applied Physiology, 123*, 501–512. https://doi.org/10.1152/japplphysiol.00855.2016

Khodabukus, A., Madden, L., Prabhu, N. K., Koves, T. R., Jackman, C. P., Muoio, D. M., & Bursac, N. (2019). Electrical stimulation increases hypertrophy and metabolic flux in tissue-engineered human skeletal muscle. *Biomaterials, 198*, 259–269. https://doi.org/10.1016/j.biomaterials.2018.08.058

Kim, S. H., Seo, Y. B., Yeon, Y. K., Lee, Y. J., Park, H. S., Sultan, M. T., Lee, J. M., Lee, J. S., Lee, O. J., Hong, H., Lee, H., Ajiteru, O., Suh, Y. J., Song, S. H., Lee, K. H., & Park, C. H. (2020). 4D-bioprinted silk hydrogels for tissue engineering. *Biomaterials, 260*, 120281. https://doi.org/10.1016/j.biomaterials.2020.120281

Kirillova, A., Maxson, R., Stoychev, G., Gomillion, C. T., & Ionov, L. (2017). 4D biofabrication using shape-morphing hydrogels. *Advanced Materials, 29*, 1703443. https://doi.org/10.1002/adma.201703443

Lee, Y. B., Jeon, O., Lee, S. J., Ding, A., Wells, D., & Alsberg, E. (2021). Induction of four-dimensional spatiotemporal geometric transformations in high cell density tissues via shape-changing hydrogels. *Advanced Functional Materials, 31*, 2010104. https://doi.org/10.1002/adfm.202010104

Li, Y.-C., Zhang, Y. S., Akpek, A., Shin, S. R., & Khademhosseini, A. (2017). 4D bioprinting: The next-generation technology for biofabrication enabled by stimuli-responsive materials. *Biofabrication, 9*, 012001. http://stacks.iop.org/1758-5090/9/i=1/a=012001.

Miao, S., Cui, H., Esworthy, T., Mahadik, B., Lee, S., Zhou, X., Hann, S. Y., Fisher, J. P., & Zhang, L. G. (2020). 4D self-morphing culture substrate for modulating cell differentiation. *Advanced Science, 7*, 1902403. https://doi.org/10.1002/advs.201902403

Miao, S., Cui, H., Nowicki, M., Xia, L., Zhou, X., Lee, S.-J., Zhu, W., Sarkar, K., Zhang, Z., & Zhang, L. G. (2018). Stereolithographic 4D bioprinting of multiresponsive architectures for neural engineering. *Advanced Biosystems, 2*, 1800101. https://doi.org/10.1002/adbi.201800101

Normand, C., Kaye, D. M., Povsic, T. J., & Dickstein, K. (2019). Beyond pharmacological treatment: An insight into therapies that target specific aspects of heart failure pathophysiology. *The Lancet, 393*, 1045–1055. https://doi.org/10.1016/S0140-6736(18)32216-5

Shadrin, I. Y., Allen, B. W., Qian, Y., Jackman, C. P., Carlson, A. L., Juhas, M. E., & Bursac, N. (2017). Cardiopatch platform enables maturation and scale-up of human pluripotent stem cell-derived engineered heart tissues. *Nature Communications, 8*, 1825. https://doi.org/10.1038/s41467-017-01946-x

Vos, T., Lim, S. S., Abbafati, C., Abbas, K. M., Abbasi, M., Abbasifard, M., Abbasi-Kangevari, M., Abbastabar, H., Abd-Allah, F., Abdelalim, A., Abdollahi, M., Abdollahpour, I., Abolhassani, H., Aboyans, V., Abrams, E. M., Abreu, L. G., Abrigo, M. R. M., Abu-Raddad, L. J., Abushouk, A. I., … Murray, C. J. L. (2020). Global burden of 369 diseases and injuries in 204 countries and territories, 1990–2019: A systematic analysis for the global burden of disease study 2019. *The Lancet, 396*, 1204–1222. https://doi.org/10.1016/S0140-6736(20)30925-9

Wang, Y., Cui, H., Wang, Y., Xu, C., Esworthy, T. J., Hann, S. Y., Boehm, M., Shen, Y.-L., Mei, D., & Zhang, L. G. (2021). 4D printed cardiac construct with aligned myofibers and adjustable curvature for myocardial regeneration. *ACS Applied Materials & Interfaces, 13*, 12746–12758. https://doi.org/10.1021/acsami.0c17610

Yang, G. H., Kim, W., Kim, J., & Kim, G. (2021). A skeleton muscle model using GelMA-based cell-aligned bioink processed with an electric-field assisted 3D/4D bioprinting. *Theranostics, 11*, 48–63. https://doi.org/10.7150/thno.50794

CHAPTER

Perspectives of 3D and 4D bioprinting

11

Taken together, the previous chapters demonstrated the tremendous potential of 4D bioprinting for biomedical applications. It is clear by now that progress in 4D bioprinting will contribute to the advancement of several research fields, including tissue engineering, biosensors, drug delivery, bioelectronics, and biorobotics. Attempting to keep a careful balance between science and fiction, this chapter will discuss potential technical developments, as well as applications envisioned by researchers involved in 4D bioprinting. We first describe promising mathematical methods, then we discuss emergent strategies in 3D bioprinting, and, to save the best for last, we conclude this chapter and the book by peeking at the endless field of potential applications.

1. Mathematical modeling

The importance of mathematical modeling in the 4D printing of inanimate smart materials has been argued in Chapter 2. A reliable approach to programmable matter created by additive manufacturing takes into account the material properties and the geometric details of the multimaterial structure. Therefore, the team led by Skylar Tibbits initiated Project Cyborg, a design platform augmented with the Maya Nucleus (Autodesk, CA, USA) dynamics solver capable of simulating the evolution of designed objects (Raviv et al., 2014). There is a need for similar software specifically tailored for 4D bioprinting.

In 3D bioprinting, mathematical models serve to optimize the printing process (Chapter 6) and predict the bioprinted structure's final state (Chapter 8). In 4D bioprinting, theoretical models also seek to provide information on the change in shape and/or function expected upon stimulation. Modeling has the potential to save time and money because fewer experiments will be required for the fine-tuning of the bioprinting process.

1.1 Analytic models of stimulus responsiveness

Mathematical and computational models of 4D bioprinting establish connections between (i) printing parameters, (ii) bioink properties, (iii) stimulus properties, and (iv)

Towards 4D Bioprinting. https://doi.org/10.1016/B978-0-12-818653-4.00003-6

final construct characteristics. To this end, two different strategies have been proposed: the forward problem approach, in which the final construct is unknown, and the inverse problem approach, in which the printing parameters are unknown. The first is investigative because it starts from a given print path and explores possible outcomes depending on printer settings and material properties, whereas the second is application-oriented since it starts from the desired outcome and aims at finding the proper print path and printing parameters for given material properties (Ashammakhi et al., 2018). Momeni and Ni formulated three fundamental laws of 4D printing and proved that the shape-morphing behavior of most multimaterial 4D-printed structures can be described by a biexponential function (Momeni & Ni, 2020). Their theoretical framework is expected to provide valuable insights into 4D bioprinting, too, at least in the context of tissue constructs whose shape changes are driven by smart materials. Future research will elucidate whether biologically driven shape transitions follow the same laws or extensions will be necessary.

In their study of 4D-printed bilayers of botanical inspiration, Gladman et al. solved both the forward and inverse design problems concerning shape changes induced by immersion in water (Gladman et al., 2016). They generalized the classical theory of bimetallic strip bending to account for the anisotropy of their composite hydrogel ink, as well as the specific patterning of each layer. The shape morphing originated from the elastic and swelling anisotropies of the composite hydrogel made of a soft acrylamide hydrogel loaded with stiff cellulose fibrils. In the course of extrusion, the fibrils adopted a preferential orientation due to shear-induced alignment along the filament length (parallel to the printing path), as shown in Fig. 11.1, *panels* A, C, and D. Therefore, the longitudinal swelling strain, $\alpha_{\parallel}$, of the printed filaments was about 4 times smaller than their transverse swelling strain $\alpha_{\perp} \approx 40\%$ (Fig. 11.1A, *middle*).

The model allowed for the independent control of both the mean and Gaussian curvatures. Planar structures comprising concentric circles became roughly conical upon water immersion, except for the apex region, which had a positive Gaussian curvature. On the other hand, an orthogonal bilayer lattice turned into a saddle-like surface with vanishing mean curvature and almost uniform negative Gaussian curvature (Fig. 11.1A, *right*). The mathematical model of Gladman et al. can also be used in 4D bioprinting to analyze the shape morphing of tissue constructs comprising bilayers of anisotropic hydrogel filaments.

1.2 Optimization of self-folding

The study of self-folding structures is a vivid field of research mainly motivated by the ease of fabrication of flat objects. Their automatic folding can lead to intricate 3D geometries. Such a fabrication strategy is especially appealing in tissue engineering because anatomical surfaces are typically irregular and patient-specific. A self-folding engineered cartilage patch, for instance, might provide a perfect fit for a given patient. Nevertheless, most 3D surfaces are not flattenable and, therefore, cannot result from the self-folding of rigid subunits connected by joints. Theoretical

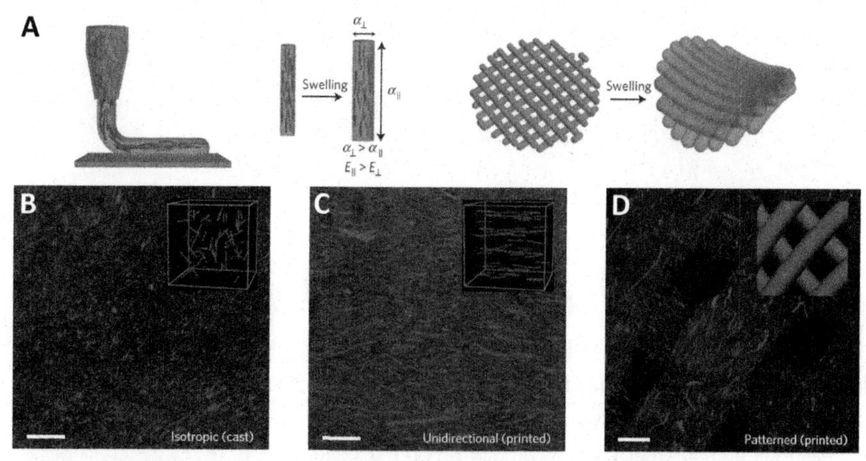

FIGURE 11.1 Biomimetic 4D printing of anisotropic hydrogel structures.

(A) *Left*: diagram of the extrusion process showing the alignment of the embedded cellulo se fibrils (the extent of alignment depends on nozzle diameter and printing speed); *midd le*: since the fibers are oriented mainly along the extruded filament, the swelling strain is smallest and the stiffness is highest along the longitudinal direction; *right*: the bilayer be comes saddle-shaped upon water immersion because the transverse swelling of each fil ament is more pronounced than the longitudinal one. (B–D) confocal microscopy imagi ng of cellulose fibrils stained blue with Calcofluor White in a representative cast specimen (B), unidirectionally printed sheet (C), and patterned printed bilayer (D); scale bars, 200 μm.

Reprinted with permission from Springer Nature from Gladman, A. S., Matsumoto, E. A., Nuzzo, R. G., Mahadevan, L. & Lewis, J. A. (2016). Biomimetic 4D printing. Nature Materials, 15, *413–418. https://doi.org/10.1038/nmat4544*

studies of the ancient Japanese arts origami and kirigami provided insights into shape optimization aimed at approximating a given surface with a flattenable one (Choi et al., 2019; Dudte et al., 2016; Kwok et al., 2015). Origami is the art of paper folding to create decorative 3D objects, whereas kirigami involves cutting and folding paper into ornamental shapes. Dudte et al. investigated the inverse problem of origami—that of designing patterns to create target shapes (Dudte et al., 2016). They developed constrained optimization algorithms to design folding patterns suitable to drape complex 3D surfaces. Fig. 11.2 depicts calculated and actual origami constructs, which conform closely to diverse target shapes. The model also quantifies the trade-off between the accuracy of using a folded structure to approximate a smooth surface and the effort put into creating finer folds. Moreover, a physical simulation allowed for testing whether the constructed patterns were rigid-foldable (i.e., suitable for thick origami) or not.

Applying constrained optimization, Choi et al. solved the inverse problem of kirigami tesselations, too (Choi et al., 2019). They devised an algorithm to design a partially cut flat sheet to morph into an open structure that roughly

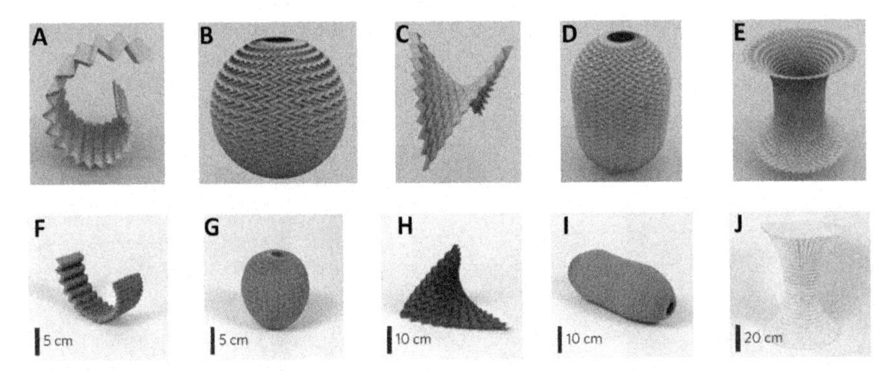

FIGURE 11.2 Calculated origami tesselations and their paper counterparts.

Digital models of the calculated origami tesselations are represented on the *top* (A–E), whereas their actual paper versions are shown on the *bottom* (F–I). (A, F) generalized cylinder (logarithmic spiral); (B, G) *sphere*; (C, H) *hyperbolic paraboloid*; (D, I) *cylinder* with positively curved caps; (E, J) *cylinder* with negatively curved caps.

Reproduced with permission from Springer Nature from Dudte, L. H., Vouga, E., Tachi, T. & Mahadevan, L. (2016). Programming curvature using origami tessellations. Nature Materials, 15, *583–588. https://doi.org/10.1038/nmat4540*

conforms to the desired target shape. To show the practical relevance of their approach, they created kirigami patterns by polydimethylsiloxane (PDMS) molding and demonstrated their deployment approximating diverse 3D surfaces. PDMS is a gas-permeable, biologically inert substance, which is extensively used in contact with live cells. In particular, it is one of the most appreciated materials for building tissue-on-chip devices. Hence, it seems feasible to build complex origami and kirigami structures to guide the self-assembly of engineered tissues.

Origami constructs comprising live cells have been built before the birth of 4D printing (Jamal et al., 2013; Kuribayashi-Shigetomi et al., 2012). Jamal et al. used conventional photolithography to assemble micropatterned poly(ethylene glycol) (PEG) bilayers of differing molecular weights (700 and 4000 Da). Placing them in aqueous media resulted in spontaneous folding because PEG 4000 had a higher swelling ratio than PEG 700. With tissue engineering applications in their mind, these authors encapsulated β-TC-6 cells into PEG bilayers—a 16 μm thick PEG 700 layer covered by a 65 μm thick cell-laden PEG 4000 layer. For comparison, they also encapsulated β-TC-6 cells into similarly patterned PEG 4000 monolayers, which remained planar. The cells remained viable, proliferated, and produced insulin in both constructs, but, remarkably, after 3 weeks in culture, the bio-origami structures secreted significantly more insulin than the planar ones. Kuribayashi-Shigetomi et al. (2012) fabricated biological origami structures by culturing cells on microtiles. These will be discussed briefly in Section 3 from the practical point of view. We will see that their folding is accomplished by cells cultured on biocompatible microtiles placed next to each other. In their simplest version, cells also serve as hinges—an elegant solution, which, however, turned out to be the Achilles' heel of cell origami, resulting in poor reproducibility. From a theoretical perspective, cell origami is a nice proof-of-concept system, suggesting that the mathematics of origami might pave the way to a myriad of applications of 4D bioprinting. Printed active composites can be used as smart hinges to accomplish reliable folding, but they are not the only option (see Chapter 2). High-resolution multi-material 3D printing is already available at reasonable costs, and it is expected to get better and cheaper in time. Thus, rigid origami and/or kirigami systems are likely to become valuable tools for building and shaping engineered tissues.

1.3 Machine learning

Progress in 3D/4D bioprinting is expected to stem also from artificial intelligence—the science devoted to making machines think and act like human beings (Ng et al., 2020; Yu & Jiang, 2020). A subfield of artificial intelligence, machine learning seeks to enable computers to perform tasks without explicit programming. A machine learning program accesses data and learns automatically, without human intervention. Machine learning based on artificial neural networks is known as deep learning. The history of additive manufacturing of inanimate objects demonstrates that machine learning is an effective tool for product quality improvement and cost reduction. It is increasingly used in 3D printing for process optimization and product characterization including material property prediction, dimensional accuracy assessment, and manufacturing defect detection (Yu & Jiang, 2020).

Machine learning was found an enabling technique also in 4D printing. For example, Hamel et al. conceived a machine-learning approach for designing 3D-printed active composite structures of desired shape-shifting response (Hamel et al., 2019). To this end, they used an evolutionary algorithm (EA) and the finite element method. An EA mimics natural selection by letting populations of potential solutions compete with each other until the fittest one prevails. In the study of Hamel et al., the EA produced candidate solutions, whereas the finite element method was used to rank them. A 2D model of a cantilever beam was partitioned into 5 rows of 100 square voxels of the same size but two different types: active and passive. Using the finite element method, Hamel et al. simulated the behavior of active voxels in terms of thermal expansion, and the output of the optimization process was the distribution of active voxels that ensured the smallest mean squared error (deviation) of the simulated shape compared to the target shape.

In 3D bioprinting, machine learning is seen as a potential source of progress, as well (Yu & Jiang, 2020). Machine learning is especially suited for the optimization of complex processes (which is the case in bioprinting), but it is based on computational algorithms that need to be trained using a vast amount of data. Indeed, machine learning is a data-driven approach, capable of generating useful results even in the absence of detailed knowledge of the phenomena involved in the investigated problem. Early studies of recurrent neural networks and support vector networks produced encouraging results and motivated the development of new algorithms, eventually leading to the widespread use of machine learning in fields ranging from face recognition to the prediction of customer preferences based on previous choices (Hamel et al., 2019). In the context of 3D bioprinting, deep learning has been defined as "a collection of algorithms where models composed of multiple processing layers learn abstract representations from raw data collected from layer-by-layer fabrication approaches to fabricate 3D tissue constructs comprising of living cells, biomaterials, and growth factors" (Ng et al., 2020).

Despite remarkable achievements in computer vision and robotics, machine learning remains largely unexplored in 3D bioprinting, perhaps due to insufficient data, but this problem is going to vanish in the near future. A few applications have been reported in the literature, and several others have been proposed for enhancing the bioprinting workflow (Ng et al., 2020). In the preprinting phase, deep learning algorithms might help improve the resolution of medical images. Also, it might be possible to expedite magnetic resonance imaging by training a neural network to generate a high-resolution image from a small number of 2D slices. In the printing phase, deep learning models can be trained to predict key outcomes (e.g., print duration or cell viability). They can also serve for bioink optimization: instead of time-consuming and expensive experiments, machine learning could be used to develop bioink compositions that boost cell proliferation and ECM deposition (Yu & Jiang, 2020). Hence, machine learning has the potential to complement the information provided by mathematical formulas of printing resolution expressed as a function of printing parameters (see Chapter 6). In the postprinting phase, tissue maturation can be promoted via machine learning algorithms trained to optimize cell culture medium formulations needed for co-culturing several cell types. Moreover, deep learning can be applied to optimize bioprinted tissue maturation by providing biomimetic conditions (Ng et al., 2020). For example, the use of generative adversarial networks has been proposed for creating unexpected designs for tissue engineering scaffolds to sustain cell growth and tissue maturation (Yu & Jiang, 2020).

Machine learning is expected to benefit also 4D bioprinting by providing better control over the additive manufacturing process as well as predicting the printout's response to stimulation. As discussed in Chapter 2, 4D printing was inspired by protein folding. Just as a chain of amino acids folds on its own, a 4D (bio)printed structure can change its shape and work as expected. The London-based artificial intelligence company DeepMind (see https://www.deepmind.com/about) has recently solved the problem of protein structure prediction (Jumper et al., 2021). AlphaFold, their open-source artificial intelligence software, has since been used to predict the 3D structures of virtually all known proteins - about one million structures made publicly available in a database maintained by the European Molecular Biology Laboratory's European Bioinformatics Institute (Callaway, 2022). The success of AlphaFold suggests that artificial intelligence will play a central role also in the progress of 4D bioprinting.

2. Emergent bioprinting techniques and materials

Chapter 5 is far from being a comprehensive account of the bioprinting strategies explored to date. Since none of them is able to replicate the geometric and compositional complexity of native tissues and organs, new techniques are being developed and old ones are being combined in innovative ways (Dalton et al., 2020; Ji & Guvendiren, 2021).

Low throughput, a major limitation of most bioprinting techniques, is surmounted by volumetric bioprinting, in which a rotating transparent cylinder filled with photosensitive material is illuminated by a sequence of 2D images (Kelly et al., 2019). Each patterned light beam traverses the material from a different angle, and the superposition of successive exposures hardens the material according to the predesigned 3D geometry. This technique is known as computed axial lithography (CAL). Within minutes, it can build centimeter-sized constructs of unrivaled surface smoothness and feature sizes in the range of hundreds or even tens of micrometers. Besides the pixel size of the projected images, the resolution of CAL depends on the diffusion of the chemical species involved in the photopolymerization process.

According to its creators, CAL was inspired by the backprojection algorithms employed in computed tomography, as well as by algorithms that govern intensity-modulated radiation therapy of cancer patients (Kelly et al., 2019). CAL relies on a digital light projector to illuminate the photopolymer precursor solution. Free radicals generated by the incoming light are deactivated by oxygen, so the material starts to crosslink as soon as oxygen is depleted locally; this happens only if the energy density of the curing radiation exceeds a certain threshold. Thus, oxygen inhibition is overcome only in regions exposed from several angles, causing the simultaneous hardening of the entire structure as opposed to the sequential addition of voxels. The uncured resin is removed by solvent rinsing at slightly elevated temperatures meant to lower the viscosity of the resin. Subsequent light curing of the entire construct improves its mechanical properties.

Bernal et al. brought CAL into the realm of regenerative medicine (Bernal et al., 2019). They prepared a cell-friendly bioresin, termed gelRESIN, by dissolving 10% (w/v) GelMA in phosphate-buffered saline and adding lithium phenyl-2,4,6-trimethylbenzoyl-phosphinate (LAP) as a photoinitiator at a relatively low final concentration, of 0.037% (w/v). As a light source, they collimated the output of six laser diodes, of 405 nm wavelength, achieving a total power of 6.4 W, and coupled the resulting beam into an optical fiber. The output of the fiber was expanded and pointed onto a digital micromirror device, which projected patterned light onto the cylindrical glass vial filled with gelRESIN. Volumetric bioprinting of living tissue constructs could be performed with a minor impact on cell viability: more than 85% of the encapsulated chondroprogenitor cells were viable postprinting. Furthermore, cell viability remained high during 1 week in culture, roughly at the same level as in tissue constructs built by hydrogel casting, extrusion bioprinting, and digital light processing (DLP). Within a volumetric bioprinted trabecular bone model, mesenchymal stem cells evolved as expected during 1 week of culture in osteogenic medium. The construct could be seeded with endothelial cells, which exhibited angiogenic sprouting on day 3, while the nearby mesenchymal stem cells overtook the role of pericytes (i.e., stabilized the capillary network precursors) (Bernal et al., 2019). Thus, it seems safe to conclude that the vast majority of cells remain fully functional in a structure built by volumetric bioprinting.

Another notable advantage of volumetric bioprinting is its ability to print overhanging and/or free-floating parts. To demonstrate this feature, Bernal et al. printed a ball-and-cage fluidic valve suitable to enforce a unidirectional fluid flow, with potential applications in microfluidics or soft robotics. Other techniques would require temporary support structures for such a feat.

Although CAL has not been employed for 4D bioprinting, it is promising because it shares several traits with stereolithography and DLP—two of the most fruitful methods for printing stimulus-responsive tissue constructs. For instance, the biocompatible resin could be loaded with magnetic nanoparticles, and the vial could be placed in a magnetic field during volumetric bioprinting. Excessive light absorption might pose limits on the achievable construct size, but novel nanoparticle coating methods might mitigate this drawback.

Oxygen inhibition is a key factor in another development in photopolymerization-based bioprinting. Less radical than CAL, continuous liquid interface production (CLIP) is a variation of DLP. In CLIP, the bottom of the vat is made of a UV transparent, oxygen-permeable window—Teflon AF 2400, an amorphous fluoropolymer (Tumbleston et al., 2015). In contact with air or pure oxygen, a 10–100 μm thick dead zone forms above the window, in which the resin remains liquid because oxygen quenches the excited photoinitiator molecules or combines with the free radicals formed by their photocleaving. A DLP projector illuminates the vat from below, projecting a continuous sequence of UV images. As opposed to bottom-up stereolithography or DLP (in which photopolymerization, hardened part lifting, and resin renewal are discrete steps taking a few seconds each), in CLIP, the hardened part is progressively pulled out of the resin bath; fresh resin streaming in from the sides of the part is cured at a steady rate. Consequently, the printing speed in CLIP is up to 8 mm/min, which is two orders of magnitude higher than in stereolithography. It is only limited by the curing rate and viscosity of the photosensitive prepolymer solution and does not depend on the slicing thickness of the 3D model. At a fine model slicing and high refresh rate of the projected images, CLIP results in 3D objects of exceptionally smooth surface and feature size below 0.1 mm (Tumbleston et al., 2015).

Fast bioprinting is achieved also in fluid-supported liquid interface polymerization (FLIP), a recently developed continuous DLP technique that relies on buoyancy provided by a support fluid immiscible with the polymer precursor solution (Fig. 11.3) (Beh et al., 2021). The support fluid, Fluorinert FC-40 (perfluorinated oil), is about 80% denser than the precursor solution.

The print bed is a surface-treated glass plate mounted on the bottom of a polypropylene tank. In the beginning, the projector is focused on the print bed covered by a thin layer of prepolymer solution, inducing photopolymerization. The resulting polymer adheres to the print bed, and the printing is carried on by continually

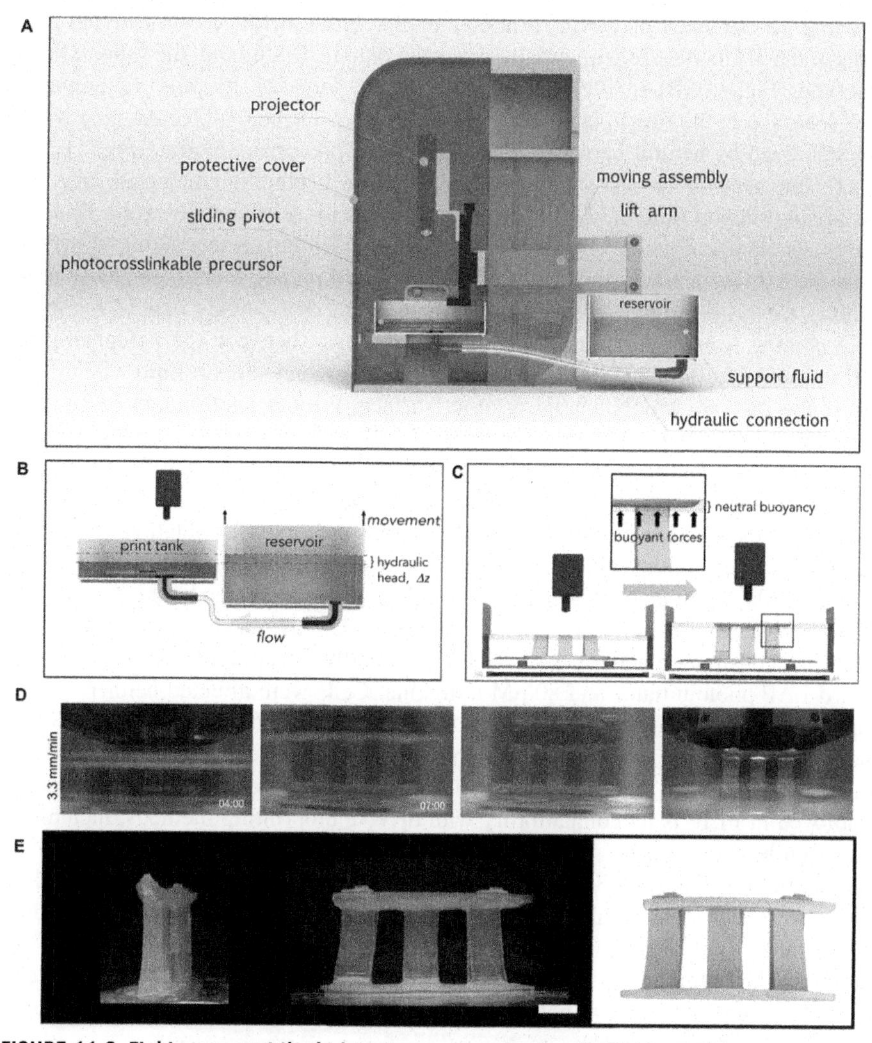

FIGURE 11.3 Fluid-supported liquid interface polymerization (FLIP) bioprinting.

(A) Schematic representation of an FLIP bioprinter; (B) scheme of the hydraulic system responsible for lifting the level of the support fluid under the prepolymer layer; (C) drawin gs of two instants of the printing process and an inset representing the role of buoyancy in stabilizing overhanging parts; (D) successive snapshots of the actual printing process; (E) digital photographs of the printed structure (left and center) corresponding to the digi tal model of the Marina Bay Sands Hotel from Singapore (right); scale bar = 10 mm.

Reprinted from Beh, C. W., Yew, D. S., Chai, R. J., Chin, S. Y., Seow, Y. & Hoon, S. S. (2021). A fluid-supported 3D hydrogel bioprinting method. Biomaterials, 276, *121034. https://doi.org/10.1016/j.biomaterials.2021. 121034, with permission from Elsevier.*

flooding the hardened part with fresh precursor solution, while cross-sectional patterns of the 3D model are sequentially projected onto it. To this end, the support fluid reservoir (Fig. 11.3B) is lifted in tandem with the projector, keeping the prepolymer's surface in the focal plane of the projector (Fig. 11.3A,C). Overhanging parts are stabilized by neutral buoyancy provided by the precursor solution (Fig. 11.3C, inset), but need to be crosslinked properly before becoming immersed into the ever-rising support fluid. (Otherwise, they are unable to withstand the stronger buoyancy). As more and more support fluid is transferred from its reservoir into the print tank, the hardened structure is rinsed of uncrosslinked prepolymer, thereby also preventing the print-through phenomenon—inadvertent crosslinking caused by stray light coming from superior layers. Since the interface between the hardened part and uncured prepolymer is steadily lifted by the inflowing support fluid, the higher the frame rate, the smoother the printout surface. Print resolution can be further improved by dissolving 0.64 mM tartrazine (a yellow food dye, E102) in the prepolymer solution meant to reduce light penetration depth. This option, however, limits the print speed to about 1.2 mm/min (Beh et al., 2021).

Despite extensive tests conducted using various precursor formulations and print parameters, the strengths and limitations of FLIP are not fully unveiled. It is known that live cells can be incorporated into hydrogel constructs created by this technique. For example, mouse fibroblasts (2.8×10^7 cells/mL) have been encapsulated in a hydrogel construct made of 10% GelMA and 10% gelatin in the presence of 1 mM LAP photoinitiator and 80 μM tartrazine. Cells were affected neither by the support fluid nor by the photoinitiator, resulting in cell viability of over 95% in the proximity of the cell culture medium. Cells deeply buried in the hydrogel died due to hypoxia and lack of nutrients. Another advantage of the upright setup employed in FLIP is its compatibility with diverse bioprinting methods, including extrusion-based and cell spheroid-based techniques. Beh et al. printed fugitive biomaterial ink (Pluronic F127) filaments into cell-laden hydrogel blocks. The filaments were deposited by microextrusion onto the free surface of the construct and trapped in it by subsequent light curing. Then, the block was cooled down to 4 °C to liquefy and eliminate the fugitive ink. The resulting channels were penetrated by cell culture medium and helped to nourish cells within roughly 200 μm from the lumen (Beh et al., 2021).

Tartrazine was used as a photoabsorber in a stereolithography apparatus for tissue engineering (SLATE) to create submillimetric perfusable channels within a biocompatible hydrogel (20% by weight PEGDA of 6 kDa molecular weight) (Grigoryan et al., 2019). The SLATE is an open-source, bottom-up DLP instrument, described in detail in the freely available supplementary material of the article by Grigoryanet al. It was used to build intricate vascular architectures akin to the pulmonary capillary system, biomimetic valves, and implantable cellularized constructs. This approach, however, did not allow for placing the cells right next to the vascular walls; instead, fibrin gel loaded with hepatocyte aggregates was placed on top of a thin, prevascularized PEGDA layer. FLIP, on the other hand, enabled Beh et al. to print a free-standing (more precisely, floating) network of millimeter-sized hydrogel tubes (Beh et al., 2021). The question is, are we able to wrap it with cells?

One option for assembling a multicellular structure around a prefabricated vascular tree is to manipulate cells or cell clusters remotely by acoustic or magnetic forces. Guo et al. generated 3D trapping nodes in a microfluidic chamber by superimposing two mutually perpendicular standing surface acoustic waves (Guo et al., 2016). To this end, they placed a rectangular PDMS chamber on a lithium niobate (LiNbO3) piezoelectric substrate and deposited two pairs of interdigital transducers next to the sides of the chamber. To move the trapping nodes parallel to the substrate, they changed the phase angle of each pair of transducers; to move them perpendicularly to the substrate, they adjusted the input acoustic power. Cells captured in these nodes were manipulated along three axes to create complex structures at single-cell resolution. The recent development of circular, slanted-finger interdigital transducers allowed for the acoustic assembly of live cells, as well as the dynamic reconfiguration of the multicellular construct, by simply changing the phase, amplitude, and frequency of the multitone excitation signal (Kang et al., 2020). This technology, however, is limited in terms of throughput, geometric complexity, and achievable construct size. Further developments might lead to acoustic tweezers that create a trapping node arrangement that matches the configuration of the vascular tree, enabling the user to cover the vessel walls with successive layers of cells.

Magnetic levitational assembly might also be capable of "dressing up" a vascular tree with tissue building blocks. Although we are not yet there, tissue constructs have already been built by magnetic levitational bioassembly of tissue spheroids both under normal gravity (Parfenov et al., 2020) and microgravity (Parfenov Vladislav et al., 2020).

Magnetic levitation was pioneered by the research group led by Whitesides to manipulate diamagnetic particles, including live cells (Winkleman et al., 2004), and to facilitate the 3D self-assembly of diamagnetic objects (Ilievski et al., 2011). Water is diamagnetic, as are most organic molecules, as well as certain plastics and metals. Diamagnets are repelled by magnetic fields, so they tend to reside at or close to the field minimum. In fields created by common magnets, however, the magnetic force exerted on most diamagnetic materials is dwarfed by gravity. Therefore, it is preferable to suspend diamagnetic objects in a paramagnetic liquid medium, which is attracted toward regions of high magnetic field. Hence, it displaces the diamagnetic object, pushing it toward regions of low magnetic field. Levitation occurs when the weight of the object is balanced by buoyancy and the magnetic force:

$$\rho_o V \vec{g} - \rho_m V \vec{g} + \frac{(\chi_m - \chi_o)}{\mu_0} V \left(\vec{B} \cdot \nabla \right) \vec{B} = 0 \tag{11.1}$$

where V is the volume of the object (irrelevant, because it cancels out), $\vec{g}$ is the gravitational acceleration (9.81 m s^{-2} in magnitude), $\vec{B}$ is the magnetic flux density, $\mu_0 = 1.26 \times 10^{-6}$ m kg s^{-2} A^{-2} is the magnetic permeability of vacuum, and, finally, $\rho_o(\rho_m)$ and $\chi_o(\chi_m)$ are the density and magnetic susceptibility of the object (medium), respectively. The more paramagnetic is the medium (the larger is χ_m), and the more diamagnetic is the object (the more negative is χ_o), the stronger is the magnetic force that draws the object toward the field minimum. An object less dense than the medium finds its equilibrium closer to the bottom magnet and is lifted by the magnetic force, whereas an object denser than the medium levitates closer to the top magnet while being pushed down by the magnetic force (Fig. 11.4).

Experiments were conducted on millimeter-scale polymethylmethacrylate (PMMA) spheres ($\rho_o = 1.19$ g cm^{-3}), immersed in 1.3 M aqueous solution of $MnCl_2$ ($\rho_m = 1.1$ g cm^{-3}, $\chi_m = 7 \times 10^{-4}$). The cuvette was placed between a pair of NdFeB magnets, fixed in an aluminum cage, with similar poles facing each other. The result of COMSOL calculations of the magnetic field distribution is visualized in the *left panel* of Fig. 11.4, whereas the other panels show the portion of the solution where PMMA spheres added from the top reach mechanical equilibrium. A planar close-packed sheet emerges up to a certain number of spheres (nine in Fig. 11.4); then, as more spheres are added, lateral magnetic forces push them toward the center, giving rise to disordered clusters. Nevertheless, as the cluster grows, its center of mass remains at the same level.

A biocompatible paramagnetic medium was prepared by Winkleman et al. using a contrast agent for magnetic resonance imaging (MRI) approved by the Food and Drug Administration, in which gadolinium (Gd^{3+}) ions are chelated with diethylenetriaminepentaacetic acid (DTPA) (Winkleman et al., 2004). A 40 mM $Gd \cdot DTPA$

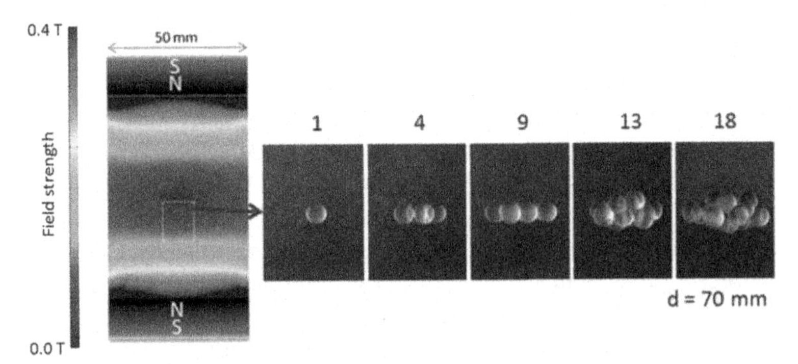

FIGURE 11.4 Self-assembly of diamagnetic hydrogel spheres in a paramagnetic solution placed between two magnets.

The shape of the magnetic trap is shown on the *left*. Photographs of self-assembled clusters of levitating polymethylmethacrylate (PMMA) spheres are depicted for different numbers of spheres (specified above each picture). In this experiment, the distance between the pair of magnets was d = 70 mm.

Republished with permission of The Royal Society of Chemistry from Ilievski, F., Mirica, K. A., Ellerbee, A. K. & Whitesides, G. M. (2011). Templated self-assembly in three dimensions using magnetic levitation. Soft Matter, *7, 9113–9118. https://doi.org/10.1039/C1SM05962A; permission conveyed through Copyright Clearance Center, Inc.*

solution was slightly hypotonic compared to the cytoplasm and provided sufficient magnetic force to trap NIH 3T3 fibroblasts. Cells cultured for 2 days in Dulbecco's Modified Eagle Medium (DMEM) supplemented with 40 mM $Gd \cdot DTPA$ remained viable but did not attach to polystyrene. Yeast cells (*Saccharomyces cerevisiae*) remained viable as well, but their proliferation rate dropped sixfold in the presence of 40 mM $Gd \cdot DTPA$, suggesting that gadolinium compounds might interfere with normal cell function at concentrations larger than a few mM. In intravenous administration of gadolinium-based contrast agents, a commonly used dose is 0.15 mmol/kg body mass, and efforts are being made to ensure more consistent imaging by establishing the dose according to blood volume. In both dosing methods, a typical concentration within the blood is about 2–3 mM, but it decreases gradually as the contrast agent diffuses beyond the vascular space into the extracellular fluid from most organs (Liu et al., 2019).

Parfenov et al. devised a biofabrication approach based on magnetic levitation in a culture medium augmented with 0.8 mM gadobutrol—a macrocyclic neutral gadolinium complex. Gadobutrol is a well-tolerated MRI contrast agent with the trade name "Gadovist 1.0." In an ultrahigh magnetic field (of 19 T flux density) generated by a Bitter magnet, aggregates of SW1353 chondrosarcoma cells have undergone magnetically driven bioassembly within 40 min (Parfenov et al., 2020). Remarkably, the initial tack between the spheroids occurred right away and tissue fusion commenced in less than an hour. A sturdy tissue construct formed within 3 h. In this context, mild magnetic forces keep the spheroids next to each other, in steady contact, which is known to favor fast fusion (Shafiee et al., 2021). What is even more important, the tissue spheroids employed by Parfenov et al. were label-free; that is, they were not loaded with magnetic nanoparticles, which would be undesirable for in vivo tissue repair or replacement.

On the downside, magnetic levitational biofabrication under normal gravity requires expensive electromagnets. To circumvent this barrier, another work by the same research group explored the magnetic bioassembly of cartilage constructs aboard the International Space Station (Parfenov Vladislav et al., 2020). Human chondrocytes were grown in DMEM, and aggregates of about 8000 cells each were fabricated by the reaggregation of a cell suspension in nonadherent micro-molds. Live aggregates were embedded in a thermoreversible hydrogel—a copolymer of poly(*N*-isopropylacrylamide) and poly(ethylene glycol) (PNIPAAm-PEG)—to prevent their premature fusion and/or spreading on the internal surface of the cuvette. PNIPAAm-PEG is liquid at 4 °C, when diluted 1.6-fold with cell culture medium, it displays a sol–gel transition at 27 °C. Hence, aggregates trapped in this hydrogel for the duration of the travel could be released by cooling down the cuvette to 17 °C (Fig. 11.5).

The cylindrical system of magnets shown in Fig. 11.5A creates a nonuniform magnetic field for trapping diamagnetic particles in the field minimum (Fig. 11.5B). During the flight to orbit, chondrocyte aggregates were kept in sterile cuvettes, immobilized in a PNIPAAm-PEG hydrogel, as shown in Fig. 11.5C, *panel* (i). To start the experiment, the P1 button was pushed to add gadobutrol dissolved in

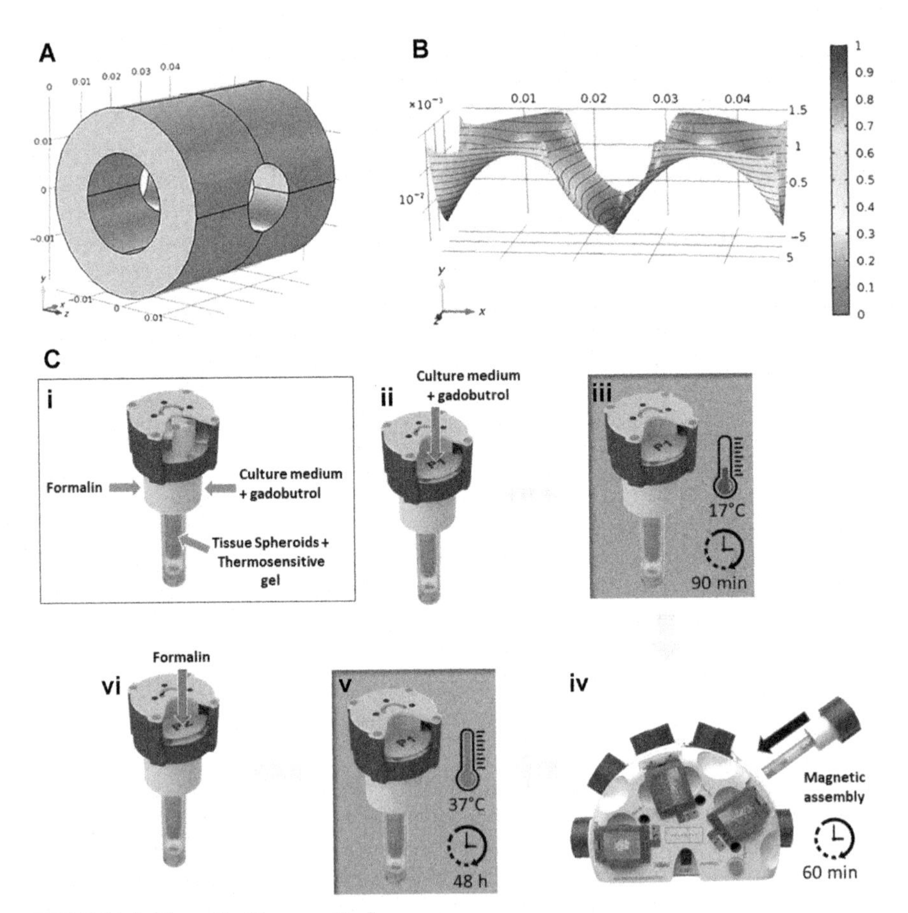

FIGURE 11.5 Magnetic bioassembly in space.

(A) NdFeB magnet and (B) plot of the magnetic flux density according to COMSOL calcula tions performed for the magnet depicted in *panel* A; (C) stages of a magnetic bioassembly experiment conducted under weightlessness conditions: (i) cuvettes with cell aggregates embedded in thermoreversible hydrogel; (ii) addition of gadobutrol and cell culture medi um; (iii) liquefaction of the hydrogel; (iv) insertion of the cuvette into the magnetic bioa ssembler; (v) maintenance of physiological temperature to facilitate the fusion of cell agg regates; and (vi) extraction of the cuvette and formalin injection for fixing the tissue constr uct until subsequent analysis.

Adapted from Parfenov Vladislav, A., Khesuani Yusef, D., Petrov Stanislav, V., Karalkin Pavel, A., Koudan Elizaveta, V., Nezhurina Elizaveta, K., Pereira Frederico, D. A. S., Krokhmal Alisa, A., Gryadunova Anna, A., Bulanova Elena, A., Vakhrushev Igor, V., Babichenko Igor, I., Kasyanov, V., Petrov Oleg, F., Vasiliev Mikhail, M., Brakke, K., Belousov Sergei, I., Grigoriev Timofei, E., Osidak Egor, O., … Mironov Vladimir, A. (2020). Magnetic levitational bioassembly of 3D tissue construct in space. Science Advances, 6, eaba4174. *https://doi.org/10.1126/sciadv.aba4174*

cell culture medium—*panel* (ii). Then, the cuvette was cooled down to liquefy the hydrogel (iii) and inserted into the orifice of a cylindrical magnet from the bioassembler—*panel* (iv). After an hour, the cuvette was warmed up to physiological temperature to ensure proper conditions for chondrocyte spheroid fusion during the next 2 days (v). Finally, the cuvette was extracted and the P2 button was pressed to fix the tissue construct by adding formalin—*panel* (vi). To achieve an assembly time of about 40 min, Parfenov et al. employed a gadobutrol concentration of 10 mM; had it been 0.8 mM, the assembly time would have been about 10 h, potentially harming the cells maintained at 17 °C to keep their embedding medium in sol state (Parfenov Vladislav et al., 2020).

Besides providing a methodological advantage by the use of inexpensive permanent magnets, cartilage tissue engineering under microgravity is exciting because it might shed light on mechanisms of articular cartilage remodeling during a long space mission.

While new methods may revolutionize 3D and 4D bioprinting, refinements of well-established techniques might also lead to important breakthroughs. Stimulus-responsive objects are notoriously difficult to fabricate because they typically involve multiple materials structured at the micrometer scale. Voxel-by-voxel printing is the most adequate for this task, as shown by the success of polyjet techniques in 4D printing. Inkjet printing, however, is limited to low-viscosity inks. Therefore, Skylar-Scott et al. developed a multimaterial multinozzle 3D (MM3D) printer capable of extruding up to eight different viscoelastic inks to create soft 3D structures with voxel sizes as small as the nozzle diameter (Skylar-Scott, Mueller, et al., 2019). The MM3D printer works with print heads comprising a single nozzle, a linear set, or a planar array of nozzles. The print head is connected to multiple syringes of different content, each of which is actuated by pneumatic solenoids that enable switching frequencies of up to 50 Hz. The materials streaming from different syringes converge into a single channel carrying a voxelated, continuous filament, which is distributed to all the nozzles (i.e., each layer of the printout will be a periodic tiling). To demonstrate the capabilities of MM3D printing, the team of Lewis printed Miura origami patterns and pneumatically controlled soft robotic walkers (Skylar-Scott, Mueller, et al., 2019). In the context of bioprinting, MM3D extrusion will be used to build heterogeneous tissue constructs made of voxels of different biomaterial and/or cellular content.

To mitigate the drawbacks of one specific method of additive manufacturing, an increasing number of research groups explore ways to combine multiple bioprinting techniques in a single, complex bioprinter capable of sequential or simultaneous deposition of different components of a hierarchical tissue construct. Moreover, automated instruments are being developed to include fabrication methods that are not part of the additive manufacturing family, such as micromolding, centrifugal spinning, electrospinning, or electrospraying. Such hybrid biofabrication technologies leverage the strengths of each manufacturing method to improve resolution, throughput, construct stability, and subsequent maturation. Acellular techniques provide scaffolding and mechanical reinforcement, whereas cellular technologies create biopatterning along length scales ranging from single cell sizes to millimeters. Nanoscale biomimicry, needed, for example, to replicate the microstructure of the extracellular matrix of native tissues, remains beyond the reach of hybrid biofabrication, but such structures can be secreted by the constituent cells in the course of tissue construct maturation; the same is true also for microvasculature (Dalton et al., 2020).

These developments are paralleled by the emergence of modular, fully automated, high-throughput production platforms for creating biomaterial libraries as well as for handling raw materials and end-products. Also, automated quality assessment (i.e., quantitative analysis of biological images, or real-time monitoring of tissue construct metabolism using pH and O_2 microsensors) combined with machine learning algorithms will streamline the entire workflow (Eggert & Hutmacher, 2019). Pioneering research in these directions is mainly based on open-source robotics hardware and software (Eggert et al., 2020, 2021), but companies will most likely catch up and bring proprietary extensions.

It seems that the holy grail of tissue engineering, the task of creating tissue constructs akin to solid organs, is like a puzzle. It is certainly complex, and still unsolved, but all the pieces are finely crafted and will be put together in the near future.

3. Potential applications

One of the motivations of 4D bioprinting research is to endow implantable tissue constructs with the ability to adapt to the dynamic environment provided by the host organism (Li et al., 2017). A 4D-bioprinted construct modifies its shape and/or develops a desired functionality upon a predetermined stimulation, and, thereby, it has a better chance to mimic the dynamics of native tissues than a static structure. The examples analyzed in Chapter 10 demonstrate the benefits of shape morphing, but they also plead for investigations aimed at developing smart materials capable of more complex, repetitive shape changes. Further research is needed to closely recapitulate native organ function, such as the response of blood vessels to chemical stimuli (vasoconstriction/vasodilation), the contraction/relaxation cycles of longitudinal and circular smooth muscles responsible for the peristaltic movements of the digestive tract, or the rhythmic contractions of the cardiac muscle.

The 4D bioprinting of cell-laden shape memory scaffolds is a promising line of development. Such scaffolds were found useful, for example, in cardiac tissue engineering. Their ability to closely fit onto the target site assured robust engraftment, whereas their microarchitecture fostered cardiac myocyte alignment and maturation (Cui et al., 2020; Wang et al., 2021). Also, shape memory was crucial for the minimally invasive delivery of cellularized cardiac constructs (Montgomery et al., 2017).

Progress in multimaterial 3D printing and bioprinting will provide versatile tools for the facile and reliable fabrication of cellularized shape memory scaffolds for tissue engineering applications. The scaffolds may incorporate bioactive agents (e.g., growth factors) to be released gradually or with precise timing controlled by intrinsic or remotely applied stimuli. Also, such constructs can be doped with magnetic micro/nanoparticles, making them suitable for manipulation by magnetic fields (e.g., for magnetic levitation culture or gentle deployment) or for inducing shape transition by exposure to an external, high-frequency electromagnetic field (magnetic hyperthermia). Also, the incorporation of an array of individually addressable nanoelectronic transducers, comparable in size to bacteria, would enable the real-time monitoring of the electrophysiological activity of the engineered tissue. Three-dimensional macroporous nanoelectronic networks were already constructed for the spatiotemporal mapping of action potentials in engineered cardiac tissues (Zhou et al., 2017). Their mechanical properties are comparable to those of fibrous scaffolds fabricated by electrospinning, suggesting that they might be sandwiched between 3D-printed layers of shape memory polymers without affecting the shape morphing of the latter. Finally, such hybrid scaffolds can be loaded with cells by 3D bioprinting or conventional cell seeding.

Smart tissue constructs should not necessarily comprise stimulus-responsive materials; cells themselves are active entities, capable of conveying shape-morphing ability. Cells explore their neighborhood by exerting traction forces on adjacent cells, ECM filaments, or biocompatible materials. Despite promising first steps (Kuribayashi-Shigetomi et al., 2012), cell traction forces (CTFs) have been largely overlooked as sources of 4D behavior. The cell origami constructs built by Kuribayashi-Shigetomi et al. comprised poly(p-xylylene) polymer (parylene) microplates coated with fibronectin. Cells were cultured on a set of microplates glued next to each other onto a glass plate by a thin layer of 0.05%–0.1% gelatin. Uncovered parts of the glass surface were coated with 2-methacryloyloxyethyl phosphorylcholine polymer to prevent protein adsorption and, consequently, cell attachment. Cells adhered to the microplates and spread across them. Mechanical detachment of a microplate resulted in a translational motion until it contacted the adjacent plate, followed by a rotation around the line of contact, up to a folding angle determined by the number of cells residing on the microplates. Precise control of the folding angle was achieved when the microplates were connected by a thin, flexible joint. Reproducible folding was observed without mechanical triggers when the concentration of the underlying gelatin layer ranged from 3% to 5%. It has been rightfully argued that such constructs cannot be considered 4D bioprinted because (i) they are not 3D printed and (ii) their shape transition occurs spontaneously, in the absence of an external stimulus (An et al., 2016). When the folding of independent plates was elicited by pushing them with a glass tip mounted on a micromanipulator, the outcome of self-folding was less predictable. In that case, stimulation was present, but the requirement of programmable postprinting evolution was not fully satisfied. Four-dimensional bioprinted constructs relying on CTF are less reliable than those based on smart materials.

Nevertheless, the drawbacks of early cell origami constructs may be surmounted by employing hybrid bioprinting to create dynamic systems capable of reliable, anticipated shape transitions under the action of certain external stimuli. For example, the microplates could be created by PolyJet printing and covered by cells via inkjet or microextrusion bioprinting. Then, they could be attached to a glass substrate using reduced graphene oxide-gelatin nanocomposite hydrogels (Piao & Chen, 2016), and the shape transition could be triggered by shining near-infrared light onto the hydrogel layer. Alternatively, the microplates might be loaded with $Nd_2Fe_{14}B$ microparticles to make them suitable for magnetic manipulation. Once covered by cells, they could be immobilized on an MPC-coated glass plate by placing a permanent magnet below the cell culture dish. Departing the magnet would trigger the shape transition driven by CTF.

Engineered tissues made of cells suspended in hydrogels have a relatively low cell density: they contain 10–100 times fewer cells in their unit volume than native tissues. When high cell density is essential, it is convenient to rely on the self-assembly of microtissue building blocks, such as tissue spheroids, or functional organoids derived from self-organizing cultures of human adult stem cells or induced pluripotent stem cells. Fabrication strategies available to date include the 3D bioprinting of tissue spheroids (Ayan et al., 2020; Mironov et al., 2009), perhaps combined with melt extrusion (Dalton et al., 2020; Mekhileri et al., 2018), or the sacrificial writing into functional tissue (SWIFT) (Skylar-Scott, Uzel, et al., 2019). The final product of such a process results from the autonomous fusion of the microtissue building blocks, as discussed in Chapters 7 and 8.

Tissue fusion is a potential source of 4D behavior. A bioprinted construct is inherently dynamic, so postprinting evolution is the rule rather than the exception. To harness the self-organizing ability of cells for 4D bioprinting, we need a deeper understanding of the molecular mechanisms involved in multicellular self-assembly. Such knowledge can lead to the discovery of bioactive compounds capable of modulating tissue fusion. The control over the self-assembly of microtissues has several benefits: First, it helps finish the product just on time for implantation or transfer into a bioreactor. Second, it favors cell survival, thereby extending the shelf-life of the engineered tissue. The microtissues are properly sized for nutrient and gas transport via diffusion. As soon as fusion proceeds, certain parts of the construct might become too thick, needing a vasculature for appropriate mass transfer. Hence, it is desirable to halt tissue fusion at the right moment, when the adjacent microtissues have already established firm contacts, making the structure sturdy enough for manipulation, but they did not block access to the cell culture medium, yet. Exceptions are thin tissue constructs and those outfitted with vascular channels (created, for example, by the SWIFT technique). Third, orchestrating tissue fusion can facilitate manufacturing because certain types of cells create a complex histoarchitecture from a relatively simple bioprinted structure.

Certain applications will rely both on stimulus-responsive materials and multicellular self-organization. For example, printing a close-packed monolayer of cell aggregates over a smart substrate can result in a tubular construct. Spontaneous tissue fusion gives rise to a patch of live cells, whose thickness depends on cell aggregate size (Neagu et al., 2005). Under the proper stimulus, the substrate bends, becoming a cylinder, and the opposite edges of the rectangular patch come in contact and fuse. The resulting multicellular tube shrinks due to surface tension and detaches from the substrate, provided that the cell-substrate adhesion is weaker than cell-cell cohesion. This concept, proposed in the early years of 4D bioprinting (An et al., 2016), became feasible due to progress in smart material research. It would provide precise control over the tube geometry while circumventing the problem of mechanical support needed during the 3D bioprinting of tubular structures.

Printable nanocomposite hydrogels hold promise to provide improved extracellular matrix mimicry and drive cell differentiation. A representative example of this kind is a GelMA hydrogel reinforced with mutually parallel filaments of iron oxide nanoparticles (Tognato et al., 2019). Spherical nanoparticles of PEG-capped γ-Fe_2O_3 crystals, with an average diameter of 3 nm, formed spherical clusters of 45–60 nm in diameter. They were incorporated in GelMA prepolymer solution and placed in a uniform magnetic field of 20 mT flux density to promote their self-assembly in filaments aligned with the magnetic field lines; then, the system was cooled below the melting temperature of the hydrogel and exposed to UV light to immobilize the nanoparticle filaments. Mesenchymal stem cells cultured on top of this hydrogel aligned along the filaments. When C2C12 mouse skeletal myoblasts were incorporated into this hydrogel, they aligned along the filaments and fused to form myotubes—and all this happened in the absence of a myogenic differentiation medium. Exploiting the printability of GelMA, refinements of this technique might lead to 4D-bioprinted skeletal muscle constructs.

Further progress in the field of smart hydrogels will enable the additive manufacturing of biocompatible soft robots, soft actuators, and targeted drug delivery systems. Cell-free structures of this kind are already available. Thermo- and photosensitive hinges and micro-gripers, of tens of microns in size, have been fabricated by two-photon laser lithography from a single photoresist made of *N*-isopropylacrylamide (pNIPAAm), using *N,N'*-methylenebisacrylamide crosslinker and lithium phenyl(2,4,6-trimethylbenzoyl)phosphinate (LAP) photoinitiator (Hippler et al., 2019). The heterogeneity of the microstructures was achieved by differential light curing known as gray-tone lithography: the laser power was varied locally to create more or less cross-linked voxels. Regions with high crosslinking density were less prone to rearrangements; i.e., they were less responsive to changes in temperature than those of low crosslinking density. Hence, a bilayer beam heated above the lower critical solution temperature of the pNIPAAm hydrogel curved toward the less crosslinked layer (because this layer expelled more water and shrunk as the polymer filaments became hydrophobic). Besides global heating, bilayer hinges could be actuated locally by focusing a laser beam on them, since two-photon absorption caused local heating via photo-thermal conversion. Rings of pNIPAAm printed inside rigid tubes made of pentaerythritol triacrylate (PETA) functioned as temperature-controlled valves. A single ring modulated the flow; when a second ring was printed around a central pole, the two swollen rings stopped the flow at room temperature and enabled it at 45 °C (Hippler et al., 2019). The vast variety of soft robots fabricated to date includes those driven by electric fields, magnetic fields, pressure, and light. Their integration with miniaturized biosensors and flexible electronics will result in standalone soft objects capable of interacting with their environment via detection and actuation. The feasibility of such integration has been demonstrated by the robotic positioning of flexible electronic devices into 3D-printed human femur models (El-Atab et al., 2020). The incorporation soft sensors and electronic devices is promising also in 3D bioprinting, in the absence of stimulus responsiveness, because they might enable the real-time monitoring of the tissue construct's metabolism and maturation.

Four-dimensional bioprinting will benefit from, and contribute to, the progress of the broader research field of physical intelligence (PI), defined as "physically encoding sensing, actuation, control, memory, logic, computation, adaptation, learning and decision-making into the body of an agent" (Sitti, 2021). The intelligence of a physical agent, such as a living organism or a man-made robot, is not limited to its computational intelligence (hosted by the brain or by a computer running artificial intelligence algorithms); it is encoded also in its body. Wound healing, for example, does not rely on the central nervous system, but on local actions undertaken by cells sensing changes in their physicochemical environment. Such repair mechanisms are vital for living beings because even small injuries can endanger their lives. Artificial structures can be endowed with a similar kind of PI. Self-healing materials have been synthesized from polymers containing dynamic covalent bonds that can be triggered to repair a damaged portion by certain stimuli, such as heat or light (Li et al., 2019). Soft robots made of certain protein-based materials spontaneously repair themselves after mechanical damage via the recovery of their reversible physical bonds. Also, a self-healing body can be created by loading it with microencapsulated healing agent while its bulk material contains the catalyst needed for the polymerization of the healing agent. PI is an interdisciplinary paradigm that builds on a vast set of key enablers, including (i) self-assembly, self-growing, self-degrading, self-healing, self-adaptation, self-regulation, self-propulsion, self-powering, self-cleaning, self-replicating capabilities, (ii) multifunction, multistability, multimodality, multiphysics features, and (iii) modularity, reconfigurability, smart structuring (e.g., origami, kirigami, tensegrity), hierarchical structuring, physical programmability, as well as mechanical memory, computation, and decision making (Sitti, 2021).

A retrospective look at this book suggests that it is hard to overstate the perspectives of 4D bioprinting. New applications are expected to emerge in the near future. Machine learning will help choose the best materials and design features, as well as select the most effective additive manufacturing technique for a specific targeted outcome. Additionally, it might facilitate the development of new material formulations and printing parameters to ensure an optimal shape transition. As experimental data continue to accumulate, artificial intelligence will most likely outperform human intuition in tailoring the biofabrication workflow needed for a particular application.

References

An, J., Chua, C. K., & Mironov, V. (2016). A perspective on 4D bioprinting. *International Journal of Bioprinting, 2*, 3–5. https://doi.org/10.18063/ijb.2016.01.003

Ashammakhi, N., Ahadian, S., Zengjie, F., Suthiwanich, K., Lorestani, F., Orive, G., Ostrovidov, S., & Khademhosseini, A. (2018). Advances and future perspectives in 4D bioprinting. *Biotechnology Journal, 13*. https://doi.org/10.1002/biot.201800148. e1800148–e1800148.

Ayan, B., Heo, D. N., Zhang, Z., Dey, M., Povilianskas, A., Drapaca, C., & Ozbolat, I. T. (2020). Aspiration-assisted bioprinting for precise positioning of biologics. *Science Advances, 6*. https://doi.org/10.1126/sciadv.aaw5111. eaaw5111.

Beh, C. W., Yew, D. S., Chai, R. J., Chin, S. Y., Seow, Y., & Hoon, S. S. (2021). A fluid-supported 3D hydrogel bioprinting method. *Biomaterials, 276*, 121034. https://doi.org/10.1016/j.biomaterials.2021.121034

Bernal, P. N., Delrot, P., Loterie, D., Li, Y., Malda, J., Moser, C., & Levato, R. (2019). Volumetric bioprinting of complex living-tissue constructs within seconds. *Advanced Materials, 31*, 1904209. https://doi.org/10.1002/adma.201904209

Callaway, E. (2022). What's next for AlphaFold and the AI protein-folding revolution. *Nature, 604*, 234–238. https://doi.org/10.1038/d41586-022-00997-5

Choi, G. P. T., Dudte, L. H., & Mahadevan, L. (2019). Programming shape using kirigami tessellations. *Nature Materials, 18*, 999–1004. https://doi.org/10.1038/s41563-019-0452-y

Cui, H., Liu, C., Esworthy, T., Huang, Y., Yu, Z., Zhou, X., San, H., Lee, S., Hann, S. Y., Boehm, M., Mohiuddin, M., Fisher, J. P., & Zhang, L. G. (2020). 4D physiologically adaptable cardiac patch: A 4-month in vivo study for the treatment of myocardial infarction. *Science Advances, 6*, eabb5067. https://doi.org/10.1126/sciadv.abb5067

Dalton, P. D., Woodfield, T. B. F., Mironov, V., & Groll, J. (2020). Advances in hybrid fabrication toward hierarchical tissue constructs. *Advanced Science, 7*, 1902953. https://doi.org/10.1002/advs.201902953

Dudte, L. H., Vouga, E., Tachi, T., & Mahadevan, L. (2016). Programming curvature using origami tessellations. *Nature Materials, 15*, 583–588. https://doi.org/10.1038/nmat4540

Eggert, S., & Hutmacher, D. W. (2019). In vitro disease models 4.0 via automation and high-throughput processing. *Biofabrication, 11*, 043002. https://doi.org/10.1088/1758-5090/ab296f

Eggert, S., Kahl, M., Bock, N., Meinert, C., Friedrich, O., & Hutmacher, D. W. (2021). An open-source technology platform to increase reproducibility and enable high-throughput production of tailorable gelatin methacryloyl (GelMA)—Based hydrogels. *Materials & Design, 204*, 109619. https://doi.org/10.1016/j.matdes.2021.109619

Eggert, S., Mieszczanek, P., Meinert, C., & Hutmacher, D. W. (2020). OpenWorkstation: A modular open-source technology for automated in vitro workflows. *HardwareX, 8*, e00152. https://doi.org/10.1016/j.ohx.2020.e00152

El-Atab, N., Mishra, R. B., Al-Modaf, F., Joharji, L., Alsharif, A. A., Alamoudi, H., Diaz, M., Qaiser, N., & Hussain, M. M. (2020). Soft actuators for soft robotic applications: A review. *Advanced Intelligent Systems, 2*, 2000128. https://doi.org/10.1002/aisy.202000128

Gladman, A. S., Matsumoto, E. A., Nuzzo, R. G., Mahadevan, L., & Lewis, J. A. (2016). Biomimetic 4D printing. *Nature Materials, 15*, 413–418. https://doi.org/10.1038/nmat4544

Grigoryan, B., Paulsen Samantha, J., Corbett Daniel, C., Sazer Daniel, W., Fortin Chelsea, L., Zaita Alexander, J., Greenfield Paul, T., Calafat Nicholas, J., Gounley John, P., Ta Anderson, H., Johansson, F., Randles, A., Rosenkrantz Jessica, E., Louis-Rosenberg Jesse, D., Galie Peter, A., Stevens Kelly, R., & Miller Jordan, S. (2019). Multivascular networks and functional intravascular topologies within biocompatible hydrogels. *Science, 364*, 458–464. https://doi.org/10.1126/science.aav9750

Guo, F., Mao, Z., Chen, Y., Xie, Z., Lata, J. P., Li, P., Ren, L., Liu, J., Yang, J., Dao, M., Suresh, S., & Huang, T. J. (2016). Three-dimensional manipulation of single cells using surface acoustic waves. *Proceedings of the National Academy of Sciences, 113*, 1522–1527. https://doi.org/10.1073/pnas.1524813113

Hamel, C. M., Roach, D. J., Long, K. N., Demoly, F., Dunn, M. L., & Qi, H. J. (2019). Machine-learning based design of active composite structures for 4D printing. *Smart Materials and Structures, 28*, 065005. https://doi.org/10.1088/1361-665x/ab1439

Hippler, M., Blasco, E., Qu, J., Tanaka, M., Barner-Kowollik, C., Wegener, M., & Bastmeyer, M. (2019). Controlling the shape of 3D microstructures by temperature and light. *Nature Communications, 10*, 232. https://doi.org/10.1038/s41467-018-08175-w

Ilievski, F., Mirica, K. A., Ellerbee, A. K., & Whitesides, G. M. (2011). Templated self-assembly in three dimensions using magnetic levitation. *Soft Matter, 7*, 9113–9118. https://doi.org/10.1039/C1SM05962A

Jamal, M., Kadam, S. S., Xiao, R., Jivan, F., Onn, T. M., Fernandes, R., Nguyen, T. D., & Gracias, D. H. (2013). Bio-origami hydrogel scaffolds composed of photocrosslinked PEG bilayers. *Advanced Healthcare Materials, 2*, 1142–1150. https://doi.org/10.1002/adhm.201200458

Ji, S., & Guvendiren, M. (2021). Complex 3D bioprinting methods. *APL Bioengineering, 5*, 011508. https://doi.org/10.1063/5.0034901

Jumper, J., Evans, R., Pritzel, A., Green, T., Figrnov, M., Ronneberger, O., et al. (2021). Highly accurate protein structure prediction with AlphaFold. *Nature, 596*, 583–589. https://doi.org/10.1038/s41586-021-03819-2

Kang, P., Tian, Z., Yang, S., Yu, W., Zhu, H., Bachman, H., Zhao, S., Zhang, P., Wang, Z., Zhong, R., & Huang, T. J. (2020). Acoustic tweezers based on circular, slanted-finger interdigital transducers for dynamic manipulation of micro-objects. *Lab on a Chip, 20*, 987–994. https://doi.org/10.1039/C9LC01124B

Kelly, B. E., Bhattacharya, I., Heidari, H., Shusteff, M., Spadaccini, C. M., & Taylor, H. K. (2019). Volumetric additive manufacturing via tomographic reconstruction. *Science, 363*, 1075–1079. https://doi.org/10.1126/science.aau7114

Kuribayashi-Shigetomi, K., Onoe, H., & Takeuchi, S. (2012). Cell origami: Self-folding of three-dimensional cell-laden microstructures driven by cell traction force. *PLoS One, 7*, e51085. https://doi.org/10.1371/journal.pone.0051085

Kwok, T.-H., Wang, C. C. L., Deng, D., Zhang, Y., & Chen, Y. (2015). Four-dimensional printing for freeform surfaces: Design optimization of origami and kirigami structures. *Journal of Mechanical Design, 137*. https://doi.org/10.1115/1.4031023

Li, L., Scheiger, J. M., & Levkin, P. A. (2019). Design and applications of photoresponsive hydrogels. *Advanced Materials, 31*, 1807333. https://doi.org/10.1002/adma.201807333

Liu, C.-Y., Lai, S., & Lima, J. A. C. (2019). MRI gadolinium dosing on basis of blood volume. *Magnetic Resonance in Medicine, 81*, 1157–1164. https://doi.org/10.1002/mrm.27454

Li, Y.-C., Zhang, Y. S., Akpek, A., Shin, S. R., & Khademhosseini, A. (2017). 4D bioprinting: The next-generation technology for biofabrication enabled by stimuli-responsive materials. *Biofabrication, 9*, 012001. http://stacks.iop.org/1758-5090/9/i=1/a=012001.

Mekhileri, N. V., Lim, K. S., Brown, G. C. J., Mutreja, I., Schon, B. S., Hooper, G. J., & Woodfield, T. B. F. (2018). Automated 3D bioassembly of micro-tissues for biofabrication of hybrid tissue engineered constructs. *Biofabrication, 10*, 024103. http://stacks.iop.org/1758-5090/10/i=2/a=024103.

Mironov, V., Visconti, R. P., Kasyanov, V., Forgacs, G., Drake, C. J., & Markwald, R. R. (2009). Organ printing: Tissue spheroids as building blocks. *Biomaterials, 30*, 2164–2174. http://www.sciencedirect.com/science/article/pii/S0142961209000052.

Momeni, F., & Ni, J. (2020). *Laws of 4D printing. Engineering*. https://doi.org/10.1016/j.eng.2020.01.015

Montgomery, M., Ahadian, S., Davenport Huyer, L., Lo Rito, M., Civitarese, R. A., Vanderlaan, R. D., Wu, J., Reis, L. A., Momen, A., Akbari, S., Pahnke, A., Li, R.-K., Caldarone, C. A., & Radisic, M. (2017). Flexible shape-memory scaffold for minimally invasive delivery of functional tissues. *Nature Materials, 16*, 1038–1046. https://doi.org/10.1038/nmat4956

Neagu, A., Jakab, K., Jamison, R., & Forgacs, G. (2005). Role of physical mechanisms in biological self-organization. *Physical Review Letters, 95*, 178104. http://<GotoISI>://000232724400079.

Ng, W. L., Chan, A., Ong, Y. S., & Chua, C. K. (2020). Deep learning for fabrication and maturation of 3D bioprinted tissues and organs. *Virtual and Physical Prototyping, 15*, 340–358. https://doi.org/10.1080/17452759.2020.1771741

Parfenov Vladislav, A., Khesuani Yusef, D., Petrov Stanislav, V., Karalkin Pavel, A., Koudan Elizaveta, V., Nezhurina Elizaveta, K., Pereira Frederico, D. A. S., Krokhmal Alisa, A., Gryadunova Anna, A., Bulanova Elena, A., Vakhrushev Igor, V., Babichenko Igor, I., Kasyanov, V., Petrov Oleg, F., Vasiliev Mikhail, M., Brakke, K., Belousov Sergei, I., Grigoriev Timofei, E., Osidak Egor, O., … Mironov Vladimir, A. (2020). Magnetic levitational bioassembly of 3D tissue construct in space. *Science Advances, 6*, eaba4174. https://doi.org/10.1126/sciadv.aba4174

Parfenov, V. A., Mironov, V. A., van Kampen, K. A., Karalkin, P. A., Koudan, E. V., Pereira, F. D. A. S., Petrov, S. V., Nezhurina, E. K., Petrov, O. F., Myasnikov, M. I., Walboomers, F. X., Engelkamp, H., Christianen, P., Khesuani, Y. D., Moroni, L., & Mota, C. (2020). Scaffold-free and label-free biofabrication technology using levitational assembly in a high magnetic field. *Biofabrication, 12*, 045022. https://doi.org/10.1088/1758-5090/ab7554

Piao, Y., & Chen, B. (2016). One-pot synthesis and characterization of reduced graphene oxide–gelatin nanocomposite hydrogels. *RSC Advances, 6*, 6171–6181. https://doi.org/10.1039/C5RA20674J

Raviv, D., Zhao, W., McKnelly, C., Papadopoulou, A., Kadambi, A., Shi, B., Hirsch, S., Dikovsky, D., Zyracki, M., Olguin, C., Raskar, R., & Tibbits, S. (2014). Active printed materials for complex self-evolving deformations. *Scientific Reports, 4*, 7422. https://doi.org/10.1038/srep07422

Shafiee, A., Kassis, J., Atala, A., & Ghadiri, E. (2021). Acceleration of tissue maturation by mechanotransduction-based bioprinting. *Physical Review Research, 3*, 013008. https://doi.org/10.1103/PhysRevResearch.3.013008

Sitti, M. (2021). Physical intelligence as a new paradigm. *Extreme Mechanics Letters, 46*, 101340. https://doi.org/10.1016/j.eml.2021.101340

Skylar-Scott, M. A., Mueller, J., Visser, C. W., & Lewis, J. A. (2019). Voxelated soft matter via multimaterial multinozzle 3D printing. *Nature, 575*, 330–335. https://doi.org/10.1038/s41586-019-1736-8

Skylar-Scott, M. A., Uzel, S. G. M., Nam, L. L., Ahrens, J. H., Truby, R. L., Damaraju, S., & Lewis, J. A. (2019). Biomanufacturing of organ-specific tissues with high cellular density and embedded vascular channels. *Science Advances, 5*, eaaw2459. https://doi.org/10.1126/sciadv.aaw2459

Tognato, R., Armiento, A. R., Bonfrate, V., Levato, R., Malda, J., Alini, M., Eglin, D., Giancane, G., & Serra, T. (2019). A stimuli-responsive nanocomposite for 3D anisotropic cell-guidance and magnetic soft robotics. *Advanced Functional Materials, 29*, 1804647. https://doi.org/10.1002/adfm.201804647

Tumbleston, J. R., Shirvanyants, D., Ermoshkin, N., Janusziewicz, R., Johnson, A. R., Kelly, D., Chen, K., Pinschmidt, R., Rolland, J. P., Ermoshkin, A., Samulski, E. T., & DeSimone, J. M. (2015). Continuous liquid interface production of 3D objects. *Science, 347*, 1349–1352. https://doi.org/10.1126/science.aaa2397

Wang, Y., Cui, H., Wang, Y., Xu, C., Esworthy, T. J., Hann, S. Y., Boehm, M., Shen, Y.-L., Mei, D., & Zhang, L. G. (2021). 4D printed cardiac construct with aligned myofibers and adjustable curvature for myocardial regeneration. *ACS Applied Materials & Interfaces, 13*, 12746–12758. https://doi.org/10.1021/acsami.0c17610

Winkleman, A., Gudiksen, K. L., Ryan, D., Whitesides, G. M., Greenfield, D., & Prentiss, M. (2004). A magnetic trap for living cells suspended in a paramagnetic buffer. *Applied Physics Letters, 85*, 2411–2413. https://doi.org/10.1063/1.1794372

Yu, C., & Jiang, J. (2020). A perspective on using machine learning in 3D bioprinting. *International Journal of Bioprinting, 6*. https://doi.org/10.18063/ijb.v6i1.253, 253–253.

Zhou, W., Dai, X., & Lieber, C. M. (2017). Advances in nanowire bioelectronics. *Reports on Progress in Physics, 80*, 016701. https://doi.org/10.1088/0034-4885/80/1/016701

Index

Note: 'Page numbers followed by "*f*" indicate figures and "*t*" indicate tables.'

S

T

V